CHEMICAL KINETICS AND REACTOR DESIGN

Chemical Kinetics AND *Reactor Design*

A. R. COOPER

Senior Lecturer, Department of Chemical Engineering,
The University of Aston in Birmingham, England

G. V. JEFFREYS

Professor and Head of Department, Department of Chemical Engineering,
The University of Aston in Birmingham, England

OLIVER & BOYD · EDINBURGH

OLIVER AND BOYD
Tweeddale Court
14 High Street
Edinburgh EH1 1YL
A Division of Longman Group Limited

ISBN 0 05 002114 1
First published 1971

Set in 10/12 Times New Roman 569 series
and printed in Great Britain by Page Bros. (Norwich) Ltd., Norwich

CONTENTS

PREFACE

In the past decade, a large number of texts on the design and analysis of chemical reactors have been published, and most of these have been welcome additions to the chemical engineering literature. Therefore, one is tempted to enquire why yet another textbook on chemical reactor design and analysis should appear. The answer to this question is that the present book is intended to be one of a series of coordinate texts covering the whole range of chemical engineering operations, and that the authors have attempted to present an undergraduate course in chemical reaction kinetics and and the design and an analysis of chemical reactors, that will prepare students for further postgraduate study or industrial experience in this field. In this respect it is hoped that this book will initiate an interest in this topic so that students will be able and encouraged to pursue the subject further in other works and in the chemical engineering journals.

In Chapter 2 of this text book, overall mass and energy balances of chemical reaction processes are considered in order to assess the feasibility of a chemical process. Following this, in Chapter 3, the methods employed for analysing laboratory kinetic experiments are discussed so that appropriate rate equations can be developed, while in the remaining chapters the ways in which these equations are applied in the selection and specification of chemical reactors are then explained Single phase reactions are first treated and are followed by the analysis of two phase reaction systems involving simultaneous chemical reaction and mass transfer processes.

Throughout the text, worked examples of practical chemical reactions which have industrial importance are presented, and it is hoped thereby that a working concept of the numerical magnitudes of reactor size and the relevant physical parameters will be imparted. At the end of the chapters discussing fundamental concepts, a wide range of supporting problems is presented. It is anticipated that the text should serve the needs of undergraduate chemical engineering courses in the introduction of chemical reactor concepts, while some of the mathematical techniques may, at present, be more appropriate to postgraduate work. A comprehensive treatise is not intended but it is hoped that students will be well prepared to undertake with confidence further detailed work in the chemical reactor field.

The authors wish to acknowledge the helpful discussions with many former colleagues and students, in particular, Dr E L Smith and Dr H Moore of Unilever Research Laboratories, and Mr B Buxton and Mr C. J. Woodcock, who have conducted supporting tutorial classes at the University of Aston in Birmingham. The assistance and constructive comments given by Dr D. E. Creasy in the preparation of many of the figures is also acknowledged.

1

INTRODUCTION TO CHEMICAL REACTION ENGINEERING

The functions of the chemical engineer in the field of chemical reaction engineering are to specify the size and geometry of a vessel for the production of a given amount of a particular chemical, and to appraise the performance of existing reactors. In the latter context, confirmation that a reactor is operating in the best way to meet its duty may be required, or means of increasing the throughput, or some other variation of duty, may need to be explored. In chemical reactor design and analysis it is necessary to establish that the reaction vessel is capable of withstanding the applied pressure or vacuum at the reaction temperature and that it possesses sufficient ancillary equipment to maintain the process operating conditions at their desired values.

The chemical reaction may be made to take place batchwise in a closed vessel or the reactants may be made to flow continuously through a vessel of specific shape with continuous discharge of the products and residual reactants from the exit. In either case it is essential that the chemical reaction is carried out at the most suitable combination of temperature and pressure for the maximum conversion of reactants by the chosen chemical route to the required product. Thus incorrect selection of processing conditions may lead to either low conversion of reactant or alternatively to its conversion by side reactions to unwanted products. In addition it is possible for the desired reaction product to undergo reaction itself and thereby diminish the reactor performance; such wastage of product can be restricted by suitable selection of reactor type and processing conditions.

While high conversion yields are desirable and are very often obtainable, there are certain reactions of industrial importance which have only very low yields. Application of thermodynamic principles to a reaction system enables the maximum possible yield—the equilibrium yield—to be evaluated for a set of proposed processing conditions, thereby enabling the exploration of feasible conditions for carrying out the envisaged reactions economically. Such a thermodynamic analysis may conclude with either the specification of a set of suitable operating conditions for high conversion or alternatively that a low conversion must be tolerated in effecting the reaction under consideration. In this latter case, it is part of the chemical reaction engineer's task to consider the physical separation of residual reactant from the reactor exit mixture with the objective of returning it to the reactor for further processing in the system. Such recycling enables the poor yields obtained on a single passage through the reactor to be overcome. The physical separation of products from residual reactants would be required in any case in order to obtain the product in reasonably pure form. The physical separation processes are themselves not perfect and some reactant is almost certain to be lost during the separation. However, the recovery of reactant is such that a single passage system having for example a 25% conversion may be engineered to a recycled one having an overall yield in excess of 90%. The actual overall operating yield is decided on economic grounds, considering the larger reactor needed for the additional throughput, the increasing capital and operating costs of the separation process as its effectiveness increases, and the value of the raw material.

The thermodynamic analysis also enables an assessment to be made of the energy changes which occur during a chemical reaction. The energy changes are usually appreciable and may take the form of either heat release or heat absorption. If the temperature of the reaction mixture is to be maintained despite these changes heat must be transferred either from the system to a surrounding coolant such as water or to the system from a heat transfer medium such as condensing steam. The rate of heat release or absorption at different times in a batch reactor and at different positions in certain continuous reactors must be assessed, and provision made to meet the greatest expected heat transfer rate.

The use of thermodynamics to establish the equilibrium yield has been introduced above. Whether or not this yield is actually attainable in a reactor of acceptable size depends on the rate at which the chemical reaction proceeds, i.e. on the reaction kinetics. In fact the selection of operating conditions based solely on a thermodynamic analysis is valid only when the reaction rate is fast at the conditions selected thereby. For slow reactions, the operating conditions may have to be selected by a compromise between the thermodynamic yield and the possibility of approaching this yield in a reactor whose size is not impracticably large. The reaction kinetic data are thus a vital feature of chemical reactor studies and are basic to the determination of the size of reactor needed to achieve the specified throughput and conversion. When slow reactions are encountered it is usual to explore the possibility of increasing the rate by the introduction of a catalyst into the reaction mixture. An effective catalyst will usually alter the detailed reaction steps in converting a reactant to product but will achieve the overall result in a shorter time than is possible in its absence It is important to realise that the catalyst, while increasing the rate of approach to equilibrium, does not change the equilibrium position.

Chemical reactions may be carried out with the reactants and products as solids, liquids or gases. Many reactions take place with gases only or with a single liquid phase and are then described as homogeneous reactions. In these cases the reactants are able to contact one another readily and problems of mass transfer do not have to be considered. In contrast heterogeneous reactions may occur between a gas and a liquid, a solid and a liquid, a gas and a solid, or between components dissolved in immiscible liquid phases; in these cases the mass transfer processes involved in bringing the reactants together may exert a controlling influence over the complete reaction mechanism and must therefore be studied. Batch reactors are usually used for homogeneous liquid phase reactions particularly when the throughput is relatively small. This manner of operation would otherwise be avoided because of the labour charges incurred in repeated reactor loading and discharging. A batch reactor may be operated adiabatically if the vessel contents are able to absorb or supply the heat of reaction without their temperature changing to such an extent that the desirable reaction no longer takes place. Alternatively, if it is essential to maintain the temperature constant within carefully defined limits, a variable rate cooling or heating system capable of countering all thermal changes must be installed. In practice, since the initial rate of heat change is very high, the installed heat transfer equipment is able only to partly counter this in the initial stages of the reaction, but takes over control as the reaction rate subsequently decreases. These situations are illustrated in Fig. 1.1 for the case of reaction heat release and a cooling system.

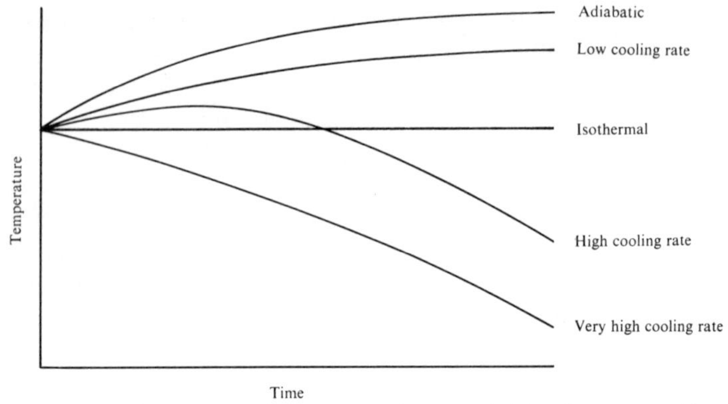

FIG. 1.1. Temperature time progress for a cooled batch reactor with exothermic reaction

Many continuous flow reactors are classifiable into either "well stirred" or "plug flow". The well stirred vessel usually processes homogenous liquid phase reactions while homogeneous gas phase reactions are normally effected in plug flow systems. The basic difference between the two types of reactor (Fig. 1.2) is that in the

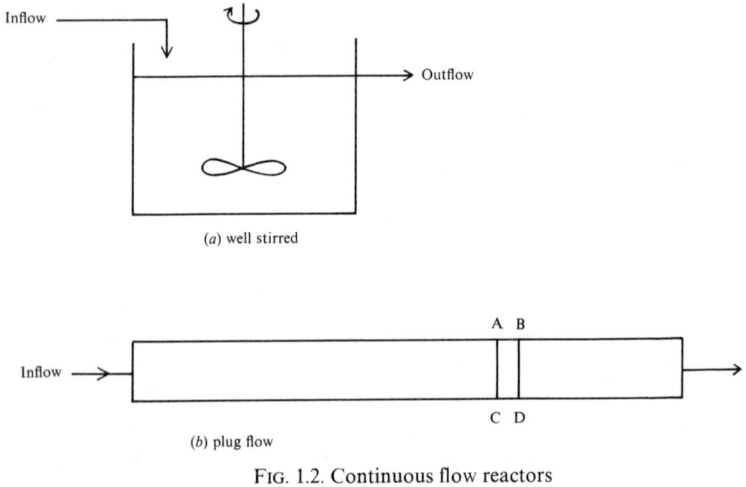

FIG. 1.2. Continuous flow reactors

well stirred system, incoming reactants are thoroughly mixed with the products very soon after entry to the vessel, whereas in the plug flow case the reaction proceeds as the reactants in a plug ABCD of

Fig. 1.2 progress along the reactor tube, but no interchange of material in the plug with material in other leading or following plugs occurs. These extreme methods of continuous operation have a profound effect on the performance of a reactor and comparisons between them have been the subject of much study. Practical continuous reactors deviate from these two idealisations, and techniques exist to allow for such deviations. By its very nature, a well stirred reactor must operate isothermally but the actual temperature at which the reaction takes place is determined by the reaction energy changes and the rate of heat transfer. A plug flow reactor operated adiabatically will have an increasing or decreasing temperature profile as the mixture progresses through the reactor; it is possible to have approximate isothermal conditions or a predetermined temperature profile by careful and detailed engineering of the heat transfer arrangements.

Continuous heterogeneous reactions in which a gas is to react with a liquid may be effected in an absorption column with countercurrent flow of the two phases In heterogenous gas—solid reactions. the solid may be preformed pellets and remain as a fixed bed with the gas passing through it in a manner deviating slightly from plug flow; if the solid is in the form of a powder with an appropriate particle size distribution, the bed of solid may become supported by the frictional forces of the rising gas stream. In this latter case of a heterogeneous reaction, gas bubbles rising through the supported solid exert a stirring action on the solid particles which may as a result exist as a well stirred bed; the reactor is then described as a fluidised bed reactor.

In the assessment of all types of reactor the fluid mechanics of reactant flow must be considered in order to specify the pumping or compression requirements for the fluids. Considerations of hydrodynamics and their influence on mass transfer rates are of particular importance for the heterogeneous systems.

The aim of the various chapters which follow is to elaborate on the different aspects of chemical reaction engineering outlined above, and as far as space allows, to consider the combination of them to evolve a coherent analysis of reacting systems.

2

CHEMICAL REACTOR THERMODYNAMICS

2.1 Introduction

As stated in the first chapter, thermodynamics enables an analysis of two very important aspects of chemical reactors to be made. Arising from the first law of thermodynamics is the energy balance which takes into account enthalpy of feed and product streams and the heats of reaction so that the duty of auxiliary heat transfer equipment can be specified. Application of the second law of thermodynamics enables the position of reaction equilibrium to be predicted for the available feedstock and any selected combination of temperature and pressure.

In the following sections these laws are considered both for closed systems and for open flow systems. Batch reactors are closed systems while continuous well stirred and tubular reactors are open systems. In fact it will be seen that in many instances conclusions made for open systems are valid for the same reaction in a closed system because the numerical differences are negligible.

2.2 Basic Application of the First Law

All chemical reactions proceed with the absorption or liberation of energy. Generally this manifests itself in the form of heat, and less frequently in the form of electricity or mechanical work. In so far as chemical reaction engineering is concerned the major form of energy change appears as heat absorbed or released during the course of the chemical reaction. However, an engine performing work can be a chemical reactor in which the chemical reaction need not necessarily be straightforward combustion. In addition it should be pointed out that a fuel cell is, in fact, a chemical reactor from which energy is released in the form of electricity. However, as stated above, the chemical reaction engineer is concerned with heat changes most frequently and therefore in the following text energy changes will be considered as heat changes. Thus, the amount of heat released or absorbed depends on the quantity and the type of chemical species reacting and on the products formed. For a particular chemical reaction, the net heat change depends on the temperature and the pressure of the reactants and products, and these factors must be considered in the thermodynamic assessment of the reaction process.

The first law of thermodynamics may be expressed as:

$$q - w = \sum E_2 - \sum E_1 \tag{2.1}$$

where q represents all heat energy input from the surroundings and w represents all work energy output to the surroundings. The total energy of the substances being considered changes from $\sum E_1$ to $\sum E_2$ as a result of chemical reaction and energy interchanges with the surroundings. The total energy includes internal energy, potential energy and kinetic energy.

2.2.1. Open flow reactor

The first law is applied to the open flow reactor (Fig. 2.1) and the corresponding equations for closed systems are then readily deduced in section 2.2.2 as simplifications of the flow equations.

Consider a mass "m" of reactants entering the system at position 1 and displacing an equal mass of products at position 2. The region between positions 1 and 2 is termed the control surface and an

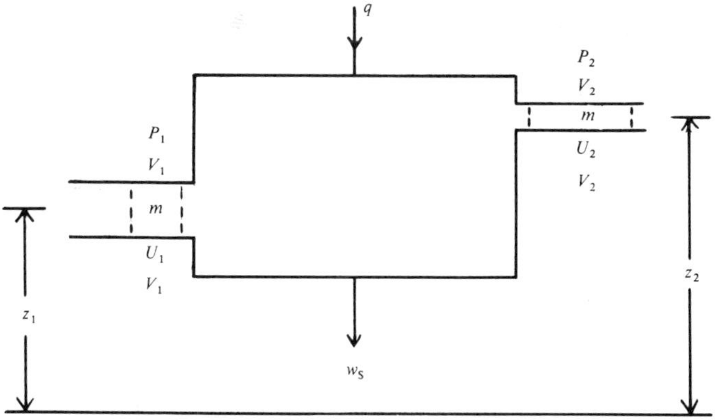

FIG. 2.1. Flow system energy balance

energy balance is made over this region in space. For these positions, let:

P_1 and P_2 be the pressures
V_1 and V_2 be the volumes of the mass "m"
U_1 and U_2 be the internal energies of the mass "m"
z_1 and z_2 the heights above a datum plane
v_1 and v_2 the velocities.

Hence

$$\sum E_1 = U_1 + m(gz_1 + \tfrac{1}{2}v_1^2) \tag{2.2}$$

and

$$\sum E_2 = U_2 + m(gz_2 + \tfrac{1}{2}v_2^2). \tag{2.3}$$

In general the work term "w" of equation (2.1) must include

(*i*) "Shaft work" w_s on any pump or turbine, etc., situated between the confines of the control surface.
(*ii*) Electrical work; this will be assumed zero in the present analysis.
(*iii*) Work done on the system to introduce the feed stream. This may be considered as the displacement of a volume V_1 against a fixed pressure P_1, i.e. the work is P_1V_1.
(*iv*) Work done on the surroundings by the exit stream which is similarly P_2V_2,

i.e.

$$w = w_s + P_2V_2 - P_1V_1. \tag{2.4}$$

Substitution of (2.2), (2.3) and (2.4) into (2.1) gives:

$$q - [w_s + P_2V_2 - P_1V_1] = U_2 + m(gz_2 + \tfrac{1}{2}v_2^2) - U_1 - m(gz_1 + \tfrac{1}{2}v_1^2) \tag{2.5}$$

or

$$q - w_s = (U_2 + P_2V_2) - (U_1 + P_1V_1) + mg(z_2 - z_1) + \tfrac{1}{2}m(v_2^2 - v_1^2). \tag{2.6}$$

Normally for a chemical reactor w_s would be zero and the difference between the potential energy and the kinetic energy at inlet and outlet would be numerically small compared with the remaining terms. Equation (2.6) would in these circumstances reduce to:

$$q = (U_2 + P_2V_2) - (U_1 + P_1V_1). \tag{2.7}$$

The group $(U + PV)$ is called enthalpy and is given the symbol "H"

i.e.

$$q = H_2 - H_1. \tag{2.8}$$

In position 1, the enthalpy refers to reactants at their feed temperature T_1. If the enthalpy is expressed relative to the elements at a datum temperature T_0, then:

$$H_1 = (\Delta H_f^0)_R + \int_{T_0}^{T_1} (\sum mC_p)_R \, dT \tag{2.9}$$

where $(\Delta H_f^0)_R$ is the enthalpy of formation of the reactants from their elements at T_0 and the integral gives the sensible enthalpy of the reactants above the datum temperature.

Similarly for products at temperature T_2:

$$H_2 = (\Delta H_f^0)_P + \int_{T_0}^{T_2} (\sum mC_p)_P \, dT \tag{2.10}$$

where $(\Delta H_f^0)_P$ is the enthalpy of formation of products from their elements at T_0 and the integral gives the sensible enthalpy of the products above T_0,

i.e.

$$q = (\Delta H_f^0)_P - (\Delta H_f^0)_R + \int_{T_0}^{T_2} (\sum mC_p)_P \, dT - \int_{T_0}^{T_1} (\sum mC_p)_R \, dT. \tag{2.11}$$

The terms $(\Delta H_f^0)_P$ and $(\Delta H_f^0)_R$ are both measured at the same datum temperature and their difference is the enthalpy change on reaction or heat of reaction at this temperature. Denoting the difference by, $\Delta H_{T_0}^0$:

$$q = \Delta H_{T_0}^0 + \int_{T_0}^{T_2} (\sum mC_p)_P \, dT - \int_{T_0}^{T_1} (\sum mC_p)_R \, dT. \tag{2.12}$$

If T_1 and T_2 are each T_0, the integrals in equation (2.12) are both zero and:

$$q = \Delta H_{T_0}^0 \equiv (\Delta H_f^0)_P - (\Delta H_f^0)_R. \tag{2.13}$$

This equation defines the heat of reaction at any chosen temperature T_0 as the heat "q" which has to be transferred to or from the surroundings in order to effect the isothermal conversion at T_0 of reactants to products.

Thus for an isothermal flow reactor, if the enthalpy of the products is less than that of the reactants, heat will have to be transferred to the surroundings, i.e. the difference $\Delta H^0_{T_0}$ will be negative, q will be negative and the reaction is described as exothermic. Hence an exothermic reaction would be written:

$$H_2 + \tfrac{1}{2}O_2 = H_2O(g) \quad \Delta H^0_{298} = -57800 \text{ cal(g mole)}^{-1}. \tag{2.14}$$

The suffix 298 is introduced to show the temperature at which this $\Delta H^0_{T_0}$ is measured, and the value -57800 cal refers to the chemical reaction as written with each reactant and product expressed in g moles.

Conversely if the product enthalpy exceeds that of the reactants, heat will have to be transferred from the surroundings, $\Delta H^0_{T_0}$ and q will be positive and the reaction is described as endothermic. For example, the reverse of the reaction of equation (2.14) would be written:

$$H_2O(g) = H_2 + \tfrac{1}{2}O_2 \quad \Delta H^0_{298} = +57800 \text{ cal(g mole)}^{-1}. \tag{2.15}$$

2.2.2 Closed reactor

For a closed system equation (2.6) would not have the PV, v^2 or z terms; the w_s term would not normally be relevant although electrical work might have to be included. In the absence of any form of work, the closed system equation corresponding to equation (2.6) is:

$$q = U_2 - U_1 \tag{2.16}$$

and the equivalent of (2.12) is:

$$q = (\Delta U^0_f)_P - (\Delta U^0_f)_R + \int_{T_0}^{T_2} (\textstyle\sum mC_v)_P \, dT - \int_{T_0}^{T_1} (\textstyle\sum mC_v)_R \, dT \tag{2.17}$$

where $(\Delta U^0_f)_R$ is the internal energy of reactants relative to the elements at T_0; similarly $(\Delta H^0_f)_P$ for products. The integrals give the internal energies of the reactants and products above the datum.

The heat of reaction in a closed system is defined with T_2 and T_1 each T_0,

i.e.

$$q = (\Delta U_f^0)_P - (\Delta U_f^0)_R = \Delta U_{T_0}^0. \tag{2.18}$$

Again there is subdivision into exothermic and endothermic reactions.

2.2.3 Estimation of heat effects

The previous section has shown the distinction between the internal energy and enthalpy changes in a reaction. The relation between these two changes will now be considered. Thus experimental observations are generally made in a bomb calorimeter which is a closed system, and therefore yield values of ΔU. However the data published in the literature are normally ΔH values deduced from experimentally determined values of ΔU. Hence the normally tabulated ΔH values can be applied directly to flow processes whereas if a closed system is to be considered the tabulated values must be adjusted.

The definition of enthalpy

$$H = U + PV \tag{2.19}$$

considered at fixed temperature and pressure gives:

$$\Delta H_{p,T} = \Delta U_{p,T} + P(\Delta V)_T. \tag{2.20}$$

While $\Delta H_{p,T}$ is the heat of reaction at constant pressure, $\Delta U_{p,T}$ is not the heat of reaction at constant volume. However, $\Delta U_{p,T}$ can be shown to be equal to $\Delta U_{v,T}$ if either:

(*i*) a gaseous system, approaching ideal behaviour is under consideration, because then internal energy is a function only of pressure;

(*ii*) extremely high pressures are not developed in the bomb calorimeter experiments.

For normal practical purposes it is assumed that the two heats of reaction of interest are connected by the relationship:

$$\Delta H_{p,T} = \Delta U_{v,T} + P(\Delta V)_T. \tag{2.21}$$

For condensed phases, ΔV will be zero. Also for gaseous reactions without change in the number of moles ΔV will be zero. In these cases $\Delta H_{p,T}$ and $\Delta U_{v,T}$ are equal within the restrictions of (*i*) and (*ii*) above. Thus since batch reactions are almost invariably concerned with condensed phase systems it is permissible to use the tabulated ΔH data to obtain a heat of reaction without any adjustment.

However for gaseous reactions or heterogeneous reactions involving gases:

$$P(\Delta V)_T = \Delta nRT. \tag{2.22}$$

For example the oxidation of benzene can be effected entirely in the gas phase or with the reactant benzene and product water in the liquid phase. In these two cases the value of Δn is given below:

$$C_6H_6(g) + 7\tfrac{1}{2}O_2 = 6CO_2 + 3H_2O(g) \tag{2.23}$$

$$\Delta n = 9 - 8\tfrac{1}{2} = +\tfrac{1}{2} \tag{2.24}$$

$$C_6H_6(l) + 7\tfrac{1}{2}O_2 = 6CO_2 + 3H_2O(l) \tag{2.25}$$

$$\Delta n = 6 - 7\tfrac{1}{2} = -1\tfrac{1}{2}. \tag{2.26}$$

Hence in the heterogeneous case, the connection between the two heats of reaction at 298°K in cal(g mole)$^{-1}$ is:

$$\Delta H_{p,T} = \Delta U_{v,T} - 1{\cdot}5 \times 1{\cdot}986 \times 298 = \Delta U_{v,T} - 886 \tag{2.27}$$

and the correction term 886 cal is small compared with the value of $\Delta H_{p,T} = -839\,580$ cal obtained from the thermodynamic tables.

To summarise this section, it is important to emphasise that the tabulated heat of formation data found in the literature will have been developed from closed bomb calorimeter experiments. However, the reverse correction in order to apply published data to the study of batch reactions is unlikely to be necessary since such reactors almost invariably operate with a condensed phase.

2.3 Estimation of Heat of Reaction

2.3.1 From standard heats of formation

Standard heats of reaction may be readily estimated using equation (2.13) if the standard heats of formation ΔH_f^0 of the various compounds taking part in the reaction are available. Then

$$\Delta H^0 = (\Delta H_f^0)_P - (\Delta H_f^0)_R. \tag{2.13}$$

TABLE 2.1

Thermodynamic Data for Elements and Inorganic Compounds

	ΔH_f^0	ΔH_v	ΔH_c	ΔG_f^0	S_{298}^0	α	β	γ	
Air						6·557	1·477	−0·2148	
Ammonia (g)	−11·04	5·581		−3·976	46·01	6·086	8·812	−1·506	NH_3
Carbon (Diamond)	+0·4532				0·583				C
Carbon (Graphite)	0		−94·052	0	1·361				C
Carbon Dioxide	−94·052			−94·26	51·06	6·393	10·100	−3·405	CO_2
Carbon Disulphide	+27·55		−263·52						CS_2
Carbon Monoxide	−26·416		−67·636	−32·81	47·30	6·342	1·836	−0·280	CO
Chlorine	0	4·878		0	53·29	7·575	2·424	−0·965	Cl_2
Hydrogen (g)	0	0·216	−68·317	0	31·21	6·947	−0·200	0·481	H_2
Hydrogen Chloride (g)	−22·06	3·860		−22·77	44·62	6·732	0·4325	0·370	HCl
Hydrogen Fluoride (g)	−64·2	1·800			4·47				
Hydrogen Sulphide	−4·815	4·463		−7·89	49·15	6·385	5·704	−1·210	H_2S
Nitric Acid (1)	−41·404			−19·10					HNO_3
Nitric Oxide	+21·60	3·292		+20·719	50·34	7·020	−0·370	2·546	NO
Nitrogen	0	1·333		0	45·77	6·449	1·412	−0·0807	N_2
Nitrogen Dioxide	+8·041			+12·390	57·47				NO_2
Nitrous Oxide	+19·49	3·956		+24·76	52·58	6·529	10·515	−3·571	N_2O
Oxygen	0	1·630		0	49·00	6·0954	3·2533	−1·0171	O_2
Sulphur Dioxide	−70·96	5·950		−71·79	59·40	7·116	9·512	3·511	SO_2
Sulphur Trioxide (g)	−94·45	9·990		−88·52	61·24	6·077	23·537	−0·687	SO_3
Sulphur Trioxide (1)	−104·80				31·7				
Water (g)	−57·798	9·717		−54·636	45·106	7·219	2·374	0·267	H_2O
Water Trioxide (1)	−68·317			−56·690	16·716				

ΔH_f^0 Standard heat of formation at 298°K kcal (g mole)$^{-1}$. ΔH_v Enthalpy change on vaporisation at the boiling point kcal (g mole)$^{-1}$. ΔH_c Heat of combustion kcal (g mole)$^{-1}$. ΔG_f^0 Standard free energy of formation at 298°K kcal (g mole)$^{-1}$. S_{298}^0 Standard entropy at 298°K relative to a zero value at 0°K cal (g mole)$^{-1}$ deg K^{-1} α, β and γ constants for specific heat in the form:

$$C_p = \alpha + \beta \times 10^{-3}T + \gamma \times 10^{-6}T^2 \text{ cal (g mole)}^{-1} \text{ deg K}^{-1}$$

with T measured in °K.

	ΔH_f^0	ΔH_v	ΔH_c	ΔG_f^0	S_{298}^0	α	β	γ	
Acetaldehyde (g)	−39·76	6·500	−284·98	−31·96		7·422	29·029	−8·742	CH_3CHO
Acetic acid (1)	−116·4	9·526	−208·34	−93·80	38·2				CH_3COOH
Acetone (1)	−59·32	7·100	−427·79		47·9	5·371	49·227	−15·182	$(CH_3)_2CO$
Acetylene (g)	+54·194	4·270	−310·615	+50·00	48·0	7·331	12·622	−3·889	C_2H_2
Benzene (g)	+19·82	7·350	789·08	30·989	64·34	−0·409	77·621	−26·429	C_6H_6
Benzene (l)	+11·72		780·98	29·756	41·30				C_6H_6
n-butane (g)	−30·15	5·350	−687·982	−4·10	74·12	4·453	72·270	−22·214	$n\text{-}C_4H_{10}$
i-butane (g)		5·089	−686·342			3·332	75·214	−23·734	$i\text{-}C_4H_{10}$
Carbon tetrachloride	−25·50	7·170	−92·01		73·95				CCl_4
Diethyl Ether		6·220	−652·59			−28·43	338·7	−593	$(C_2H_5)_2O$
Ethane	−20·236	3·517	−372·82	−7·86	54·85	2·247	38·201	−11·049	C_2H_6
Ethylene	+12·496	3·237	−337·23	+16·282	52·45	2·830	28·601	−8·726	C_2H_4
Ethyl Alcohol (g)	−56·240	9·220	−326·70	−40·30	67·4	6·990	39·741	−11·926	C_2H_5OH
Ethyl Alcohol (l)	−66·356			−41·77	38·4				C_2H_5OH
Ethylene oxide (g)	−12·19	6·101		−2·79		−1·12	4·925	−23·89	$(CH_2)_2O$
Formaldehyde (g)	−27·7		−134·67		52·26	4·498	13·953	−3·73	HCHO
Formic acid (1)	−97·8	5·510	−64·57		30·82				HCOOH
Methane	−17·889	1·955	−212·798	−12·14	44·50	3·381	18·044	−4·30	CH_4
Methyl Alcohol (g)	−48·100	8·430	−173·65	−38·700	56·80	4·398	24·274	−6·855	CH_3OH
Methyl Alcohol (l)	−57·036			−39·750	30·30				CH_3OH
Methylamine	−6·7								CH_3NH_2
Methyl Chloride	−19·58	5·150	−182·81			3·563	22·998	7·571	CH_3Cl
Propane	−24·82	4·487	−530·605	−5·614	64·51	2·410	57·195	−17·533	C_3H_8
Propylene	+4·879	4·405	−491·987	+14·99	63·8	2·253	45·116	−13·740	C_3H_6
Propyne			−463·109		59·30	6·334	30·990	−9·457	C_3H_4
Toluene (g)	+11·950	8·000		+29·228	76·42	0·576	93·493	−31·227	$C_6H_5CH_3$
Toluene (l)	+2·870			+27·282	52·48				$C_6H_5CH_3$

ΔH_f^0 Standard heat of formation at 298°K kcal (g mole)$^{-1}$. ΔH_v Enthalpy change on vaporisation at the boiling point kcal (g mole)$^{-1}$. ΔH_c Heat of combustion kcal (g mole)$^{-1}$. ΔG_f^0 Standard free energy of formation at 298°K kcal (g mole)$^{-1}$. S_{298}^0 Standard entropy at 298°K relative to a zero value at 0°K cal (g mole)$^{-1}$ deg K^{-1} α, β and γ constants for the specific heat in the form:

$$C_p = \alpha + \beta \times 10^{-3} T + \gamma \times 10^{-6} T^2 \text{ cal (g mole)}^{-1} \text{ deg K}^{-1}$$

with T measured in °K.

The necessary standard heats of formation can be obtained from tables of thermodynamic data such as those found in Perry (1950) and Weast (1966). These standard heats of formation are special cases of formation reactions of compounds from their elements at 298°K (or some other specified temperature) with the elements in their standard states at 1 atm. pressure.

Selected thermodynamic data for elements and inorganic compounds are reproduced in Table 2.1 and for organic compounds in Table 2.2.

Example 1. Calculation of the standard heat of reaction at 298°K for the hydration of acetylene:

$$C_2H_2 + H_2O(l) \rightarrow CH_3CHO(g) \quad \text{(I)}$$

The required tabulated values of ΔH_f^0 from Tables 2.1 and 2.2 are:

$$C_2H_2 + 54194 \text{ cal(g mole)}^{-1}$$
$$H_2O(l) - 68317 \text{ cal(g mole)}^{-1}$$
$$CH_3CHO(g) - 39760 \text{ cal(g mole)}^{-1}$$

and the heat of reaction is deduced as:

$$\Delta H_{298}^0 = (-39760) - (+54194 - 68317)$$
$$= -35637 \text{ cal(g mole)}^{-1}. \quad \text{(II)}$$

2.3.2 From heats of combustion

Direct measurement of a heat of formation is not always possible. For instance a mixture of hydrogen, carbon and oxygen in a bomb calorimeter would yield many other organic compounds than acetaldehyde. The heats of formation of such compounds are therefore deduced from the results of combustion experiments and frequently the heat of combustion only is tabulated. Thus in the case of acetaldehyde the available heat of combustion is -284980 cal(g mole)$^{-1}$.

Example 2. This data is used to deduce the heat of formation of acetaldehyde

$$CH_3CHO(g) + 2\tfrac{1}{2}O_2 \rightarrow 2CO_2 + 2H_2O(l)\ \Delta H_{298}^0 = -284980 \quad \text{(I)}$$

and the following data for the products is also required:

$$(\Delta H_f^0)_{298} CO_2 = -94052$$
$$(\Delta H_f^0)\ H_2O(l) = -68317.$$

Hence if ΔH_f^0 is the required heat of formation of acetaldehyde:

$$(2 \times -68317 + 2 \times -94052) - (\Delta H_f^0) = -284980 \qquad \text{(II)}$$

from which

$$\Delta H_f^0 = -39758. \qquad \text{(III)}$$

A value of heat of formation obtained in this way could then be utilised as in section 2.3.1 to deduce the heat of reaction for a reaction involving acetaldehyde.

2.3.3 Effect of temperature on the heat of reaction

Since most reactions are effected commercially at elevated temperatures it may be convenient to estimate the heat of reaction at the appropriate temperature. Having obtained such a heat of reaction it is important to realise that the value is relevant to reactants and products all at the stated temperature. In practice reactants may enter at temperature T_1, products leave at temperature T_2 and reaction take place at all temperatures between T_1 and T_2; it would not then be relevant to evaluate the heat of reaction at any single elevated temperature.

When the heat of reaction is required at a single elevated temperature it is calculated from Kirchhoff's Law which is a particular application of the independence on route of enthalpy change (or internal energy change). This fact is expressed by considering reactants

(*i*) to be converted to products at T_0 and the products then heated to the elevated temperature T and
(*ii*) to be heated to T and then reacted at T.

The enthalpy changes involved in these two routes ABC and ADC (Fig. 2.2) from reactants at T_0 to products at T are then equated to give:

$$\Delta H_{T_0}^0 + \int_{T_0}^{T} \sum (mC_p)_{\mathrm{P}}\, dT = \int_{T_0}^{T} \sum (mC_p)_{\mathrm{R}}\, dT + \Delta H_T^0 \qquad (2.28)$$

$$\Delta H_T^0 = \Delta H_{T_0}^0 + \int_{T_0}^{T} \left[\sum (mC_p)_P - \sum (mC_p)_R\right] dT \qquad (2.29)$$

where m is the number of modes of reactant or product.

C_p is the temperature dependent molar heat capacity of reactant

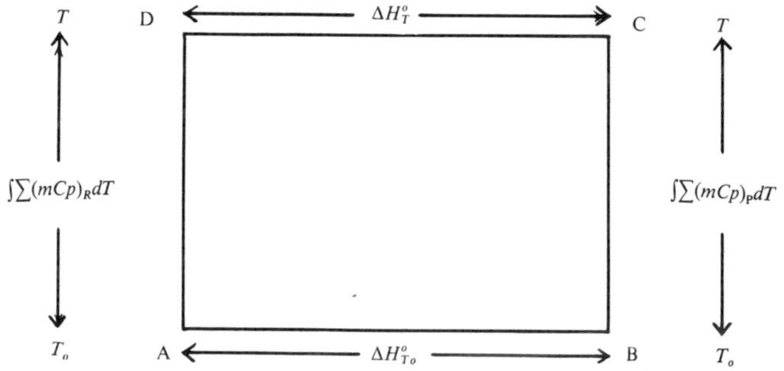

FIG. 2.2. Variation of reaction enthalpy change with temperature

or product and $\Delta H^0{}_{T_0}$ is the known heat of reaction at temperature T_0.

Since the heat capacity of each component is frequently expressed in the form:

$$C_p = \alpha + \beta T + \gamma T^2$$

the term $\Sigma(mC_p)_P - \Sigma(mC_p)_R$
may be developed to:

$$\Sigma[(m\alpha)_P - (m\alpha)_R] + T\Sigma[(m\beta)_P - (m\beta)_R] + T^2\Sigma[(m\gamma)_P - (m\gamma)_R]$$

in which the three summations are abbreviated respectively to $\Delta\alpha$, $\Delta\beta$, and $\Delta\gamma$ to give the following form for integration:

$$\Delta H^0_T = \Delta H^0_{T_0} + \int_{T_0}^{T} (\Delta\alpha + T\Delta\beta + T^2\Delta\gamma)\,\mathrm{d}T \qquad (2.30)$$

i.e.

$$\Delta H^0_T = \Delta H^0_{T_0} + \Delta\alpha(T - T_0) + \tfrac{1}{2}\Delta\beta\,(T^2 - T_0^2) + \tfrac{1}{3}\Delta\gamma\,(T^3 - T_0^3) \qquad (2.31)$$

which is the form of equation frequently employed in the present context. For use in equilibrium calculations it is normally developed by re-arrangement as follows:

$$\Delta H^0_T = [\Delta H^0_{T_0} - \Delta\alpha T_0 - \tfrac{1}{2}\Delta\beta T_0^2 - \tfrac{1}{3}\Delta\gamma T_0^3] + \Delta\alpha T + \tfrac{1}{2}\Delta\beta T^2 + \tfrac{1}{3}\Delta\gamma T^3. \qquad (2.32)$$

The group:

$$\Delta H^0_{T_0} - \Delta\alpha T_0 - \tfrac{1}{2}\Delta\beta\, T_0^2 - \tfrac{1}{3}\Delta\gamma\, T_0^3$$

is a constant often denoted as ΔH_0^0; it is the hypothetical value of ΔH^0 at 0°K if the specific heat data were valid to such temperatures. Hence equation (2.32) becomes:

$$\Delta H_T^0 = \Delta H_0^0 + \Delta\alpha T + \tfrac{1}{2}\Delta\beta\, T^2 + \tfrac{1}{3}\Delta\gamma\, T^3 \qquad (2.33)$$

and is used in this form in section 2.5.1.

Equation (2.29) may be simplified in two ways. First, if specific heat is not a significant function of temperature, the ΣmC_p terms may be taken outside the integration sign and the following result obtained:

$$\Delta H_T^0 = \Delta H_{T_0}^0 + [\Sigma(mC_p)_P - \Sigma(mC_p)_R]\,(T - T_0). \qquad (2.34)$$

Secondly, mean specific heats "$\bar{C}_p$" for individual components are frequently tabulated and since these are calculated from:

$$\bar{C}_p = \frac{\int_{T_0}^{T} C_p dT}{T - T_0} \qquad (2.35)$$

the integrals in (2.29) may be replaced to give:

$$\Delta H_T^0 = \Delta H_{T_0}^0 + [\Sigma(m\bar{C}_p)_P - \Sigma(m\bar{C}_p)_R]\,(T - T_0) \qquad (2.36)$$

where the summations require $\bar{C}_p$ values for the various reactants and products relevant to the temperature range T_0 to T.

Example 3. The application of these equations is illustrated in this example by considering the oxidation of sulphur dioxide. A general expression is developed for the heat of this reaction as a function of temperature and the actual value at a series of temperatures is then tabulated to show the extent of the variation.

Thermodynamic tables contain the data:

Standard heat of formation $(\Delta H_f^0)_{298}$:

SO_2: -70960 cal (g mole)$^{-1}$
SO_3: -94450 cal (g mole)$^{-1}$
O_2: zero

Specific heat as a function of T°K:

[cal(g mole)$^{-1}$ degK^{-1}]

$SO_2: 7{\cdot}116 + 9{\cdot}512 \times 10^{-3}\, T + 3{\cdot}511 \times 10^{-6} T^2$
$SO_3: 6{\cdot}077 + 23{\cdot}537 \times 10^{-3}\, T - 0{\cdot}687 \times 10^{-6} T^2$
$O_2: 6{\cdot}148 + 3{\cdot}102 \times 10^{-3}\, T - 0{\cdot}923 \times 10^{-6} T^2$

From the standard heats of formation, the heat of the reaction:

$$SO_2 + \tfrac{1}{2}O_2 \rightarrow SO_3 \tag{I}$$

is:

$$\Delta H^0_{298} = (-94450) - (-70960) = -23490. \tag{II}$$

From the specific heat data:

$$\Delta\alpha = 6{\cdot}077 - (7{\cdot}116 + \tfrac{1}{2} \times 6{\cdot}148) = -4{\cdot}113 \tag{III}$$

$$\Delta\beta = [23{\cdot}537 - (9{\cdot}512 + \tfrac{1}{2} \times 3{\cdot}102)] \times 10^{-3}$$
$$= 12{\cdot}474 \times 10^{-3} \tag{IV}$$

$$\Delta\gamma = [-0{\cdot}687 - (3{\cdot}511 - \tfrac{1}{2} \times 0{\cdot}923)] \times 10^{-6}$$
$$= -3{\cdot}7365 \times 10^{-6} \tag{V}$$

Equation (2.31) is now used with $\Delta H^0_{298} = -23490$ (i.e. $T_0 = 298$) to give:

$$\Delta H^0_T = -23490 - 4{\cdot}113(T - 298) + \frac{12{\cdot}474 \times 10^{-3}}{2}(T^2 - 298^2)$$
$$- \frac{3{\cdot}7365 \times 10^{-6}}{3}(T^3 - 298^3) \tag{VI}$$

or

$$\Delta H^0_T = -22785 - 4{\cdot}113T + 6{\cdot}237 \times 10^{-3}\, T^2$$
$$- 1{\cdot}245 \times 10^{-6}\, T^3. \tag{VII}$$

The term -22785 in this result is to be identified with ΔH^0_0 of equation (2.33).

Equation (VII) applied to a series of temperatures shows a variation of about 6% over a 500°K range:

T°K	ΔH^0_T
473	−23467
573	−23328
673	−23108
773	−22813
873	−22451
973	−22029

A value of -23467 cal (g mole)$^{-1}$ for instance would be relevant to an energy balance in which sulphur dioxide and oxygen each at 473°K entered a reactor to give product sulphur trioxide also at

473°K. Two energy balances are considered below in which the reactants and products are not at the same temperature; in the first of these, the necessary heat transfer to achieve a specified product temperature is deduced while in the second one, an expression for the adiabatic reaction temperature is developed and used in section 2.6 in conjunction with an equilibrium expression as a function of temperature.

Example 4. A stream of gas of composition: SO_2, 12%; O_2, 9%; and N_2, 79% enters a contact converter at 400°C and is required to leave in equilibrium at 500°C. The equilibrium fractional conversion at this temperature is 0·911 and so based on 100 moles of feed and hence 10·94 moles of SO_2 having reacted the exit gas is:

SO_2	1·06 moles
O_2	3·53 moles
SO_3	10·94 moles
N_2	79·0 moles

The specific heat data of Example 3 are used together with that for nitrogen:

$$C_p = 6{\cdot}449 + 1{\cdot}412 \times 10^{-3}\,T - 0{\cdot}0807 \times 10^{-6}\,T^2 \text{ cal (g mole)}^{-1}.$$

Hence using equation (2.12) with a datum temperature of 298°K, the various terms are:

$$\int_{298}^{673} (\Sigma m C_p)_{\mathrm{R}}\, dT = 12\left[7{\cdot}116(673-298) + \frac{9{\cdot}512}{2}(673^2 - 298^2) \times 10^{-3} + \frac{3{\cdot}511}{3}(673^3 - 298^3) \times 10^{-6}\right]$$

$$+\ 9\left[6{\cdot}148 \times 375 + \frac{3{\cdot}102}{2} \times 364{\cdot}5 - \frac{0{\cdot}923}{3} \times 278{\cdot}5\right]$$

$$+\ 79\left[6{\cdot}449 \times 375 + \frac{1{\cdot}412}{2} \times 364{\cdot}5 - \frac{0{\cdot}0807}{3} \times 278{\cdot}5\right]$$

$$= 56800 + 25050 + 211000 = 292850$$

$$\int_{298}^{773} (\Sigma m C_p)_{\mathrm{P}}\, dT = 1{\cdot}06[7{\cdot}116(773-298) + 4{\cdot}756(773^2 - 298^2) \times 10^{-3} + 1{\cdot}170(773^3 - 298^3) \times 10^{-6}]$$

$$+ 3{\cdot}53\,[6{\cdot}148 \times 475 + 1{\cdot}551 \times 509 - 0{\cdot}308 \times 437{\cdot}5]$$
$$+ 79\,[6{\cdot}449 \times 475 + 0{\cdot}706 \times 509 - 0{\cdot}0269 \times 437{\cdot}5]$$
$$+ 10{\cdot}94\,[6{\cdot}077 \times 475 + 11{\cdot}768 \times 509 - 0{\cdot}229 \times 437{\cdot}5]$$
$$= 6700 + 12620 + 269\,500 + 96\,100 = 384920.$$

For the number of SO_2 moles reacted:

$$\Delta H^0_{298} = -23490 \times 10{\cdot}94 = -257000.$$

Hence on applying equation (2.12):

$$q = -\,257\,000 + 384920 - 292850$$

and $$q = -164930,$$

i.e. 164930 cal have to be removed from the reactor in order that 500°C is not exceeded

Example 5. The above type of converter would, in practice, be operated adiabatically and the method of calculation of the exit conditions for this manner of operation is now considered. Applying equation (2.12), q is zero and the feed enthalpy has the value 292 850 cal as in Example 4. The enthalpy of the products must be expressed in terms of the unknown outlet temperature T and conversion f. Since the gas at exit consists of:

SO_2	$12(1 - f)$ moles
O_2	$(9 - 0{\cdot}5f)$ moles
N_2	79 moles
SO_3	$12f$ moles

the enthalpy of the products 'H_P' is:

$$H_P = \int_{298}^{T} (\Sigma mC_p)_P\, dT$$
$$= 12(1 - f)\,[7{\cdot}116(T - 298) + 4{\cdot}756(T^2 - 298^2) \times 10^{-3} + 1{\cdot}170(T^3 - 298^3) \times 10^{-6}]$$
$$+ (9 - 0{\cdot}5f)\,[6{\cdot}148(T - 298) + 1{\cdot}551(T^2 - 298^2) - 0{\cdot}308(T^3 - 298^3) \times 10^{-6}]$$
$$+ 79\,[6{\cdot}449(T - 298) + 0{\cdot}706(T^2 - 298^2) - 0{\cdot}0269(T^3 - 298^3) \times 10^{-6}]$$
$$+ 12f\,[6{\cdot}077(T - 298) + 11{\cdot}768(T^2 - 298^2) - 0{\cdot}229(T^3 - 298^3) \times 10^{-6}] \qquad \text{(I)}$$

and

$$\Delta H^0_{298} = -23490 \times 12f. \tag{II}$$

These terms substituted into equation (2.12) give the following equation in T and f:

$$0 = -23490 \times 12f + H_P - 292850 \tag{III}$$

wherein H_P is the function of both T and f given by equation (I).

In order to solve equation (III), a further relationship between f and T is needed and is developed from the variation of reaction equilibrium with temperature in section 2.6.

2.3.4 Effect of pressure on heat of reaction

The effect of pressure is generally small but the tendency nowadays to carry out gaseous synthesis reactions at high pressure can lead to situations where the pressure effect is important. The change in enthalpy of a component as the pressure is changed from P_1 to P_2 is shown in many thermodynamics texts as:

$$H_2 - H_1 = \int_{P_1}^{P_2} \left[V - T\left(\frac{\partial V}{\partial T}\right)_P \right] dP. \tag{2.37}$$

The change is negligibly small for the liquid and solid states and also for gases removed from the critical point. The integral can be evaluated analytically if an equation of state is known but it is more usual to make a graphical integration from plots of P–V–T data. In fact the results of such integrations as a function of reduced temperature and pressure have been plotted as in Fig. 2.3 and are used in the following example.

Example 6. A gaseous mixture containing 2 parts of hydrogen to 1 part of carbon monoxide by volume is fed at 350°C and 150 atm and reacts in a continuous reactor. Under the prevailing conditions 10% of the carbon monoxide is converted to methanol and the reaction mixture leaves the tubes at 450°C. The reaction occurs inside tubes surrounded by water so that the reactor is an integral part of a waste heat boiler. Use the following data to estimate the amount of steam that will be generated for each mole of methanol formed.

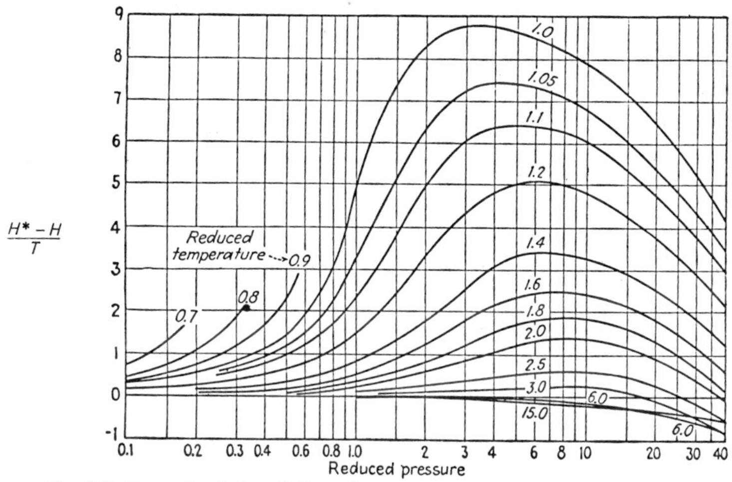

FIG. 2.3. Generalised plot of effect of temperature and pressure on enthalpy
Enthalpy reduction is in Btu lb moles^{-1} with T in ^{0}Rankine

Enthalpy change between cold water feed and steam raised:

500 CHU lb^{-1} steam

Heat of reaction at 1 atm and 25°C: – 21 500 CHU lb mole^{-1} CH_3OH

Mean specific heats:

CO	7·2 CHU lb mole^{-1}degC^{-1}
H_2	7·0 CHU lb mole^{-1}degC^{-1}
CH_3OH	15·1 CHU lb mole^{-1}degC^{-1}

Critical Properties:

	T_C°K	P_c atm
CO	133	34·5
H_2	41·3	20·8
CH_3OH	513·2	78·7

The Reduced Properties are then:

	P_R	T_R at inlet	T_R at outlet
CO	4·35	4·7	5·45
H_2	7·21	15·2	17·55
CH_3OH	1·91	—	1·41

Consider 1 lb mole of CO fed and take as the datum for enthalpy 25°C and 1 atm pressure.
Enthalpy of feed:

1 lb mole CO at 1 atm: $1 \times 7{\cdot}2\,(350 - 25) = 2340$ CHU
2 lb mole H_2 at 1 atm: $2 \times 7{\cdot}0 \times 325 = 4550$ CHU

Pressure correction for CO using $P_R = 4{\cdot}35$ and $T_R = 4{\cdot}7$ is seen from Fig. 2.3 to be very small and no correction is applied. For the H_2, the figure gives:

$$\frac{H^* - H}{T} = -0{\cdot}15 \text{ Btu lb mole}^{-1} \text{ deg R}^{-1}$$

$$= -\frac{0{\cdot}15}{1{\cdot}8} \text{ CHU lb mole}^{-1} \text{ deg R}^{-1}$$

and the enthalpy correction is:

$$-2(1{\cdot}8 \times 623) \times \left(-\frac{0{\cdot}15}{1{\cdot}8}\right) = +190 \text{ CHU.}$$

The total enthalpy of the feed is then:

$$2340 + (4550 + 190) = 7080 \text{ CHU.}$$

Enthalpy of products:

0·9 moles CO at 1 atm: $0{\cdot}9 \times 7{\cdot}2\,(450 - 25) = 2754$ CHU.
1·8 moles H_2 at 1 atm: $1{\cdot}8 \times 7{\cdot}0 \times 425 = 5355$ CHU.
0·1 moles CH_3OH at 1 atm: $0{\cdot}1 \times 15{\cdot}1 \times 425 = 642$ CHU.

Correction for pressure for CO: zero
Estimated pressure correction for H_2 (outside range of figure): 210 CHU. Fig. 2.3 gives a correction of 1·8 for $P_R = 1{\cdot}91$ and $T_R = 1{\cdot}41$ and so the correction for methanol is:

$$-0{\cdot}1\,(1{\cdot}8 \times 723) \times \frac{1{\cdot}8}{1{\cdot}8} = -130 \text{ CHU.}$$

The total enthalpy of the products is:

$$2754 + (5355 + 210) + (642 - 130) = 8831 \text{ CHU}$$

Reaction heat release:

$$0{\cdot}1 \times 21\,500 = 2150 \text{ CHU.}$$

For M lb steam generated, the heat transfer to the boiler is: 500M CHU. Hence an energy balance gives:

$$7080 + 2150 = 8831 + 500M$$

and $M = 0{\cdot}798$ lb/lb mole CO fed
For 1 lb mole methanol produced, the steam raised is therefore 7·98 lb.

If the pressure corrections are not taken into account the energy balance is:

$$6890 + 2150 = 8751 + 500M$$

and $M = 0{\cdot}578$.
This example shows that while the effect of pressure is small on the actual values of enthalpy, the quantities involved cannot be neglected. Therefore in the case of high pressure gas reactions, the effects of pressure should always be investigated.

2.4 Application of the Second Law of Thermodynamics

When the chemical reaction takes place, the products of reaction themselves react to regenerate the original reactants, and with all of the reactants and products present both the forward and reverse reactions proceed simultaneoulsy. The situation is written:

$$pA + qB \rightleftharpoons rC + sD. \tag{2.38}$$

Since the rate of the chemical reaction depends on the effective concentrations of the reacting species, a situation will develop in which the rates of the forward and backward reactions will just balance one another, and a dynamic chemical equilibrium is established. While such a state exists for all chemical reactions, in many the ultimate concentrations of either the reactants or products is so small that their equilibrium concentration cannot be detected experimentally. When this occurs the effective rate of reaction depends on only one of the reactions and is described as irreversible for the purpose of reaction kinetics. The reaction would then be written for instance as:

$$pA + qB \rightarrow rC + sD \tag{2.39}$$

or

$$rC + sD \rightarrow pA + qB \tag{2.40}$$

depending upon which process predominates. Thus in the extremes the equilibrium mixture might contain essentially C and D (equation 2.39) or A and B (equation 2.40).

The second law of thermodynamics is concerned with the prediction of the relative concentrations of A, B, C and D when any given mixture of these is allowed to come to equilibrium at selected combinations of temperature and pressure. The criteria which have been derived to predict the equilibrium position in a mixture at a given temperature are that:

(*i*) For a constant volume system of reactants, the Helmholtz Free Energy "A" shall be at a minimum, i.e. its rate of change with respect to composition shall be zero:

$$-d(U - TS)_{T,V} = 0 \tag{2.41}$$

or

$$dA_{T,V} = 0. \tag{2.42}$$

(*ii*) For a constant pressure system, the Gibbs Free Energy "G" shall be at a minimum, i.e. its rate of change with respect to composition shall be zero:

$$-d(H - TS)_{T,P} = 0 \tag{2.43}$$

or

$$dG_{T,P} = 0. \tag{2.44}$$

It is the latter of these which is the more relevant to chemical engineering operations and the two conditions should be viewed in parallel with the two heats of reaction discussed in sections 2.2.1 and 2.2.2. For the constant pressure and temperature system the criterion of equation (2.44) is developed to:

$$r\mu_C + s\mu_D = p\mu_A + q\mu_B \tag{2.45}$$

where μ_A, μ_B, μ_C, and μ_D are the chemical potentials of the four species at equilibrium. Since the chemical potentials are functions of composition, the introduction of this composition dependence enables the equilibrium position to be evaluated. The chemical potential of each component is related to that component's standard chemical potential μ^0 at unit activity and the actual activity "a" by:

$$\mu = \mu^0 + RT \text{ Ln } a. \tag{2.46}$$

It should be noted that μ^0 in this equation is a function of temperature only and makes the equilibrium constant defined below also a function of temperature only Substituting values of μ from expressions of the type given in equation (2.46) into equation (2.45) leads to:

$$-RT\,\mathrm{Ln}\,\frac{a_C^r a_D^s}{a_A^p a_B^q} = -(p\mu_A^0 + q\mu_B^0) + (r\mu_C^0 + s\mu_D^0). \qquad (2.47)$$

The μ^0 values are tabulated thermodynamic data and are listed as standard free energies of formation for individual compounds at a specified temperature (usually 298°K). The right-hand side of equation (2·47) can thus be evaluated from a knowledge of the stoichiometric numbers and tabulated data and is called the standard free energy change of the reaction, ΔG_T^0.

i.e.
$$\Delta G_T^0 \equiv (r\mu_C^0 + s\mu_D^0) - (p\mu_A^0 + q\mu_B^0). \qquad (2.48)$$

Comparing (2 47) with (2 48), ΔG_T^0 may be set equal to $-RT\,\mathrm{Ln}\,K$, so defining the equilibrium constant,

i.e.
$$-RT\,\mathrm{Ln}\,K = \Delta G_T^0. \qquad (2.49)$$

Thus K can be evaluated from tabulated data and as stated above is a function of temperature only Hence the value of K may be equated to the grouping of activities: $a_C^r a_D^s / a_A^p a_B^q$ to find the equilibrium composition It remains now to find K at any required temperature and to express the activities in terms of the corresponding partial pressures, concentrations or mole fractions before an equilibrium composition can be deduced The development of activities follows in detail and the effect of temperature is treated in section 2 5

(*i*) For perfect gases, the activity is equal to the partial pressure, and so:

$$K = \frac{p_C^r p_D^s}{p_A^p p_B^q} = K_p \qquad (2.50)$$

or since each partial pressure, concentration, and mole fraction is related by equations of the form:

$$p_A = y_A P \qquad (2.51)$$

and
$$p_A = c_A RT \qquad (2.52)$$

where P is the total pressure,

$$K = \frac{y_C^r y_D^s}{y_A^p y_B^q} P^{(r+s)-(p+q)} = \frac{y_C^r y_D^s}{y_A^p y_B^q} \,.\, P^{\Delta n} = K_y P^{\Delta n} \qquad (2.53)$$

thereby defining K_y

or
$$K = \frac{c_C^r c_D^s}{c_A^p c_B^q}(RT)^{\Delta n} = K_c(RT)^{\Delta n} \tag{2 54}$$

defining K_c

(*ii*) For imperfect gases, the activity is equal to the fugacity Assuming the Lewis and Randall rule, the fugacity in the mixture f_A is related to the partial pressure by:

$$\frac{f_A}{p_A} = \alpha_A \tag{2.55}$$

where α_A is the fugacity coefficient. Furthermore the fugacity coefficient is given by:

$$\alpha_A = \frac{f'_A}{P} \tag{2.56}$$

where f'_A is the fugacity of pure A and as such is available on plots as a function of reduced temperature and pressure (Fig 2 4)

Hence for imperfect gases, equations (2 50), (2 53) and (2 54) become:

$$K = \frac{p_C^r p_D^s}{p_A^p p_B^q} \cdot \frac{\alpha_C^r \alpha_D^s}{\alpha_A^p \alpha_B^q} = \frac{p_C^r p_D^s}{p_A^p p_B^q} K_\alpha = K_p K_\alpha \tag{2.57}$$

defining K_α.

$$K = K_\alpha \frac{y_C^r y_D^s}{y_A^p y_B^q} P^{\Delta n} = K_\alpha K_y P^{\Delta n} \tag{2.58}$$

and
$$K = K_\alpha \frac{c_C^r c_D^s}{c_A^p c_B^q}(RT)^{\Delta n} = K_\alpha K_c (RT)^{\Delta n} \tag{2.59}$$

(*iii*) For liquid phase reactions, the activities may be expressed as concentrations x_A, etc. for ideal solutions or as $\gamma_A x_A$ where γ_A is the activity coefficient of A in solution for non-ideal solutions; K_x and K_y may then be introduced for combinations of concentrations and activity coefficients.

The above discussion of equilibrium constants is illustrated below by several examples.

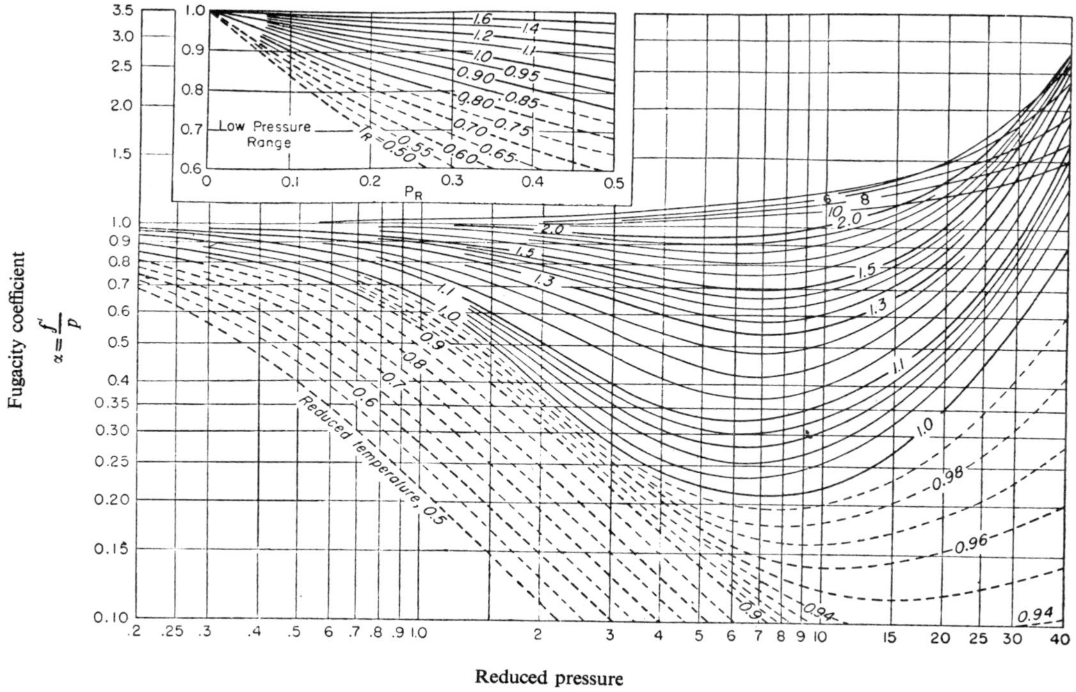

FIG. 2.4. Fugacity coefficient of pure compounds as a function of reduced temperature and pressure

Example 7. A mixture of 7·8 g of ethyl alcohol and 10 g of acetic acid are retained at a constant temperature until equilibrium is established when 2·7 g of acetic acid remain in the mixture. Estimate the equilibrium constant for this reaction and determine the equilibrium mixture formed by allowing 12 g of ethyl alcohol, 8·1 g of acetic acid, 4·0 g of ethyl acetate and 10·8 g of water to react under the above circumstances.

The esterification is represented by:

$$C_2H_5OH + CH_3COOH \rightleftharpoons CH_3COOC_2H_5 + H_2O$$

The starting mixture contains:

$$C_2H_5OH: \quad \frac{7{\cdot}8}{46} = 0{\cdot}1695 \text{ moles}$$

$$CH_3COOH: \frac{10}{60} = 0{\cdot}1667 \text{ moles}$$

At equilibrium:

$$CH_3COOH: \frac{2{\cdot}7}{60} = 0{\cdot}045 \text{ moles remain}$$

i.e. $(0{\cdot}1667 - 0{\cdot}045) = 0{\cdot}1217$ moles of acid and hence of alcohol have reacted.

$$C_2H_5OH: 0{\cdot}1695 - 0{\cdot}1217 = 0{\cdot}0478 \text{ moles remain}$$

$$CH_3COOC_2H_5 \text{ and } H_2O: 0{\cdot}1217 \text{ moles each.}$$

Expressing the equilibrium constant in terms of mole fractions gives:

$$K = \frac{(0{\cdot}1217)^2}{0{\cdot}0478 \times 0{\cdot}045} = 6{\cdot}885$$

since the total number of moles cancels out.

This value is now used to predict the equilibrium mixture from the starting mixture given.

Let x be moles of alcohol reacted at equilibrium

$$\text{Alcohol present at equilibrium: } \left(\frac{12}{46} - x\right) = 0{\cdot}261 - x$$

$$\text{Acid present at equilibrium: } \left(\frac{8{\cdot}1}{60} - x\right) = 0{\cdot}135 - x$$

Ester present at equilibrium: $\left(\frac{4\cdot0}{88} + x\right) = 0\cdot0455 + x$

Water present at equilibrium: $\left(\frac{10\cdot8}{18} + x\right) = 0\cdot6 + x$

Then

$$\frac{(0\cdot0455 + x)(0\cdot6 + x)}{(0\cdot261 - x)(0\cdot135 - x)} = 6\cdot885$$

and

$$x = 0\cdot072 \text{ or } 0\cdot5.$$

Since x must be less than 0·135, the admissable solution is 0·072.

Converting back to grams by multiplying by the appropriate molecular weights gives the composition of the equilibrium mixture as Alcohol 8·7 g; Acid 3·8 g; Ester 10·34 g and water 12·1 g.

A detailed analysis of the reaction between sulphur dioxide and oxygen will now be considered and the influence of the various variables demonstrated. The reaction will be considered at an elevated temperature for which the thermodynamic data give $K = 85$ (see Example 11 in section 2.5.1). Assuming perfect gases, equation (2.50) will be applied to the reaction:

$$SO_2 + \tfrac{1}{2}O_2 \rightleftharpoons SO_3 \tag{2.60}$$

to give:

$$85 = \frac{p_{SO_3}}{p_{SO_2} \cdot p_{O_2}^{\frac{1}{2}}} \tag{2.61}$$

A general feed of gas will be considered initially and a number of particular applications then made in Example 8 below. Let the feed have the following moles:

SO_2	S
O_2	R
SO_3	Q
Inert	N

and consider a fractional conversion "f" of SO_2. Then the moles, mole fractions and partial pressures of the participants in the reaction are shown in the table:

	Moles	Mole fractions	Partial pressures
SO_2	$S(1-f)$	$\frac{S(1-f)}{N+Q+R+S-\frac{1}{2}fS}$	$\frac{S(1-f)P}{N+Q+R+S-\frac{1}{2}fS}$
O_2	$R-\frac{1}{2}fS$	$\frac{R-\frac{1}{2}fS}{N+Q+R+S-\frac{1}{2}fS}$	$\frac{(R-\frac{1}{2}fS)P}{N+Q+R+S-\frac{1}{2}fS}$
SO_3	$Q+Sf$	$\frac{Q+Sf}{N+Q+R+S-\frac{1}{2}fS}$	$\frac{(Q+Sf)P}{N+Q+R+S-\frac{1}{2}fS}$
Inert	N		
Total	$N+Q+R+S-\frac{1}{2}fS$		

Example 8.

(*i*) Consider a stoichiometric feed of SO_2 and O_2 with no SO_3 or non-reactant at 1 atm pressure. Then $S = 2$, $R = 1$, $Q = 0 = N$, $P = 1$ and the partial pressures are:

$$p_{SO_2} = \frac{2(1-f)}{3-f} \tag{I}$$

$$p_{O_2} = \frac{1-f}{3-f} \tag{II}$$

$$p_{SO_3} = \frac{2f}{3-f}.$$

Hence

$$85 = \frac{2f}{(3-f)} \cdot \frac{(3-f)}{2(1-f)} \cdot \frac{(3-f)^{\frac{1}{2}}}{(1-f)^{\frac{1}{2}}} = \frac{f(3-f)^{\frac{1}{2}}}{(1-f)^{\frac{3}{2}}} \tag{III}$$

and the fractional conversion which satisfies this equation is 0·937.

(*ii*) Take the same feed as above but consider operation at 2 atm pressure. The equation to be solved is then

$$85 = \frac{f(3-f)^{\frac{1}{2}}}{(1-f)^{\frac{3}{2}}2^{\frac{1}{2}}} \tag{IV}$$

and the fractional conversion 0·95.

The increase in fractional conversion induced by a pressure increase is consistent with Le Chatelier's Principle.

(*iii*) The effect of a 50% excess of oxygen at 1 atm pressure will next be considered and the result compared with (*i*).

Thus keeping $S = 2$, R will be 1·5 and:

$$p_{SO_2} = \frac{2(1 - f)}{3{\cdot}5 - f} \tag{V}$$

$$p_{O_2} = \frac{1{\cdot}5 - f}{3{\cdot}5 - f} \tag{VI}$$

$$p_{SO_3} = \frac{2f}{3{\cdot}5 - f}. \tag{VII}$$

Hence

$$\frac{f(3{\cdot}5 - f)^{\frac{1}{2}}}{(1 - f)(1{\cdot}5 - f)^{\frac{1}{2}}} = 85 \tag{VIII}$$

giving $f = 0{\cdot}975$ and demonstrating the enhancement of the SO_2's conversion by the excess oxygen.

(*iv*) Finally the influence of an inert gas will be shown by considering the feed to a practical contact plant at 1 atm. The composition of gas after burning sulphur in air is:

$$SO_2\ 12\%;\quad O_2\ 9\%;\quad N_2\ 79\%$$

Consider 100 moles of this mixture as the basis for partial pressure and note that for comparison purposes the oxygen is in 50% excess as in (*iii*), i.e. $S = 12$, $R = 9$ and $N = 79$

$$p_{SO_2} = \frac{12(1 - f)}{100 - 6f} \tag{IX}$$

$$p_{O_2} = \frac{9 - 6f}{100 - 6f} \tag{X}$$

$$p_{SO_3} = \frac{12f}{100 - 6f} \tag{XI}$$

and

$$\frac{f(100 - 6f)^{\frac{1}{2}}}{(1 - f)(9 - 6f)^{\frac{1}{2}}} = 85 \tag{XII}$$

giving $f = 0{\cdot}941$.

The presence of nitrogen has reduced the fractional conversion despite the excess oxygen and in general an inert gas may be considered to have the same effect as a reduction in pressure.

The general expression for this reaction may be written as:

$$\frac{(Q + Sf)(N + Q + R + S - \frac{1}{2}fS)^{\frac{1}{2}}}{S(1 - f)(R - \frac{1}{2}fS)^{\frac{1}{2}} P^{\frac{1}{2}}} = K. \tag{2.62}$$

In none of the examples given above has the feed contained any SO_3. However, a practical plant may bring gases to equilibrium at one temperature and then feed the remaining reactants and products to a second converter at a different temperature; the estimation of the output from the second converter would utilise equation (2.62).

Since a high yield is obtainable at atmospheric pressure, sulphur dioxide converters are unlikely to utilise high pressure operation, and hence no deviation from perfect gas behaviour had to be considered in the above illustrations. In the case of ammonia production, the yield is negligible at normal pressures and the gases do not have ideal behaviour at the elevated pressures required to ensure satisfactory yield. The influence of non-ideal behaviour is now illustrated.

Example 9. A stoichiometric mixture of nitrogen and hydrogen is to be brought to equilibrium at 600 atm pressure. For the reaction:

$$\tfrac{1}{2}N_2 + \tfrac{3}{2}H_2 \rightleftharpoons NH_3$$

at 773°K the equilibrium constant, K, from thermodynamic data is 0·0034 atm.

The fractional conversion to ammonia will be considered allowing for imperfect gases and then compared with the result obtained on the assumption of ideal gases.

For the operating conditions the reduced properties and hence the fugacity coefficients from Fig. 2.4 are:

	T_R	P_R	Fugacity coefficient
N_2	6·1	17·9	1·37
H_2	18·8	28·9	1·20
NH_3	1·91	5·41	0·90

Using equation (2.57):

$$K_\alpha = \frac{0{\cdot}90}{(1{\cdot}37)^{\frac{1}{2}}\,(1{\cdot}20)^{\frac{3}{2}}} = 0{\cdot}59$$

and since $\Delta n = 1 - \frac{1}{2} - \frac{3}{2} = -1$

$$K_y = \frac{K}{K_\alpha P^{\Delta n}} = \frac{0{\cdot}0034 \times 600}{0{\cdot}59} = 3{\cdot}5$$

and

$$K_p = \frac{K}{K_\alpha} = 0{\cdot}0058.$$

Let the initial mixture contain 1 mole of nitrogen and let 'f' moles have reacted at equilibrium.

	N_2	H_2	NH_3
Moles in initial mixture	1	3	0
Moles at equilibrium	$1-f$	$3(1-f)$	$2f$
Moles fractions	$\frac{1-f}{4-2f}$	$\frac{3(1-f)}{2(2-f)}$	$\frac{f}{2-f}$
Partial pressures	$\frac{(1-f)P}{2(2-f)}$	$\frac{3(1-f)P}{2(2-f)}$	$\frac{fP}{2-f}$

Using mole fractions and partial pressures the following expressions are obtained:

$$\frac{4f(2-f)}{3\sqrt{3}(1-f)^2} = 3{\cdot}5$$

and

$$\frac{4f(2-f)}{3\sqrt{3}(1-f)^2 P} = 0{\cdot}0058.$$

On substitution of $P = 600$ these are identical and give a solution of $f = 0{\cdot}69$ and $52{\cdot}7\%$ NH_3 in the equilibrium gas mixture.

If ideal gases are assumed, then $K_\alpha = 1$ and $K_y = 2{\cdot}065$.

Hence
$$\frac{4f(2-f)}{3\sqrt{3}(1-f)^2} = 2{\cdot}065$$

and the equilibrium fractional conversion would then be 0·58 and the outlet ammonia concentration 40·9%. In this particular case, the fact that there is deviation from ideal behaviour leads to an increase in fractional conversion.

In many industrial processes two or more reactions involving common molecules are brought to equilibrium simultaneously. The conversions achieved in the several reactions are solved by setting up an equilibrium equation for each of the reactions involved. The procedure for two reactions is considered in the next example.

Example 10. The feed to a reactor contains A and B in the ratio 3:1 and the reactions occurring are:

$$\mathrm{A + B \rightleftharpoons C + D} \qquad K_p = 0{\cdot}15$$

$$\mathrm{A + C \rightleftharpoons 2E} \qquad K_p = 1{\cdot}60$$

Consider a feed of 4 moles of mixture and at equilibrium let f_1 moles of A have reacted by the first reaction, and f_2 by the second. Then the moles of the various components in the mixture are:

A	$3 - f_1 - f_2$
B	$1 - f_1$
C	$f_1 - f_2$
D	f_1
E	$2f_2$
Total	4 (constant)

Hence:

$$0{\cdot}15 = \frac{f_1(f_1 - f_2)}{(1 - f_1)(3 - f_1 - f_2)}$$

and

$$1{\cdot}6 = \frac{(2f_2)^2}{(f_1 - f_2)(3 - f_1 - f_2)}.$$

Elimination of f_2 between these equations gives:

$$385f_1^4 + 24f_1^3 + 342f_1^2 - 432f_1 + 81 = 0$$

from which $f_1 = 0{\cdot}6$ and f_2 by back substitution is 0·4.
The gas mixture has the equilibrium composition:

	Moles	Mole %
A	2·0	50
B	0·4	10
C	0·2	5
D	0·6	15
E	0·8	20

2.5 Effect of Temperature on the Equilibrium Constant

The effect of temperature on equilibrium constant has been developed extensively from specific heat data but more recently spectroscopic measurements have become increasingly available. The two methods utilising specific heat data are treated in sections 2.5.1 and 2.5.2 and are followed by the use of spectroscopic data in 2.5.3.

2.5.1 DEVELOPMENT OF ΔG_T^0 FROM ΔH_T^0

Differentiation of equation (2.47) at constant pressure and using the thermodynamic relationship:

$$\frac{\partial}{\partial T}\left(\frac{\mu^0}{T}\right)_P = -\frac{h^0}{T^2} \tag{2.63}$$

to eliminate the chemical potentials or molar standard free energies of formation (μ^0) gives:

$$\frac{d}{dT}(\text{Ln } K)_P = \frac{1}{RT^2}[-p\bar{h}_A^0 - q\bar{h}_B^0 + r\bar{h}_C^0 + s\bar{h}_D^0] \tag{2.64}$$

and on denoting:

$$\Delta H_T^0 = r\bar{h}_C^0 + s\bar{h}_D^0 - p\bar{h}_A^0 - q\bar{h}_B^0 \tag{2.65}$$

this becomes:

$$\frac{d}{dT}(\text{Ln } K)_P = \frac{\Delta H_T^0}{RT^2} \tag{2.66}$$

and then from equation (2.49)

$$\frac{d}{dT}\left(\frac{\Delta G_T^0}{T}\right)_P = -\frac{\Delta H_T^0}{T^2}. \tag{2.67}$$

It has been shown in section 2.3.3 that the heat of reaction is in general a function of temperature. However, it is sometimes sufficiently accurate to integrate equation (2.66) assuming ΔH_T^0 to be constant at ΔH^0 giving:

$$\text{Ln}\frac{K_2}{K_1} = \frac{\Delta H^0}{R}\left[\frac{1}{T_1} - \frac{1}{T_2}\right] \tag{2.68}$$

where K_1 is the equilibrium constant at temperature T_1
and K_2 is the equilibrium constant at temperature T_2.

Hence a knowledge of equilibrium constant at any one temperature, e.g. 298°K enables the value at any other temperature to be found.

If variation of ΔH_T^0 with temperature is to be taken into account, an expression for ΔH_T^0 as in equation (2.30) may be developed and substituted into (2.66) to yield:

$$\frac{d}{dT}(\text{Ln } K) = \frac{\Delta H_0^0}{RT^2} + \frac{\Delta\alpha}{RT} + \frac{\Delta\beta}{2R} + \frac{\Delta\gamma}{3R}T \tag{2.69}$$

which on integration and introducing C_1 as a constant gives:

$$\text{Ln } K_T = C_1 - \frac{\Delta H_0^0}{RT} + \frac{\Delta\alpha}{R}\text{Ln } T + \frac{\Delta\beta}{2R}T + \frac{\Delta\gamma}{6R}T^2. \tag{2.70}$$

The constant C_1 will usually be evaluated from a knowledge of K_{298} and equation (2.70) then becomes a general expression for K_T at any temperature within the range of applicability of the incorporated specific heat data.

A corresponding expression may be developed from equation (2.67) for ΔG_T^0 as a function of temperature.
If C_2 is the constant of integration, the result is:

$$\Delta G_T^0 = C_2 T + \Delta H_0^0 - \Delta\alpha T\,\text{Ln } T - \tfrac{1}{2}\Delta\beta T^2 - \tfrac{1}{6}\Delta\gamma T^3 \tag{2.71}$$

and the constant C_2 will be obtained from the known value of ΔG_{298}^0.

These equations are used below in example (11) to find a general expression for the equilibrium constant for the SO_2 oxidation reaction and hence the numerical value at a particular temperature. From tables of thermodynamic data, the standard free energies of formation are:

	$(\Delta G_f^0)_{298}$
SO_2	$-71\,790$ cal(g mole)$^{-1}$
SO_3	$-88\,520$ cal(g mole)$^{-1}$
O_2	0

Hence for the reaction:

$$SO_2 + \tfrac{1}{2}O_2 \rightleftharpoons SO_3 \tag{2.72}$$

$$\Delta G^0_{298} = (-88\,520) - (-71\,790) = -16\,730 \text{ cal (g mole)}^{-1} \tag{2.73}$$

and using equation (2.46):

$$\text{Ln } K_{298} = 28{\cdot}3. \tag{2.74}$$

It should be noted that had the equation been written:

$$2SO_2 + O_2 \rightleftharpoons 2SO_3 \tag{2.75}$$

ΔG^0_{298} would have been $-33\,460$ cal (g mole)$^{-1}$ and Ln K_{298} would have been 56·6. Thus the equilibrium constant for the reaction (2.75) is the square of the one for reaction (2.72) and an equilibrium composition could equally well be deduced from either:

$$\frac{p_{SO_3}}{p_{SO_2} p_{O_2}^{\frac{1}{2}}} = 1{\cdot}86 \times 10^{12} \tag{2.76}$$

or

$$\frac{p^2_{SO_3}}{p^2_{SO_2} p_{O_2}} = 3{\cdot}5 \times 10^{24}. \tag{2.77}$$

Great care must be taken to ensure that an equilibrium constant is used only in conjunction with the form of stoichiometric equation from which it was developed. This same caution applies equally to equilibrium constants at elevated temperatures and also to the development of heats of reaction as in section 2.3.

Example 11. The values of ΔH^0_0, $\Delta\alpha$, $\Delta\beta$ and $\Delta\gamma$ evaluated in earlier examples may be substituted into equation 2.71 to give:

$$\Delta G^0_T = C_2 T - 22785 + 4{\cdot}113T \text{ Ln } T - 6{\cdot}237 \times 10^{-3} T^2 + 0{\cdot}6225 \times 10^{-6} T^3. \tag{I}$$

The constant C_2 is evaluated as $-1{\cdot}309$ on substituting $\Delta G^0_{298} = -16730$ at $T = 298°K$. Hence the expression for ΔG^0_T at any temperature is:

$$\Delta G^0_T = -22785 - 1{\cdot}309T - 6{\cdot}237 \times 10^{-3} T^2 + 0{\cdot}6225 \times 10^{-6} T^3 + 4{\cdot}113T \text{ Ln } T. \tag{II}$$

Substituting $T = 749°K$ gives $\Delta G^0_{749} = -6612$ and then using equation (2.49)

$$\text{Ln } K = \frac{6612}{749 \times 1{\cdot}986} = 4{\cdot}444 \tag{III}$$

and $K = 85{\cdot}1$.

A value of $K = 85$ has been used in example 8 while illustrating the effect of feed composition on equilibrium yield.

2.5.2 Development of ΔG_T^0 from ΔH_T^0 and ΔS_T^0

For a reaction at any single temperature:

$$\Delta G_T^0 = \Delta H_T^0 - T\Delta S_T^0. \tag{2.78}$$

Just as $(\Delta H_f^0)_{298}$ is tabulated for the individual compounds, so also is their entropy S_{298}^0 relative to 0°K. Equations (2.30) and (2.31) may be adapted to give:

$$\Delta H_T^0 = \Delta H_{298}^0 + \Delta\alpha(T - 298) + \tfrac{1}{2}\Delta\beta(T^2 - 298^2) + \tfrac{1}{3}\Delta\gamma(T^3 - 298^3). \tag{2.79}$$

Parallel equations for standard entropy changes of reaction are:

$$\Delta S_T^0 = \Delta S_{298}^0 + \int_{298}^{T} \left(\frac{\Delta\alpha}{T} + \Delta\beta + T\Delta\gamma\right) dT \tag{2.80}$$

and

$$\Delta S_T^0 = \Delta S_{298}^0 + \Delta\alpha \,\mathrm{Ln}\frac{T}{298} + \Delta\beta(T - 298) + \tfrac{1}{2}\Delta\gamma(T^2 - 298^2). \tag{2.81}$$

Thus ΔH_T^0 and ΔS_T^0 can be found from equations (2.79) and (2.81) respectively and substituted into (2.78) as an alternative means of obtaining ΔG_T^0.

2.5.3 Development of ΔG_T^0 from spectroscopic data

Spectrographic measurements yield for individual compounds the 'free energy function' which is tabulated as a function of temperature. The free energy function of a substance is:

$$-\frac{G_T^0 - H_0^0}{T}$$

where G_T^0 is the free energy of the substance in its standard state at temperature T°K and H_0^0 is the enthalpy of the substance in its standard state at 0°K. Then the standard free energy of formation of the substance at temperature T°K relative to the elements at this temperature is expressed as:

$$\frac{\Delta G_f^0}{T} = \left[\frac{G_T^0 - H_0^0}{T} + \frac{(\Delta H_f^0)_0}{T}\right]_{\text{Substance}} - \left[\Sigma \frac{G_T^0 - H_0^0}{T}\right]_{\text{Elements}} \tag{2.82}$$

in which $(\Delta H_f^0)_0$ is the standard heat of formation of the substance at 0°K.

This equation may be developed to apply to the standard free energy change on reaction in terms of the products and reactants, giving:

$$\frac{\Delta G_T^0}{T} = \Sigma \left[\frac{G_T^0 - H_0^0}{T} + \frac{(\Delta H_f^0)_0}{T}\right]_P - \Sigma \left[\frac{G_T^0 - H_0^0}{T} + \frac{(\Delta H_f^0)_0}{T}\right]_R. \tag{2.83}$$

The use of this equation is illustrated in the following example.

Example 12. Consider the value of ΔG_{600}^0 for the water-gas shift reaction:

$$CO + H_2O \rightleftharpoons CO_2 + H_2. \tag{I}$$

From thermodynamic tables:

	$-\left(\frac{G_T^0 - H_0^0}{T}\right)$ cal(g mole)$^{-1}$ deg C^{-1}	$(\Delta H_f^0)_0$ kcal(g mole)$^{-1}$
CO_2	49·238	—93·969
H_2	29·203	0
CO	45·222	−27·202
H_2O	42·766	−57·107

where the free energy functions are the entries appropriate to 600°K and since all values are negative, the sign is incorporated in the column heading.

Equation (2.83) may be re-arranged to:

$$\Delta G_T^0 = T\Delta \left[\frac{G_T^0 - H_0^0}{T}\right] + (\Delta H_f^0)_0 \tag{II}$$

where the term $(\Delta H_f^0)_0$ now refers to the standard heat of reaction at 0°K. Hence substituting the above numerical values the value of ΔG^0 at 600°K in cal (g mole)$^{-1}$ is:

$$\Delta G^0_{600} = 600\,[(-49{\cdot}238 - 29{\cdot}203) - (-45{\cdot}222 - 42{\cdot}766)] + 1000\,[-93{\cdot}969 - (-27{\cdot}202 - 57{\cdot}107)] \qquad \text{(III)}$$

$$= 600\,[9{\cdot}547] - 1000 \times 9{\cdot}660 \qquad \text{(IV)}$$

$$= -3932 \text{ cal (g mole)}^{-1}. \qquad \text{(V)}$$

2.5.4 Factors Affecting Equilibrium Position

It is convenient at this stage to summarise the effects of processing conditions on the equilibrium yield. Most of the effects have been demonstrated in the various examples, but the general effect of temperature, pressure and composition is now summarised. The conclusions form the basis for Le Chatelier's Principle.

(*i*) *Temperature.* It can be seen from equation (2.68) that an increase in temperature will increase the equilibrium constant provided ΔH^0 is positive. Since an increase in equilibrium constant entails a greater yield, yield rises with temperature for endothermic reactions and falls with temperature for exothermic ones.

(*ii*) *Pressure.* At a given temperature, the equilibrium constant K is fixed and hence the mole fraction function of equation (2.53) can be written as:

$$\frac{y_C^r y_D^s}{y_A^p y_B^q} = \frac{K}{(P)^{\Delta n}}.$$

Thus the effect of pressure will depend upon whether Δn is positive, negative or zero. If there is no change in the total number of reactant and product moles, Δn is zero and pressure has no effect on the equilibrium position. If the moles increase on reaction, Δn is positive and so an increase in pressure will decrease the term $K/(P)^{\Delta n}$ and hence decrease the yield Conversely, when there is a decrease in moles, pressure increase will favour yield.

(*iii*) *Composition.* Extension of these considerations confirms that an excess of reactant will increase the yield of other reactants, while the presence of products in a feed will suppress the conversion of reactants. The presence of an inert gas has the same effect as a reduction in pressure and depends, therefore, on the sign of Δn.

2.6 Combined Consideration of Heat of Reaction and Equilibrium

In the earlier sections of this chapter, the concepts of heat of reaction and adiabatic reaction temperature on the one hand, and equilibrium position on the other, have been considered in isolation. In fact the equilibrium position and the outlet temperature are inextricably related in an adiabatically operating reactor. This is best illustrated by considering a feed of SO_2, 12%; O_2, 9% and N_2, 79% at 400°C as shown in Fig. 2.5 to the first stage of a two-stage converter. From a combination of the function of "f" in equation (XII) of Example 8 and the expression developed for equilibrium constant as a function of temperature using equation (2.49) and equation (II) of Example 11, the following equation connecting "f" and "T" can be formulated:

$$\frac{f(100-6f)^{\frac{1}{2}}}{(1-f)(9-6f)^{\frac{1}{2}}} = \exp\left[\frac{1}{R}\left\{\frac{22\,785}{T} + 1{\cdot}309 + 6{\cdot}237 \times 10^{-3}\,T - 0{\cdot}6225 \times 10^{-6}T^2 - 4{\cdot}113\,\mathrm{Ln}\,T\right\}\right]. \qquad (2.84)$$

Solution of the simultaneous equations (2.84) and (III) of Example 5 leads to the result that an outlet fractional conversion 0·694 at a temperature 872°K will be achieved. The physical situation is illustrated by Fig. 2.5; the adiabatic temperature line represents equation (III) of Example 5 and the equilibrium line (2.84). It is seen that as fractional conversion increases, temperature increases but that the increase in temperature reduces the potential equilibrium yield; a compromise between these conflicting features is found at the intersection of the two lines.

Since the fractional conversion from this first stage would not be high enough, the products would be cooled to say 400°C and then fed to the second stage The effects of the intercooler and second reaction stage are indicated by the dotted lines of Fig 2 5 The detailed calculation may be performed by extension of the above equations with the utilisation of equation (2.62) to allow for the SO_3 in the feed to the second reactor.

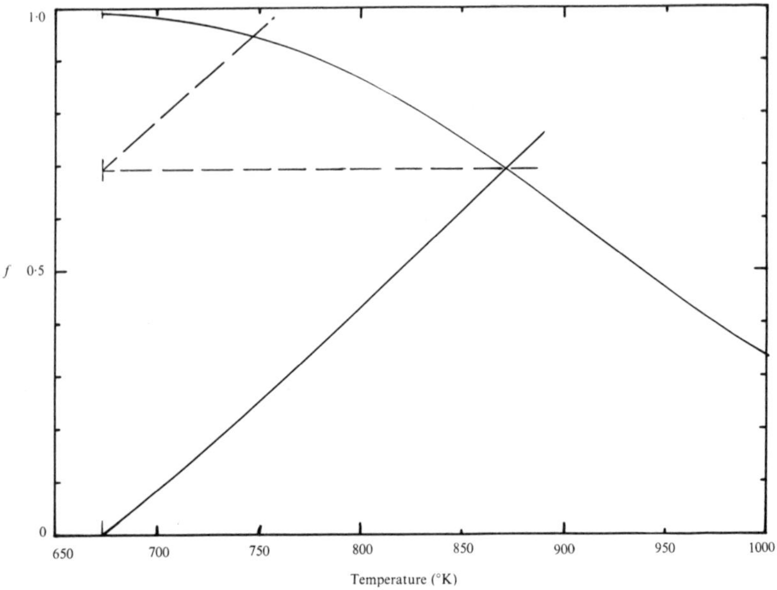

FIG. 2.5. Adiabatic operation of an SO_2 converter to the position of thermodynamic and thermal equilibrium

2.7 Possibility of a Chemical Reaction

It has been stated earlier that all chemical reactions are reversible although from a practical point of view some reactions appear not to proceed at all while others appear to go to completion. In fact these extremes are found in any given reaction over a practical range of temperature. Thus the SO_2 oxidation reaction has $\Delta G^0 = -16730$ at 298°K and an equilibrium constant of $1{\cdot}86 \times 10^{12}$; use of this equilibrium constant in equation (III) of Example 8 would give a value of f essentially unity. Conversely at 2000°K, $\Delta G^\circ = +17200$ and $K = 0{\cdot}0133$ and solving the same equation for "f" will give a value near to zero conversion.

It is possible to find reactions having acceptable reaction rates at convenient operating temperatures and at the same time having all extents of equilibrium conversion between the extremes. A negative value of ΔG_T^0 will be seen by equation (2.49) to give a value of K greater than unity while positive ΔG_T^0 values give equilibrium constants less than unity. However, when the standard free energy change is positive it does not necessarily signify that the reaction is not possible; it is sufficient only to indicate the likelihood of a low yield. As already demonstrated with the SO_2 oxidation reaction, the actual yield for a given K (i.e. given ΔG_T^0) depends on the feed composition and the operating pressure and so each case must be considered on its merits. The yield for a given K also depends on the form of the reaction equation and consideration will now be given to the following three commonly occurring forms of equation:

$$\mathrm{A + B \rightleftharpoons C + D} \tag{2.85}$$

$$\mathrm{A + B \rightleftharpoons C} \tag{2.86}$$

$$\mathrm{A + \tfrac{1}{2}B \rightleftharpoons C} \tag{2.87}$$

For a stoichiometric feed of reactants, the fractional conversions at pressure P are given respectively by:

$$K = \frac{f^2}{(1-f)^2} \text{ (independent of pressure)} \tag{2.88}$$

$$K = \frac{f(2-f)}{(1-f)^2 P} \tag{2.89}$$

and

$$K = \frac{f(3-f)^{\frac{1}{2}}}{(1-f)^{\frac{3}{2}} P^{\frac{1}{2}}}. \tag{2.90}$$

Since the equilibrium constant in each case would be evaluated from:

$$K = \exp\left[-\frac{1}{R}\left(\frac{\Delta G_T^0}{T}\right)\right] \tag{2.91}$$

it is convenient to plot the fractional conversion directly as a function of the parameter $\Delta G_T^0/T$.

The variation of fractional conversion with $\Delta G_T^0/T$ for each of these reactions at 1 atm pressure is shown in Table 2.3 and Fig. 2.6. These show clearly how the nature of the reaction influences the fractional conversion for given values of ΔG_T^0 and T.

The fractional conversions for the two reactions which are dependent upon pressure is shown in Tables 2.4 and 2.5. The results for the one reaction are given over a narrow range of pressure (tenfold) while for the other a thousandfold range is shown; the

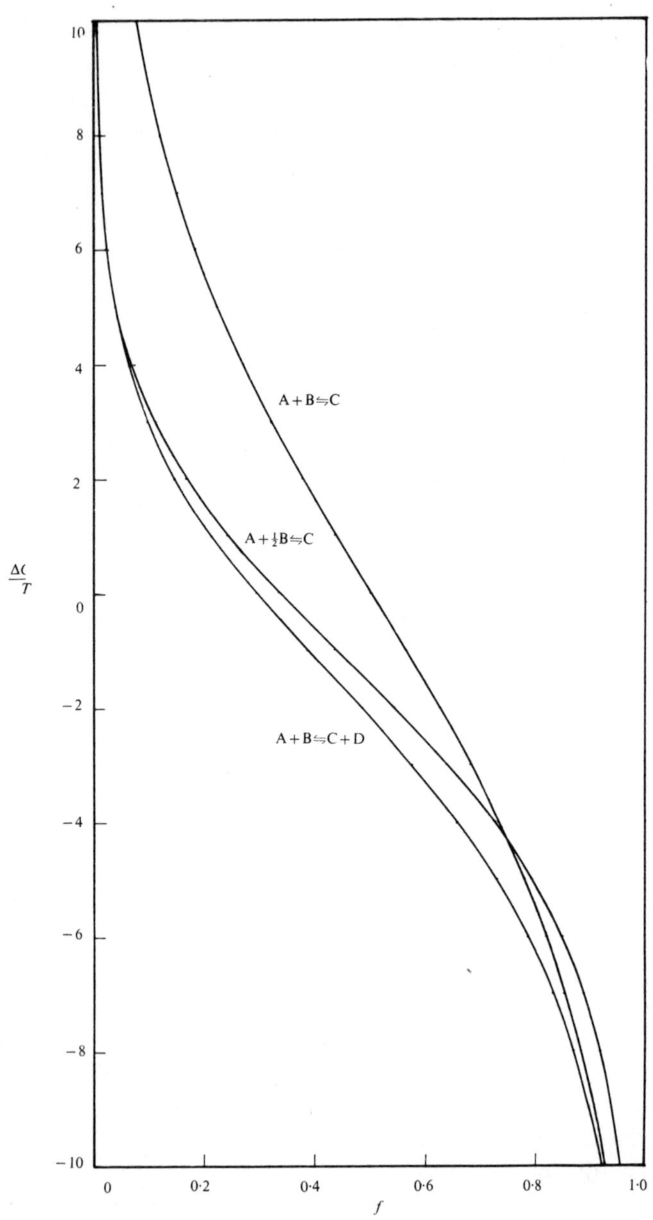

FIG. 2.6. Fractional conversions at 1 atm pressure for various reaction types

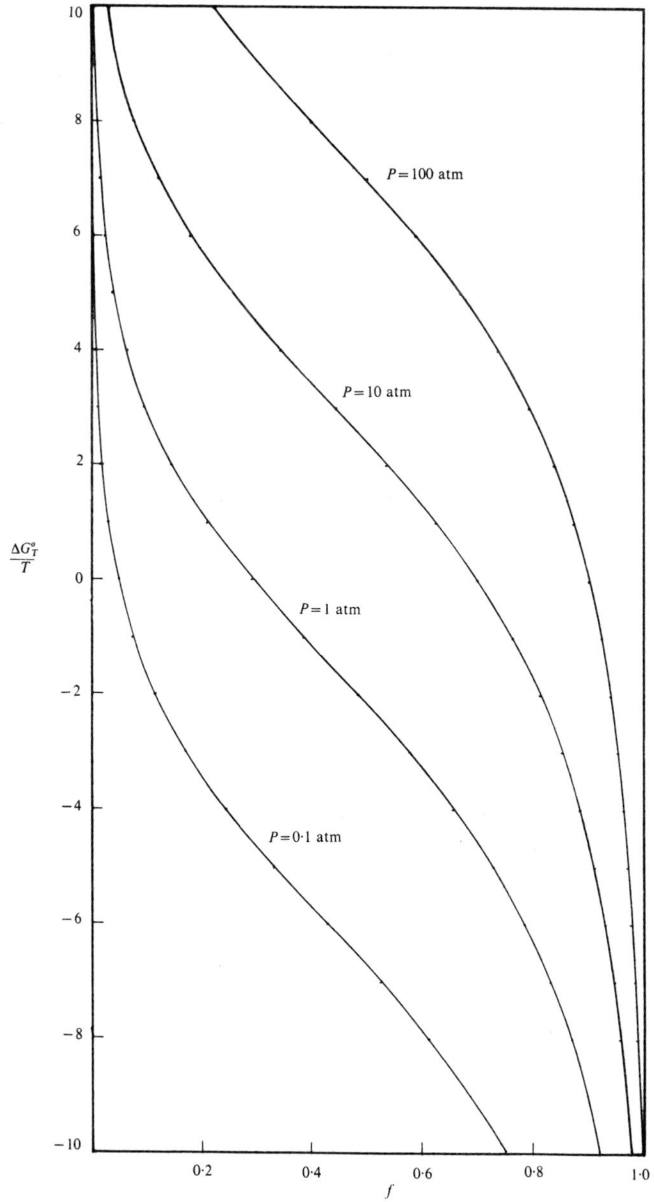

FIG. 2.7. Fractional conversions for reaction A + B ⇌ C at varying pressure

TABLE 2.3

[Fractional conversion at 1 atm pressure]

$\frac{\Delta G_T^0}{T}$	$A + B \rightleftharpoons C + D$	$A + B \rightleftharpoons C$	$A + \frac{1}{2}B \rightleftharpoons C$
−10	0·925	0·920	0·957
−8	0·882	0·868	0·918
−7	0·853	0·831	0·887
−6	0·819	0·784	0·846
−5	0·779	0·727	0·792
−4	0·732	0·657	0·723
−3	0·680	0·575	0·639
−2	0·623	0·483	0·542
−1	0·563	0·386	0·437
0	0·500	0·293	0·333
1	0·437	0·211	0·241
2	0·377	0·144	0·165
3	0·320	0·0949	0·109
4	0·268	0·0607	0·0699
5	0·221	0·0380	0·0439
6	0·181	0·0235	0·0271
7	0·146	0·0144	0·0166
8	0·118	0·00879	0·0101
10	0·0746	0·00324	0·00374

TABLE 2.4

[Fractional conversions for reaction $A + \frac{1}{2}B \leftrightharpoons C$ at varying pressure]

$\frac{\Delta G_T^0}{T}$	$P = 0·5$ atm	$P = 1·5$ atm	$P = 3$ atm	$P = 5$ atm
−10	0·946	0·962	0·970	0·975
−8	0·898	0·928	0·942	0·951
−7	0·860	0·900	0·920	0·932
−6	0·810	0·864	0·890	0·906
−5	0·746	0·815	0·850	0·872
−4	0·667	0·753	0·797	0·826
−3	0·573	0·675	0·730	0·766
−2	0·470	0·582	0·647	0·691
−1	0·365	0·479	0·551	0·601
0	0·268	0·374	0·446	0·500
1	0·187	0·276	0·342	0·394
2	0·125	0·193	0·248	0·294
3	0·0806	0·130	0·171	0·208
4	0·0508	0·0839	0·113	0·141
5	0·0315	0·0530	0·0729	0·0916
6	0·0194	0·0330	0·0458	0·0581
7	0·0118	0·0203	0·0284	0·0362
8	0·00720	0·0124	0·0174	0·0223
10	0·00265	0·00457	0·00645	0·00830

TABLE 2.5
[Fractional conversions for reaction A + B ⇌ C at varying pressure]

$\dfrac{\Delta G_T^0}{T}$	$P = 0{\cdot}1$ atm	$P = 1$ atm	$P = 10$ atm	$P = 100$ atm
−10	0·753	0·920	0·975	0·992
−8	0·611	0·868	0·958	0·987
−7	0·523	0·831	0·946	0·983
−6	0·428	0·784	0·930	0·978
−5	0·332	0·727	0·911	0·972
−4	0·244	0·657	0·885	0·963
−3	0·170	0·575	0·853	0·953
−2	0·114	0·483	0·812	0·940
−1	0·0737	0·386	0·761	0·922
0	0·0465	0·293	0·698	0·900
1	0·0289	0·211	0·623	0·872
2	0·0178	0·144	0·536	0·837
3	0·0109	0·0949	0·442	0·792
4	0·00661	0·0607	0·345	0·736
5	0·00401	0·0380	0·256	0·668
6	0·00243	0·0235	0·180	0·587
7	0·00147	0·0144	0·121	0·500
8	0·000889	0·00879	0·0787	0·400
10	0·000325	0·00324	0·0310	0·222

latter situation is also plotted in Fig. 2.7 and demonstrates the significant effect which may be induced by a suitable choice of pressure. It can be seen from Table 2.5 that even with a large positive ΔG_T^0 (say 10000 cal mole^{-1} at 1000°K) a significant fractional conversion (0·222) may be achieved by use of elevated pressure.

2.8 Use of Recycle to Increase Fractional Conversion

It has been seen that equilibrium in SO_2 oxidation can give a satisfactorily high yield of SO_2 conversion. In contrast the synthesis of ammonia from nitrogen and hydrogen gives a yield of the order of only 0·2–0·3 at temperatures for which the rate of reaction is acceptable. It is economically essential to recover the unreacted hydrogen and nitrogen and return these to the reactor inlet. Thus for a stoichiometric feed of reactants with the reactor operating at a temperature for which the fractional conversion is 0·25 the mass balance on the reactor with a recycle is as shown in Fig. 2.8 and complete conversion is achieved. The economic penalties are the pressure boosting of recycle gas and the larger reactor to cope with the greater rate of throughput.

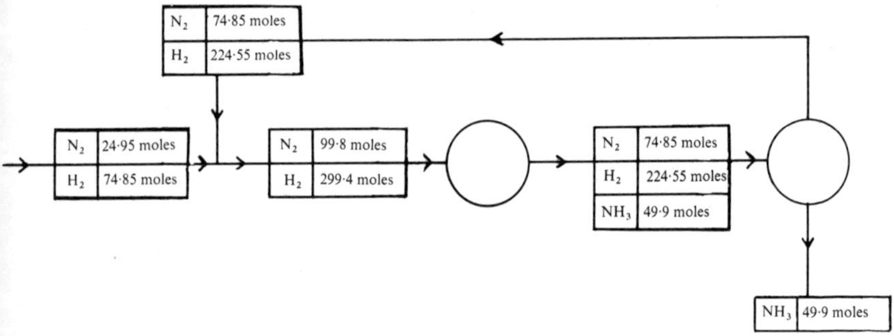

FIG. 2.8. Ammonia reactor with recycle

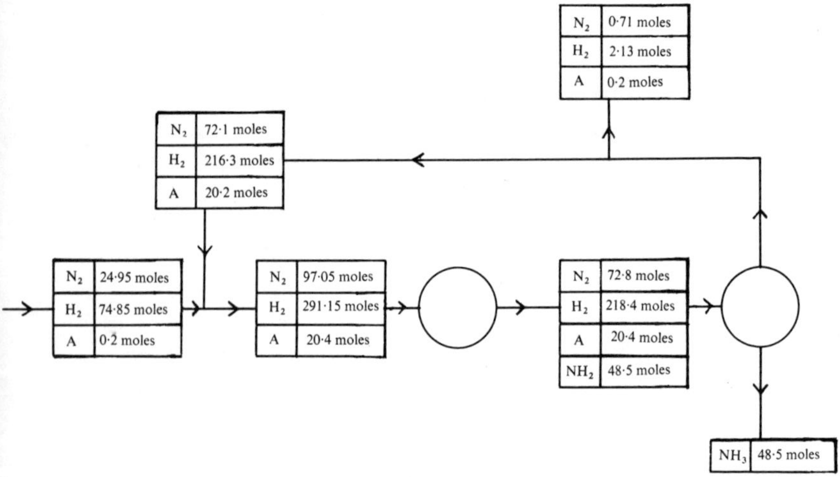

FIG. 2.9. Ammonia reactor with recycle and purge

The practical situation is more complicated in the sense that the feed nitrogen may contain a small percentage of argon which will accumulate in the reactor and would eventually occupy the whole reaction space. The input of argon must be balanced by a purge stream to prevent this accumulation and it is usual to specify the argon content which is to be tolerated at some point in the reactor system. The same complication might arise from a feedstock which does not contain stoichiometric reactants; then the excess reactant must be purged to prevent accumulation. Once a purge stream is discharged from a recycle, loss of some reactant must be tolerated, the extent of the loss being dependent upon the recycle ratio employed. The idealised balance of Fig. 2.8 has been made more realistic (Fig. 2.9) to allow for argon in the feed. The balance in this case is made on the premise of 5% argon at the reactor inlet with the reactor pass efficiency still at 25% It is arranged that the argon in the purge exactly balances that in the feed. However, the purge composition is the same as the recycle gas composition and the hydrogen and nitrogen lost with the argon purge represent an inefficiency in overall reactant conversion. For this particular set of conditions the overall efficiency is:

$$\frac{24{\cdot}95 - 0{\cdot}71}{24.95} = 97{\cdot}14\%.$$

A further example (Fig. 2.10) is now considered in which there is an excess of one reactant as well as an inert to be purged from the the system. The basis of the following development is the feed of 10 lb moles h^{-1} of air together with ethylene to yield 1·0 lb mole of ethylene oxide; the latter is completely separated and the remaining reactants recycled.

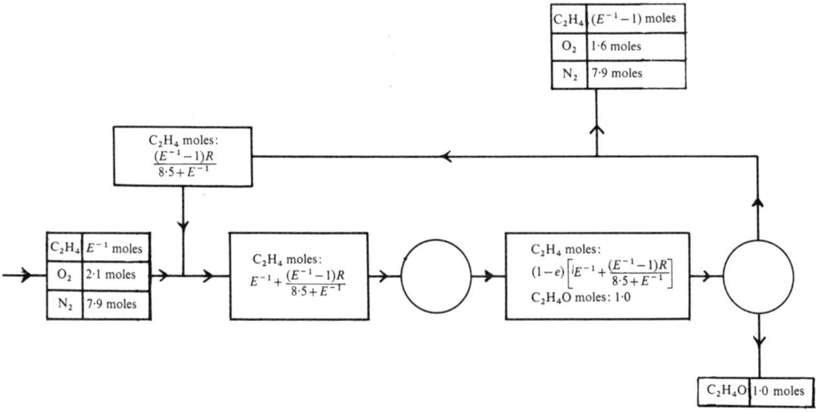

FIG. 2.10. Ethylene oxide reactor with recycle and purge of inert and excess oxygen

Let E be the overall system efficiency of C_2H_4 conversion
Let e be the pass C_2H_4 conversion efficiency
and Let R be the total gas recycle.

Then for 1·0 mole C_2H_4O, the ethylene feed is:

$$\frac{1}{E}\text{ moles}$$

and the purge ethylene:

$$\left(\frac{1}{E} - 1\right) \text{ moles.}$$

The purge must also contain 7·9 moles N_2 and 1·6 moles of excess O_2, so that its ethylene mole fraction is:

$$\frac{E^{-1} - 1}{9 \cdot 5 + (E^{-1} - 1)}.$$

Hence the C_2H_4 recycle is:

$$\frac{(E^{-1} - 1)\,R}{8 \cdot 5 + E^{-1}}.$$

The C_2H_4 entering the reactor is:

$$E^{-1} + \frac{(E^{-1} - 1)\,R}{8 \cdot 5 + E^{-1}}$$

and leaving the reactor:

$$(1 - e)\left[E^{-1} + \frac{(E^{-1} - 1)\,R}{8 \cdot 5 + E^{-1}}\right].$$

After the ethylene oxide separation, this ethylene has to balance the total C_2H_4 in the purge and recycle and so yields the equation:

$$(1 - e)\left[E^{-1} + \frac{(E^{-1} - 1)\,R}{8 \cdot 5 + E^{-1}}\right] = (E^{-1} - 1) + \frac{(E^{-1} - 1)\,R}{8 \cdot 5 + E^{-1}} \qquad (2.92)$$

from which:

$$R = \frac{(e^{-1} - E^{-1})(8 \cdot 5E + 1)}{1 - E}. \qquad (2.93)$$

A balance around the system is shown in Fig. 2.11 for the case where $e = 0 \cdot 25$ and $E = 0 \cdot 9$.

Values of R for varying overall efficiency E at fixed pass efficiency ($e = 0 \cdot 25$) are tabulated below:

TABLE 2.6

E	R (lb moles)
0·5	21·0
0·6	35·6
0·7	59·6
0·8	107
0·9	250
0·96	677
0·98	1390
0·99	2815
0·995	5665
0·999	28465

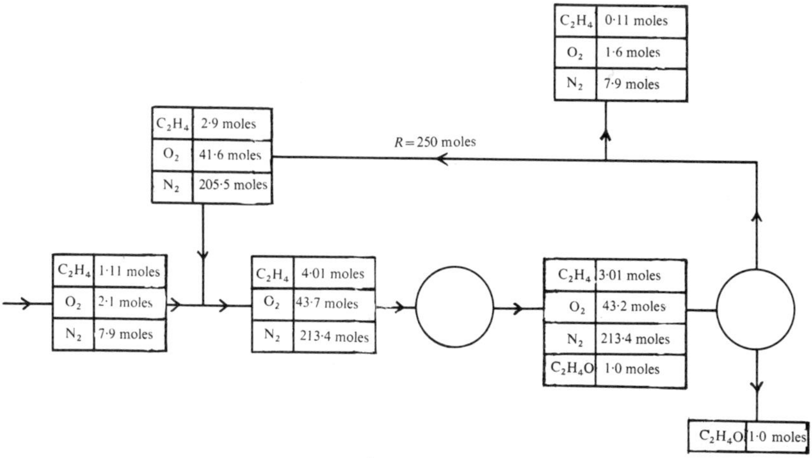

FIG. 2.11. Ethylene oxide reactor mass balance with particular values of pass efficiency (0·25) and overall efficiency (0·9)

The results show that in general as the overall efficiency rises, the moles of recycle gas increase and that the increase becomes very pronounced as the demanded overall efficiency exceeds about 95%. Thus the operating recycle rate could be determined by an economic balance of the savings from prevention of ethylene wastage and the increasing compression costs incurred thereby. A plot of such an economic analysis would take the form of Fig. 2.12.

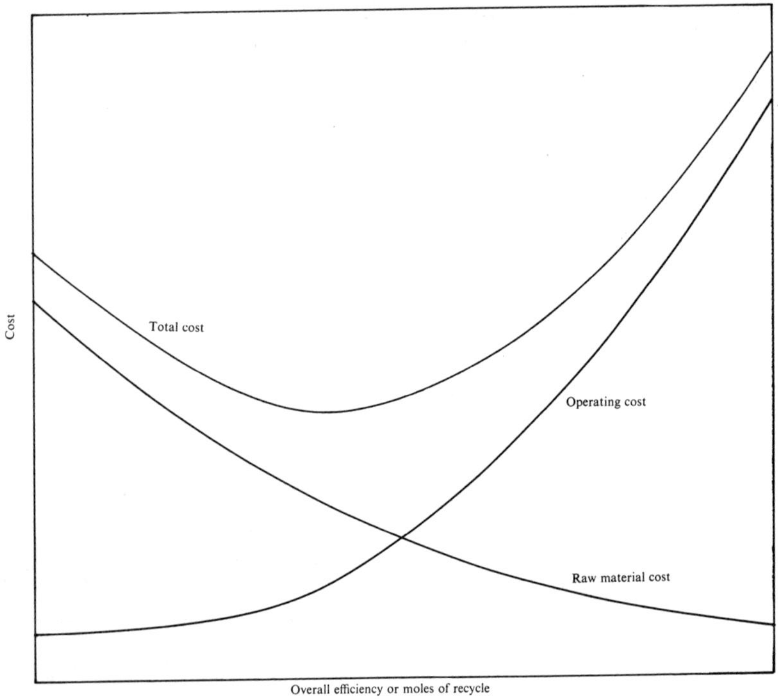

FIG. 2. 12. Selection of economic operating point

Problems

1. Find the standard heat of formation of CS_2 from the following standard heats of combustion.

$$CS_2 + 3O_2 = CO_2 + 2SO_2 \qquad \Delta H^0_{25} = -265\ \text{kcal mole}^{-1}$$
$$C + O_2 = CO_2 \qquad \Delta H^0_{25} = -94\ \text{kcal mole}^{-1}$$
$$S + O_2 = SO_2 \qquad \Delta H^0_{25} = -71\ \text{kcal mole}^{-1}$$

ANS. +29 kcal

2. Evaluate the standard heat of formation of C_3H_8 given its heat of combustion as 530·605 kcal and other standard heats of formation:

$$CO_2 = -94052\ \text{cal}$$
$$H_2O(g) = -57798\ \text{cal}$$
$$H_2O(1) = -68317\ \text{cal}$$

ANS. −24·819 kcal

3. For the reaction:

$$C_3H_8(g) + 5O_2(g) \rightarrow 3CO_2(g) + 4H_2O$$

use data from the tables to calculate the internal energy and enthalpy changes at 25°C:

(*a*) when the water is in the gaseous state
(*b*) when it is in the liquid phase.

ANS. (*a*) $\Delta H = -488{\cdot}528$ kcal; $\Delta U = -489{\cdot}120$ kcal
(*b*) $\Delta H = -530{\cdot}604$ kcal; $\Delta U = -528{\cdot}826$ kcal

4. The standard heats of formation of CO and CO_2 at 25°C are $-26{\cdot}416$ and $-94{\cdot}052$ kcal (g mole)$^{-1}$. Find the heat of combustion of CO at 827°C.

Specific heats (C_p) at T°K
CO: $6{\cdot}342 + 1{\cdot}836T \times 10^{-3} - 0{\cdot}28T^2 \times 10^{-6}$ cal (g mole)$^{-1}$ degC^{-1}
CO_2: $6{\cdot}393 + 10{\cdot}10T \times 10^{-3} - 3{\cdot}405T^2 \times 10^{-6}$ cal (g mole)$^{-1}$ degC^{-1}
O_2: $6{\cdot}095 + 3{\cdot}253T \times 10^{-3} - 1{\cdot}017T^2 \times 10^{-6}$ cal (g mole)$^{-1}$ degC^{-1}

ANS. -67446 cal (g mole)$^{-1}$ CO

5 Ethylene oxide is produced from ethylene by the following reaction:

$$C_2H_4(g) + \tfrac{1}{2}O_2(g) \rightarrow C_2H_4O(g)$$
$$\Delta H^0_{18°C} = -51\,500 \text{ Btu lb mole}^{-1}.$$

Find the hourly heat removal from a plant producing 5 tons per day of ethylene oxide. 10% excess oxygen above stiochiometric is used and 80% of the ethylene feed is converted. The oxygen enters at 150°C, the ethylene at 200°C and the products leave at 280°C. Mean specific heats above 18°C in Btu lb mole^{-1} degF^{-1}

C_2H_4: 18·0 at 200°C, 19·0 at 280°C
C_2H_4O: 21·0 at 280°C
O_2: 7·15 at 150°C, 7·3 at 280°C

ANS. 501 000 Btu

6. For the reaction:

$H_2O + CO \rightleftharpoons CO_2 + H_2$ $\quad \Delta H^0_{298} = -9840$ cal (g mole)$^{-1}$

the CO originates by passing air over coke with complete oxygen conversion. The CO/N_2 stream at 300°F is mixed with 2 moles of steam per mole of CO at 400°F and the above reaction then effected with a 98% CO conversion and an exit temperature of 900°F.

Find the final gas composition and prepare a flow sheet for the production of 2 ton moles of hydrogen in the product stream.

Use specific heat data to find the thermal load on the reactor.

7. At 400°C and a total pressure of 10 atmosphere, ammonia is dissociated to the extent of 98%. Calculate K_p, K_y and K_c for the reaction: $2NH_3 \rightleftharpoons N_2 + 3H_2$

ANS. $K_p = 9{\cdot}92 \times 10^4$ atm^2; $K_y = 990$; $K_c = 32{\cdot}55$ g mole^2 litre^{-2}

8. A gaseous mixture containing 60·0% H_2, 20·0% N_2 and the remainder inert gas is passed over a catalyst at 50 atm pressure in order to produce ammonia. Estimate the maximum conversion if K_p for the reaction: $N_2 + 3H_2 \rightleftharpoons 2NH_3$ is 0·0125 and the system is behaving ideally.

What are the equilibrium conversions of nitrogen and hydrogen for a feed containing 65% H_2, 15% N_2 and balance inerts?

ANS. $f = 0{\cdot}60$; For N_2, $f = 0{\cdot}765$ and for H_2, $f = 0{\cdot}47$

9. A gas mixture of 25 mole % CO_2 and 75% CO is mixed with the stoichiometric H_2 for the following reactions:

$$CO_2 + 3H_2 \rightleftharpoons CH_3OH + H_2O \qquad K = 2{\cdot}82 \times 10^{-6}$$
$$CO + 2H_2 \rightleftharpoons CH_3OH \qquad K = 2{\cdot}00 \times 10^{-5}$$

The conversion to equilibrium is effected in the presence of a catalyst at 300 atm pressure. Assuming the behaviour to remain ideal, calculate the fractional conversion for each reaction and hence the volume composition of the equilibrium mixture.

ANS. First reaction $f = 0{\cdot}3$; second $f = 0{\cdot}965$.

10. For the reaction:

$$2CO + 4H_2 \rightleftharpoons C_2H_5OH + H_2O$$

the feed consists of 'r' moles of H_2 per mole of CO. Find an expression for the fractional conversion of CO in the presence of excess hydrogen (i.e. for $r > 2$) at 20 atm pressure and 573°K for which $\Delta G^0 = 2126$ cal(g mole)$^{-1}$. Develop the corresponding expression for the fractional conversion of H_2 when CO is in excess.

ANS. $$\left(\frac{f}{1-f}\right)^2 \left(\frac{1+r-2f}{r-2f}\right)^4 = 9{\cdot}91 \times 10^4 \quad \text{and}$$

$$\left(\frac{f}{r(2-rf)}\right)^2 \left(\frac{1+r(1-f)}{1-f}\right)^4 = 9{\cdot}91 \times 10^4.$$

11. Assuming that ΔH is independent of temperature for the following pyrolysis reaction, develop an expression for ΔG at any temperature:

$$C_2H_6 \rightleftharpoons C_2H_4 + H_2.$$

Hence estimate the fractional conversion at 500°C and 1·5 atm pressure for a feed containing 80% C_2H_6, 15% H_2 and balance non-reactants.

	C_2H_6	C_2H_4
Standard heats of formation	−20236	+12496 cal(g mole)$^{-1}$
Standard free energies of formation at 25°C	−7860	+16282 cal(g mole)$^{-1}$

ANS. $\Delta G_T^0 = (32732 - 29T)$ cal; $f = 0{\cdot}0053$

12. For the reaction:

$$C_2H_4(g) + H_2O(g) \rightleftharpoons C_2H_5OH(g)$$

$\Delta G^0 = 1685$ cal(g mole)$^{-1}$ and $\Delta H^0 = -10400$ cal(g mole)$^{-1}$ at the operating temperature of 418°K. The hourly production rate is to be 250 kg of C_2H_5OH from a feed containing 2 moles H_2O per mole of C_2H_4. Assuming that the feed enters and the products leave at the reaction temperature and that equilibrium is attained at a pressure of 2 atm, deduce as much as possible about the reactor operation.

ANS. $f = 0{\cdot}147$; 56500 kcal h^{-1} to be removed

13. Calculate ΔG^0 and ΔH^0 for the ammonia oxidation reaction at 500°K. The reaction may be taken as:

$$4NH_3 + 5O_2 \rightarrow 4NO + 6H_2O(g).$$

Use data from the tables and take into account the variation of specific heat with temperature.

ANS − 238041 cal; − 215243 cal

14. For the reaction:

$$\tfrac{1}{2}N_2(g) + \tfrac{3}{2}H_2(g) \rightleftharpoons NH_3(g)$$

$\Delta H_{298}^0 = -11040$ cal(g mole)$^{-1}$

$\Delta S_{298}^0 = -23{\cdot}69$ e.u.

$$C_p(NH_3) = 6{\cdot}086 + 8{\cdot}812 \times 10^{-3}T - 1{\cdot}506 \times 10^{-6}T^2 \text{ cal(g mole)}^{-1} \text{ degK}^{-1}$$

$$C_p(N_2) = 6{\cdot}449 + 1{\cdot}412 \times 10^{-3}T - 0{\cdot}0807 \times 10^{-6}T^2 \text{ cal(g mole)}^{-1} \text{ degK}^{-1}$$

$$C_p(H_2) = 6{\cdot}947 - 0{\cdot}200 \times 10^{-3}T + 0{\cdot}481 \times 10^{-6}T^2 \text{ cal(g mole)}^{-1} \text{ degK}^{-1}$$

(*a*) Evaluate ΔG^0_{298}
(*b*) Develop an expression for ΔG^0_T as a function of temperature.

ANS. (*a*) -3980 cal (*b*) $12490 - 7{\cdot}559T \text{ Ln } T + 72{\cdot}049T + 4{\cdot}203 \times 10^{-3}T^2 - 0{\cdot}3635 \times 10^{-6}T^3$

15. For the species in the reaction:

$$C_3H_8(g) \rightleftharpoons C_3H_6(g) + H_2(g)$$

thermodynamic tables give the following entries for $T = 800$°K

	C_3H_6	H_2	C_3H_8
$-\dfrac{G^0_T - H^0_0}{T}$ (cal degC^{-1})	67·04	31·186	68·74
$(\Delta H^0_f)_0$ (kcal)	8·468	0	−19·482

Evaluate the equilibrium constant at this temperature

ANS. $K = 0{\cdot}0643$

16 For the reaction $UO_2(s) + 4HF(g) \rightleftharpoons UF_4(s) + 2H_2O(g)$ $\Delta H^0 = -43\,200$ cal(g mole)$^{-1}$ UO_2 and ΔG^0 at 298°K is $-31\,200$ cal(g mole)$^{-1}$ UO_2. Assuming ΔH^0 to be independent of temperature, calculate ΔG^0 at 250°C and 500°C and hence the equilibrium gas composition at these temperatures and 1 atm pressure.

ANS. $T = 250$, $p_{HF} = 0{\cdot}005$ atm; $T = 500$, $p_{HF} = 0{\cdot}128$ atm

17. Ethylene is to be converted by catalytic air oxidation to ethylene oxide. The air and ethylene are mixed in the ratio of 10:1 by volume; this mixture is combined with a recycle stream and the two streams are fed to the reactor. Of the ethylene entering the reactor, 40% is converted to ethylene oxide, 20% is converted to carbon dioxide and water, and the rest does not react. The exit gases from the reactor are treated to remove substantially all of the ethylene oxide and water, and the residue recycled. Purging of the recycle is required to avoid accumulation and hence maintain a constant feed to the reactor. Calculate the ratio of purge to recycle if not more than 8% of the ethylene fed is lost in the purge.

What will be the composition of the corresponding reactor feed gas?

ANS. 6·7; N_2 81·8 mole %, O_2 10·7, CO_2 5·5 and C_2H_4 2·1

18. An aqueous solution of HF is pumped from a feed tank through a vaporiser to a reactor in which solid UO_2 is converted to solid UF_4 according to the following equation:

$$UO_2 + 4HF \rightleftharpoons UF_4 + 2H_2O.$$

The reaction reaches equilibrium with 10% m/m HF in the exit gases. The exit gas passes to a distillation column which gives end products of pure H_2O and 38% m/m HF. The latter is continuously returned to the feed tank where it is mixed with the stoichiometric requirement of pure HF from stock.

Determine the composition of the solution in the feed tank and the reactor pass efficiency for HF.

ANS. 90·6% m/m HF and 93·5%

19. A gas stream, of composition SO_2, 12%; O_2, 8%; N_2, 80% enters a two-stage SO_2 converter at 400°C, and the SO_2 is to be 97% converted. The fractional conversion of SO_2 in the first stage is 0·8 and in the second stage 0·75.

(*a*) Find the required recycle from the exit of stage 2 to the inlet of stage 2 to achieve the required conversion. Take 100 moles SO_2 as the basis.

(*b*) What is the composition at entry to stage 2?

(*c*) Find the temperature at the exit to stage 1, given the mean specific heat values in cal(g mole)$^{-1}$ degC^{-1} as:

$$SO_2\ 10{\cdot}0;\quad O_2\ 7{\cdot}4;\quad SO_3\ 12{\cdot}3;\quad N_2\ 7{\cdot}1$$

and the exothermic heat of reaction (independent of temperature) as -22980 cal(g mole)$^{-1}$.

ANS. (*a*) 697 moles; (*b*) SO_2 1·52 mole %, SO_3 11·1, O_2 2·88 and N_2 84·5; (*c*) 700°C.

REFERENCES

PERRY, (Ed.). 1950. *Chemical engineers handbook*. McGraw-Hill, New York.

WEAST, R. C. (Ed.). 1966. *Handbook of physics and chemistry*. Chemical Rubber Publishing Co., Cleveland, Ohio.

3

KINETICS OF CHEMICAL REACTIONS

3.1 Introduction

A very important aspect of the design and analysis of a chemical reactor is the study of the rates of the chemical reactions taking place. Generally such studies will be carried out in conjunction with thermal and equilibrium measurements, which will be made to correct or confirm the thermodynamic analysis. The laboratory kinetic investigation will establish the effect of different physical conditions on the rate of conversion of reactants, and on the yield of the desired product. Fundamental studies dealing with the nature of possible reaction steps are seldom considered in chemical reaction engineering; the kinetic study will be concentrated on the development of chemical rate equations in a useful form for subsequent design or analysis.

The reaction rates may be studied on a pilot plant scale. This form of investigation has the advantage that it usually attempts to closely simulate commercial practice and the results obtained are, therefore, a good guide for subsequent scale-up design calculations. Since this type of investigation is expensive and time consuming there is a tendency nowadays to confine the laboratory work to bench scale experiments. However, irrespective of the scale of the laboratory work, the objective of the study is the provision of one or more rate equations to adequately describe the chemical reaction kinetics. The basis for interpretation and analysis of the fundamental observations of reactant conversion with time, forms the subject matter of this chapter.

3.2 Terminology in Chemical Engineering Reaction Kinetics

Before analysing chemical reaction rate equations it is necessary to introduce the following pertinent definitions:

3.2.1 LIMITING REACTANT

If in a reacting system, the reactants are not present in the proportions required by the stoichiometric equation, then the reactant which is not in excess is the limiting reactant.

For instance, the combustion of carbon is represented by the stoichiometric equation:

$$C + O_2 \rightarrow CO_2 \tag{3.1}$$

which signifies reaction between 12 mass units of carbon and 32 of oxygen to yield 44 mass units of CO_2. However, if the combustion mixture were not stoichiometric and had 12 mass units of carbon and, say 64 of oxygen, then 44 mass units of CO_2 would still be formed. The oxygen is present in excess and the carbon is the limiting reactant.

3.2.2 FRACTIONAL CONVERSION

This is the fraction of a reactant that has undergone chemical change at a particular stage of the reaction process. It is usually of practical importance to consider the fractional conversion of the limiting reactant. Thus, in the above combustion example, the range of fractional conversion of the carbon is 0–1 in both cases, while for oxygen it is also 0–1 in the stoichiometric case but 0–0·5 in the non-stoichiometric case.

3.2.2 YIELD

This concept needs to be introduced for the description of the frequent situation in which only part of a reactant is converted to the desired product, while some is unavoidably converted to unwanted side products.

Thus in the reaction system represented by the stoichiometric equations:

$$C_2H_5OH \begin{cases} \nearrow CH_3CHO + H_2 \\ \searrow C_2H_4 + H_2O \end{cases} \tag{3.2}$$

let 'a' moles of alcohol be fed to the reactor of which a total of 'x' moles react to give

'y' moles of the aldehyde

and

'z' moles of ethylene.

The total alcohol conversion is then: $\frac{x}{a}$,

the aldehyde yield is either: $\frac{y}{a}$ or $\frac{y}{x}$,

depending upon whether one chooses as basis the total alcohol fed to the system or that part of the alcohol which reacts.

and similarly the ethylene yield is either: $\frac{z}{a}$ or $\frac{z}{x}$.

Thus in discussing yield, great care must be taken to establish which basis is being used in each particular instance; in this text yield is based on the reactant feed, i.e. y/a and z/a.

3.2.4 Reaction rate

The rate of chemical reaction, sometimes called the reaction velocity, may be expressed as the rate of decrease of the concentration of any reactant or the rate of increase of the concentration of any product of the reaction. Concentration may be expressed in any appropriate units but since chemical reactions occur between molecules it is convenient to express concentration in terms of molar concentration. However, this is not essential and, from a chemical engineering point of view, it is sometimes desirable to express concentration in mass units. Whatever system of units is used, it is important to realise that the actual value of the rate depends on the particular reactant or product chosen for the expression of the rate. Thus in the reaction

$$pA + qB \rightarrow wC \tag{3.3}$$

the rate of change of A moles is expressed as: $-dn_A/dt$ the negative sign indicating that n_A decreases with time. Similarly the rates of change of B and C are expressed as $-dn_B/dt$ and $+dn_C/dt$. The relationship between these three rates is then:

$$-\frac{1}{p}\frac{dn_A}{dt} = -\frac{1}{q}\frac{dn_B}{dt} = +\frac{1}{w}\frac{dn_C}{dt} \tag{3.4}$$

and so in discussion of the rate of a chemical reaction, it is important to state clearly which reactant or product rate represents the reaction under consideration.

The rate of a chemical reaction taking place in unit volume depends on the concentrations of the substances taking part, on the absolute temperature and sometimes on the system pressure. In the case of gas phase reactions, the concentration is usually expressed as a partial pressure which may in turn be expressed as the product of mole fraction and total pressure.

3.2.5 Reaction order

The manner in which the rate of a chemical reaction is controlled by the concentration of each reactant is specified as a power exponent of that reactant's concentration and this is the reaction order with respect to that particular reactant. Thus for the reaction expressed by equation (3.3) the rate of conversion of A per unit volume of mixture is:

$$-\frac{dn_A}{dt} = k_n c_A^{\alpha} c_B^{\beta} \tag{3.5}$$

where 'α' is the reaction order with respect to A
'β' is the reaction order with respect to B
$\alpha + \beta \equiv n$ is the overall reaction order

and k_n is the specific reaction rate, section (3.2.7).

The exponents α and β may or may not be equal to their respective stoichiometric numbers p and q.

The order of a chemical reaction must at our present state of knowledge be deduced from the results of experimental observations, and may not have any apparent direct relation to the mechanism of the reaction. The deduction of reaction orders is the subject of later sections of this chapter.

The overall order of a chemical reaction need not be an integer; it may be zero or a small positive quantity but seldom, if ever, exceeds three. Furthermore the order of a reaction can vary with concentration level in liquid phase reactions and with absolute pressure in gas phase reactions.

3.2.6 Molecularity

The molecularity describes the mechanism of a chemical reaction and must be an integer. Many apparently simple chemical reactions take place through a sequence of steps each with a minimum of atomic rearrangement. Of these steps most will be very rapid but one of them is likely to be comparatively slow and will therefore have an over-riding effect on the reaction rate. This slow step is the rate determining step of the reaction sequence and the number of molecules reacting determines the molecularity of the reaction.

Consider the overall reaction:

$$2A + B \rightarrow C + D \tag{3.6}$$

proceeding by the two steps:

$$A + B \rightarrow C + E \tag{3.7}$$

and

$$A + E \rightarrow D \tag{3.8}$$

Reaction (3.7) is slow and rate determining while (3.8) is a rapid reaction. The molecularity is two because the rate determining step involves the reaction between one molecule of each of A and B, and is clearly not deducible from a knowledge of the stoichiometric equation.

3.2.7 Fundamental rate equation and specific reaction rate

Equation (3.5) expresses the rate of conversion of moles of A per unit volume of reactor space. If the reactor has a volume V, then the total rate of conversion of A is:

$$-\frac{dn_A}{dt} = k_n V c_A^{\alpha} c_A^{\beta} \text{ moles/time.} \tag{3.9}$$

This equation is applicable for variable volume and in the case of a fixed volume system, the V may be allowed within the differential sign to give:

$$-\frac{d}{dt}\left(\frac{n_A}{V}\right) = k_n c_A^{\alpha} c_B^{\beta} \tag{3.10}$$

or simply

$$-\frac{dc_A}{dt} = k_n c_A^{\alpha} c_B^{\beta} \tag{3.11}$$

Equation (3.11) is of very frequent application and the assumption of constant volume will be used throughout the present text unless otherwise stated.

In situations where volume variations are significant either equation (3.9) involving the three time dependent variables n_A, c_A and V must be used or alternatively the connection between these variables: $c_A = n_A/V$ may be used to eliminate one of them.

e.g. eliminating n_A gives:

$$-\frac{d}{dt}(c_A V) = k_n V c_A^{\alpha} c_B^{\beta} \tag{3.12}$$

or

$$c_A \frac{dV}{dt} + V\frac{dc_A}{dt} = -k_n V c_A^{\alpha} c_B^{\beta} \tag{3.13}$$

The term k_n in the above equations is known as the specific reaction rate. In other texts k_n is variously called the 'velocity constant' or 'reaction constant' or 'rate coefficient'. Its numerical value is equal to the rate of the reaction per unit volume when the concentrations of the reactants are each 1·0. The dimensions of k_n vary with the order of the reaction and with the units employed to express the concentration of the reacting species. However, there are equivalent values of k_n for different sets of units of concentration and the way in which the dimensions of k_n vary with the order of reaction may be obtained as follows.

Considering equation (3.9) or (3.11) it will be seen that the units of k_n are:

$$\frac{[\text{CONC}]}{[\text{TIME}]} \times \frac{1}{[\text{CONC}]^n} \quad \text{or} \quad \frac{1}{[\text{TIME}]\,[\text{CONC}]^{(n-1)}}.$$

where $n = \alpha + \beta$.

Thus if, for example, the chosen concentration units are g mole litre^{-1} and the time units are seconds, the following dimensions and units for k_n follow:

TABLE 3.1
Units of specific reaction rate

zero order	$n = 0$	[CONC] [TIME]$^{-1}$	g mole litre^{-1} sec^{-1}
1st order	$n = 1$	[TIME]$^{-1}$	sec^{-1}
2nd order	$n = 2$	[CONC]$^{-1}$ [TIME]$^{-1}$	litre (g mole)$^{-1}$ sec^{-1}
$\frac{1}{2}$ order	$n = 0{\cdot}5$	[CONC]$^{\frac{1}{2}}$ [TIME]$^{-1}$	(g mole)$^{\frac{1}{2}}$ litre$^{-\frac{1}{2}}$ sec^{-1}
$\frac{3}{2}$ order	$n = 1{\cdot}5$	[CONC]$^{-\frac{1}{2}}$ [TIME]$^{-1}$	litre$^{\frac{1}{2}}$ (g mole)$^{-\frac{1}{2}}$ sec^{-1}

It is thus seen that all specific reaction rates have reciprocal time units and that with the sole exception of first order reactions all have concentration units.

3.3 The First Order Forward Reaction

A first order forward reaction

$$A \rightarrow \text{Products} \tag{3.14}$$

has constant volume kinetics of the form:

$$-\frac{dc_A}{dt} = k_1 c_A . \tag{3.15}$$

With an initial concentration c_{A_0} of A and a concentration c_A at time t, rearrangement and integration of equation (3.15) gives:

$$-\int_{c_{A0}}^{c_A} \frac{dc_A}{c_A} = k_1 \int_0^t dt \tag{3.16}$$

and

$$k_1 t = \text{Ln}\,\frac{c_{A0}}{c_A} \tag{3.17}$$

or

$$c_A = c_{A0} \exp(-k_1 t). \tag{3.18}$$

It is also useful to express such results in terms of the fractional conversion "f" of the reactant A. Since the ratio c_A/c_{A0} represents the fraction of original A remaining unreacted at time t, it follows that:

$$f = 1 - \frac{c_A}{c_{A0}} \quad \text{or} \quad \frac{c_A}{c_{A0}} = 1 - f \tag{3.19}$$

giving

$$k_1 t = \text{Ln}\,\frac{1}{1-f} \tag{3.20}$$

or

$$f = 1 - \exp(-k_1 t). \tag{3.21}$$

It is apparent from these results that in this special instance of a first order reaction, the fractional conversion at a given time does not depend upon the starting concentration. By contrast, the results obtained in the following sections for reaction orders other than unity, all give fractional conversions at a given time which do depend upon the initial concentration.

Example 1. Decomposition of hydrogen peroxide solution in the presence of a catalyst gave 23·5 ml oxygen in 10 min and 37·0 ml oxygen in 20 min. Given that the decomposition reaction is first order, estimate the specific reaction rate and the quantity of oxygen evolved in 15 minutes. The decomposition of hydrogen peroxide can be presented by the reactions

$$H_2O_2 \rightarrow H_2O + O \tag{I}$$

followed by

$$O + O \rightarrow O_2 \tag{II}$$

Since the combination of the oxygen atoms is very rapid, reaction (I) is the first order rate determining step.

Let x be the volume of oxygen liberated in time t, and let y be the total volume to be liberated on complete reaction.

Then equation (3.21) can be applied to give

$$\frac{x}{y} = 1 - \exp(-k_1 t). \tag{III}$$

Application of this equation to the two known values of x and t and then dividing the resultant equations, eliminates the unknown y to give:

$$\frac{23 \cdot 5}{37 \cdot 0} = \frac{1 - \exp(-10k_1)}{1 - \exp(-20k_1)} = 0 \cdot 635. \quad \text{(IV)}$$

This particular equation may be solved as a quadratic since if $\exp(-10\,k_1)$ is given the symbol 'z', then $\exp(-20\,k_1) = z^2$ from which (IV) gives:

$$0 \cdot 635\, z^2 - z + 0 \cdot 365 = 0 \quad \text{(V)}$$

i.e.

$$z = \exp(-10\,k_1) = 0 \cdot 573 \quad \text{(VI)}$$

and

$$k_1 = 0 \cdot 055\ \text{min}^{-1}. \quad \text{(VII)}$$

Had the two times not been separated by a factor of two, the following graphical solution would have been appropriate. In this the function:

$$\frac{1 - \exp(-10\,k_1)}{1 - \exp(-20k_1)}$$

is tabulated for a set of values of k_1 and then plotted in Fig. 3.1. It is seen that $k_1 = 0 \cdot 055\ \text{min}^{-1}$ satisfies the required value of 0·635 for the ordinate.

k_1	$(1 - e^{-k_1 t_1})/(1 - e^{-k_1 t_2})$
0·15	0·815
0·10	0·73
0·075	0·68
0·05	0·625
0·025	0·565
0·01	0·52

The volume of oxygen x_{15} evolved in 15 minutes can be obtained by straight applications of equation (III) and then dividing to give:

$$\frac{x_{15}}{37 \cdot 0} = \frac{1 - \exp(-15 \times 0 \cdot 055)}{1 - \exp(-20 \times 0 \cdot 055)} \quad \text{(VIII)}$$

Hence $x_{15} = 31 \cdot 2$ ml.

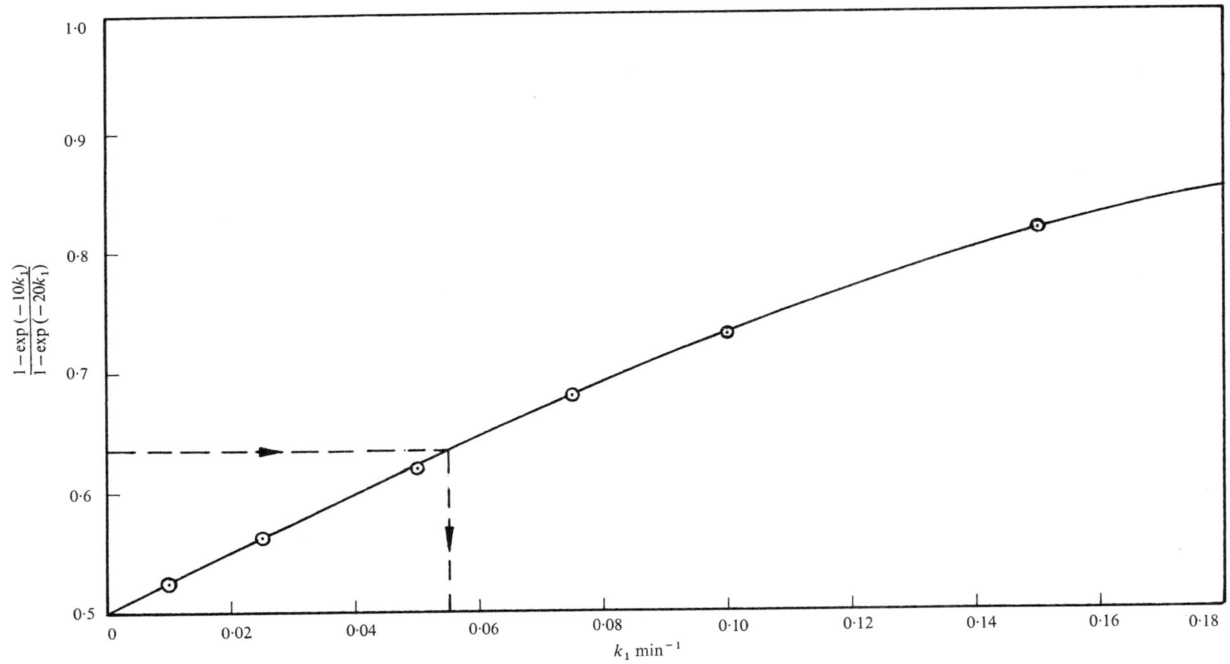

FIG. 3.1. Graphical deduction of velocity constant

3.4 The Second Order Forward Reaction

The second order forward reaction may involve the reaction of identical molecules or of different molecules and may therefore be represented as either

$$2A \rightarrow \text{Products} \tag{3.22}$$

with

$$-\frac{dc_A}{dt} = k_2 c_A^2 \tag{3.22a}$$

or

$$A + B \rightarrow \text{Products} \tag{3.23}$$

with

$$-\frac{dc_A}{dt} = -\frac{dc_B}{dt} = k_2 c_A c_B \tag{3.23a}$$

With the usual boundary conditions equation (3.22a) becomes:

$$-\int_{c_{A0}}^{c_A} \frac{dc_A}{c_A^2} = k_2 \int_0^t dt \tag{3.24}$$

or

$$\frac{1}{c_A} - \frac{1}{c_{A0}} = k_2 t$$

or in the terms of fractional conversion:

$$\frac{1}{c_{A0}} \left[\frac{f}{1-f}\right] = k_2 t. \tag{3.26}$$

This result is also applicable to the integration of equation (3.23a) for the special case where A and B have equal initial concentrations c_{A0}. However, in the general case when the initial concentrations of A and B are different, the instantaneous concentration c_B of B must be expressed in terms of the instantaneous concentration of A to permit integration. This is conveniently carried out as follows:

Let initial concentration of A be a.
Let initial concentration of B be b.

Let the change in concentration of A and B in time t be x. Then in terms of x the second order rate equation is

$$-\frac{d}{dt}(a-x) = k_2(a-x)(b-x) = \frac{dx}{dt} \tag{3.27}$$

or

$$k_2 t = \int_0^x \frac{dx}{(a-x)(b-x)} \tag{3.28}$$

which on taking partial fractions gives:

$$k_2 t = \frac{1}{a-b} \int_0^x \left[\frac{1}{b-x} - \frac{1}{a-x}\right] dx. \tag{3.29}$$

Hence

$$k_2 t = \frac{1}{(a-b)} \ln \frac{b(a-x)}{a(b-x)}. \tag{3.30}$$

Equation (3.30) is the classical second order reaction rate equation for irreversible reactions when two reactants at different initial concentrations are undergoing reaction. Since a large number of chemical reactions follow this second order law, equation (3.30) is applied extensively.

Example 2. A quantity of ethyl acetate was mixed with sodium hydroxide solution in a thermostat at 25°C. 100 cm^3 of the mixture was titrated immediately with 0·05 N HCl and required 68·2 cm^3 for neutralisation. Titration of 100 cm^3 samples of reaction mixture with the same strength acid required 49·7 cm^3 of acid after 30 minutes of reaction and 15·6 cm^3 at the completion of reaction. The second order specific rate constant is calculated below utilising equation (3.30).

The reaction is:

$$CH_3COOC_2H_5 + NaOH \rightarrow CH_3COONa + C_2H_5OH$$

Let b be the initial NaOH concentration

Let a be the initial Ester concentration

and Let x be the change in concentration of both NaOH and ester after 30 minutes.

Then $b = 0{\cdot}05 \times \dfrac{68{\cdot}2}{100} = 0{\cdot}0340$ g mole litre^{-1}.

At 30 minutes, NaOH is $0{\cdot}05 \times \dfrac{49{\cdot}7}{100} = 0{\cdot}0249$ g mole litre^{-1}.

Hence $x = 0{\cdot}0340 - 0{\cdot}0249 = 0{\cdot}0091$ g mole litre^{-1}.

At completion of reaction, NaOH is $0{\cdot}05 \times \dfrac{15{\cdot}6}{100} = 0{\cdot}0078$ g mole litre^{-1}.

Hence the overall NaOH concentration change is:

$$0{\cdot}0340 - 0{\cdot}0078 = 0{\cdot}0262 \text{ g mole litre}^{-1}.$$

Since no ester remains and the two reagents have had the same change in concentration,

$$a = 0{\cdot}0262 \text{ g mole litre}^{-1}.$$

Hence on substitution into equation (3.30):

$$k = \frac{2{\cdot}303}{30(0{\cdot}0340 - 0{\cdot}0261)} \operatorname{Log} \frac{0{\cdot}0262 \times 0{\cdot}0249}{0{\cdot}0340 \times 0{\cdot}0171}$$
$$= 0{\cdot}49 \text{ min}^{-1} (\text{g mole}^{-1}) \text{ litre}$$

3.5 General Order Forward Reaction

For a forward reaction involving only one type of molecule, the kinetic equation is:

$$-\frac{dc_A}{dt} = k_n c_A^n \tag{3.31}$$

which on integration leads to:

$$k_n t = \frac{1}{n-1}\left[\frac{1}{c_A^{n-1}} - \frac{1}{c_{A0}^{n-1}}\right] \tag{3.32}$$

or to

$$k_n t = \frac{1}{(n-1)\,c_{A0}^{n-1}}\left[\frac{1}{(1-f)^{n-1}} - 1\right]. \tag{3.33}$$

Equations (3.32) and (3.33) apply to any fractional or integral value of n except 1·0 and can readily be reduced to the results of equations (3.25) and (3.26) for the particular case when $n = 2$. The special case when $n = 1$ is covered in section 3.3.

In the general case of an nth order reaction involving more than one type of molecule, one is faced with integrations of the form

$$\int \frac{dx}{(a - x)^{\alpha}(b - x)^{\beta}}$$

and if α and β are not integers, recourse to numerical methods will be necessary.

3.6 Determination of Orders for Forward Reactions

The results of chemical kinetic experiments always take the form of a tabulation or plot of progress of reaction against time. The reaction progress may be a change in concentration (or some equivalent measurement) of a reactant or a product or of both. The observations will be sufficient to enable mass balances to be performed at each time observation so that a full knowledge of the reaction mixture composition may be deduced. From this composition-time history it is necessary to first deduce the reaction order with respect to each reactant and when the results of this analysis are satisfactory, to evaluate the rate constant. Methods relevant to forward reactions follow.

3.6.1 Integration Method

The experimental results are substituted into the integrated form of the zero, first, second and other order equations:

$$\text{for zero order: } k_0 = \frac{c_{A0}}{t} f = \frac{c_{A0} - c_A}{t}$$

$$\text{for first order: } k_1 = \frac{1}{t} \operatorname{Ln} \frac{1}{1 - f} = \frac{1}{t} \operatorname{Ln} \frac{c_{A0}}{c_A}$$

and for second order with identical reacting molecules:

$$k_2 = \frac{1}{tc_{A0}} \left[\frac{f}{1 - f}\right] = \frac{1}{t}\left[\frac{1}{c_A} - \frac{1}{c_{A0}}\right].$$

A value of the rate constant k_n is evaluated for each time observation, and will be essentially constant for the correct order but will change progressively for incorrect orders. A typical analysis of results for a reaction of the form:

$$A \rightarrow \text{Products}$$

is shown in Table 3.2 for an initial concentration of A of 0·01 mole litre^{-1}.

TABLE 3.2

Deduction of reaction order from integrated kinetic equations

t min	f	k_0 mole litre^{-1} min^{-1}	k_1 min^{-1}	k_2 litre mole^{-1} min^{-1}
1	0·049	$4·9 \times 10^{-4}$	$5·1 \times 10^{-2}$	5·15
2	0·094	$4·6 \times 10^{-4}$	$4·9 \times 10^{-2}$	5·07
5	0·206	$4·1 \times 10^{-4}$	$4·6 \times 10^{-2}$	5·19
10	0·342	$3·4 \times 10^{-4}$	$4·2 \times 10^{-2}$	5·20
15	0·438	$2·9 \times 10^{-4}$	$3·8 \times 10^{-2}$	5·19
25	0·565	$2·3 \times 10^{-4}$	$3·3 \times 10^{-2}$	5·20

On the basis of these results, the hypothesis of second order would be accepted and the other two orders rejected.

This is a trial and error method and, since it is extremely dependent on the accuracy of the experimental observations, the result should be confirmed by one of the other methods. Furthermore, the trial and error nature of the method renders it unsatisfactory if the order turns out to be fractional.

3.6.2 HALF-LIFE METHOD

This method consists of determining the time taken for the initial concentration of a reactant to be reduced by a factor of two; it must be emphasised that it is not half of the time for complete reaction.

Let the half-life period be τ.

$$\text{At this time } c_A = c_{A0}/2 \tag{3.34}$$

Applying this to a first order reaction using equation (3.17) gives:

$$\tau = \frac{1}{k_1} \text{Ln} \frac{c_{A0}}{c_{A0}/2} = \frac{\text{Ln}2}{k_1} \tag{3.35}$$

and to a second order reaction with a single reactant:

$$\tau = \frac{1}{k_2} \left[\frac{1}{c_{A0}/2} - \frac{1}{c_{A0}} \right] = \frac{1}{k_2 c_{A0}}. \tag{3.36}$$

In general for an nth order reaction with a single reactant:

$$\tau = \frac{2^{n-1} - 1}{(n-1) k_n c_{A0}^{n-1}} \tag{3.37}$$

i.e.

$$\tau \propto (c_{A0})^{-(n-1)}. \tag{3.38}$$

Experiments are carried out at two or more different initial concentrations. First, if the half-life is independent of initial concentration, then the reaction is first order. If the half-life varies with initial concentration, a plot of half-life against initial concentration will give a straight line of slope $-(n-1)$. Alternatively the results τ_1 and τ_2 of two experiments with initial concentrations c_1 and c_2 may be divided to yield the expression:

$$\frac{\tau_1}{\tau_2} = \left(\frac{c_2}{c_1}\right)^{n-1} \tag{3.39}$$

from which

$$n = 1 + \frac{\mathrm{Log}\,\dfrac{\tau_1}{\tau_2}}{\mathrm{Log}\,\dfrac{c_2}{c_1}}. \tag{3.40}$$

Example 3. The following results have been published for the decomposition of ammonium cyanate solutions:

	Expt 1	Expt 2	Expt 3
Initial NH_4CNO concentration (g mole litre^{-1})	0·05	0·1	0·2
Half-life (h)	37·0	19·2	9·5

Applying equation (3.40) to the three possible combinations of these data leads to the following three estimates of reaction order:

Expts 1 and 2 $n = 1{\cdot}950$
Expts 1 and 3 $n = 1{\cdot}982$
Expts 2 and 3 $n = 2{\cdot}015$

It is reasonably concluded that the reaction order is 2·0.

3.6.3 Variation of initial slopes

Pairs of experiments are performed with different initial concentrations and the resulting different initial reaction rates are deduced by measurement of the initial slope of the concentration time plot. The limited accuracy of measurement of such initial slopes is a serious restriction of this method.

If initial concentrations c_1 and c_2 lead to initial rate values (dc_1/dt) and (dc_2/dt) respectively, then:

$$-\left(\frac{dc_1}{dt}\right) = k_n c_1^n \tag{3.41}$$

and

$$-\left(\frac{dc_2}{dt}\right) = k_n c_2^n. \tag{3.42}$$

Taking logarithms, subtracting and rearranging leads to the elimination of k_n and to the following equation for n:

$$n = \frac{\text{Log}\,(dc_1/dt) - \text{Log}\,(dc_2/dt)}{\text{Log}\,c_1 - \text{Log}\,c_2}. \tag{3.43}$$

3.6.4 Isolation method

This technique is applied to situations where the rate of reaction is governed by the concentration of more than one reactant. In making the experimental observations of conversion with time, it is arranged that all of the reactants except one are present in excess. Hence in the course of the reaction the concentration of the one reactant will change appreciably while the concentrations of those present in excess will remain substantially constant. Thus a kinetic equation of the form:

$$-\frac{dc_A}{dt} = K_n c_A^\alpha c_B^\beta c_C^\gamma \tag{3.44}$$

would be reduced in such experimentation to

$$-\frac{dc_A}{dt} = k_s c_A^\alpha \tag{3.45}$$

where B and C are present in excess and

$$k_s \equiv k_n c_B^\beta c_C^\gamma$$

is called a pseudo-rate constant. Using the techniques already described the order 'α' is determinable from the observations. Further similar experiments are performed keeping A and C in excess to determine β, and with A and B in excess to determine γ.

3.6.5 Reference curve method

Theoretical plots may be prepared of fractional conversion against a dimensionless time for various nominated reaction orders. These theoretical plots thus have reaction order as a parameter but the use of a dimensionless time is seen below to render them independent of the numerical magnitude of the rate constant. Actual observed experimental data may then be plotted on the same coordinates and the position of this superimposed data allows an estimate of the reaction order to be made. The method has the advantage of utilising a whole range of experimental observations and of giving a good estimate of fractional orders.

Considering the general forward equation

$$\frac{1}{c_t^{n-1}} - \frac{1}{c_0^{n-1}} = (n-1)\,k_n t \tag{3.32}$$

and letting $t_{0\cdot9}$ be the time at which the initial concentration has fallen by a factor of ten to $c_{0\cdot9}$ (i.e. 90% conversion) we have:

$$\frac{1}{c_{0\cdot9}^{n-1}} - \frac{1}{c_0^{n-1}} = (n-1)\,k_n t_{0\cdot9}. \tag{3.46}$$

Division of these equations gives:

$$\frac{t}{t_{0\cdot9}} = \frac{\dfrac{1}{c_t^{n-1}} - \dfrac{1}{c_0^{n-1}}}{\dfrac{1}{c_{0\cdot9}^{n-1}} - \dfrac{1}{c_0^{n-1}}} = \frac{\left(\dfrac{c_0}{c_t}\right)^{n-1} - 1}{\left(\dfrac{c_0}{c_{0\cdot9}}\right)^{n-1} - 1}. \tag{3.47}$$

Now the above definition of $c_{0\cdot9}$ means that

$$\frac{c_{0\cdot9}}{c_0} = \frac{1}{10} \tag{3.48}$$

and using the definition of fractional conversion:

$$\frac{c_t}{c_0} = 1 - f \tag{3.19}$$

equation (3.47) is converted to (3.49) as required for direct plotting of the reference curves

$$\frac{t}{t_{0\cdot9}} = \frac{\left(\dfrac{1}{1-f}\right)^{n-1} - 1}{(10)^{n-1} - 1} \quad \text{for} \quad n \neq 1. \tag{3.49}$$

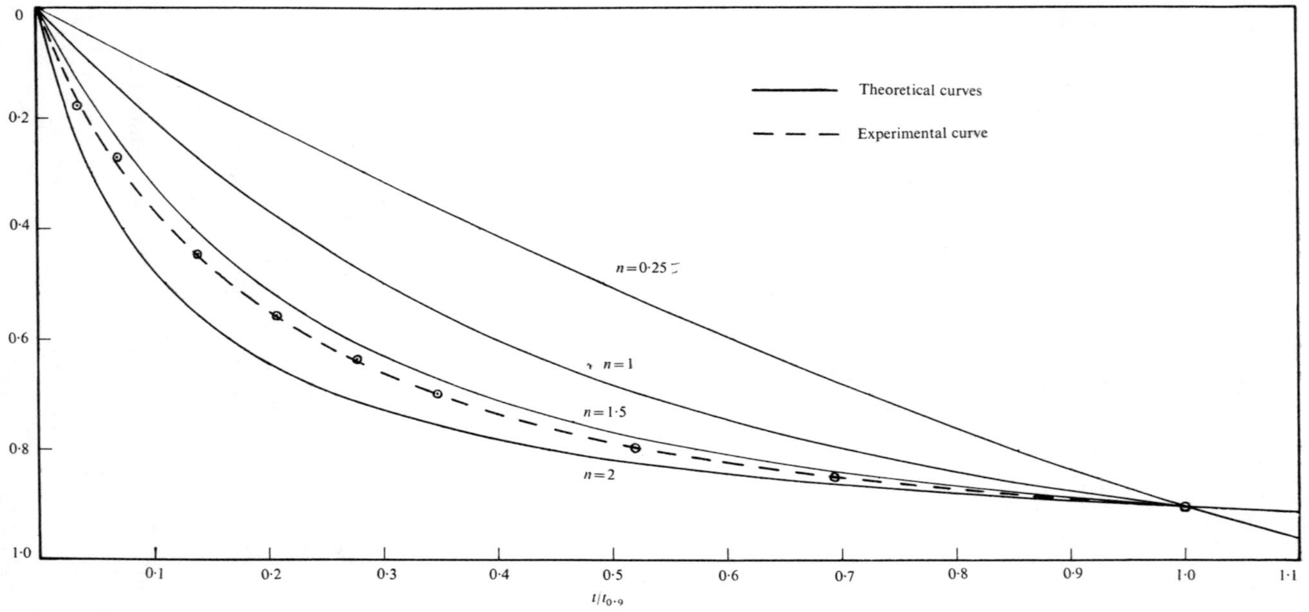

FIG. 3.2. Reference curves

Similar treatment of equation (3.20) for first order reactions leads to

$$\frac{t}{t_{0\cdot 9}} = \frac{\mathrm{Ln}\,\dfrac{1}{1-f}}{\mathrm{Ln}\,10} \qquad \text{for} \quad n = 1 \tag{3.50}$$

Applying these equations to the cases of $n = \frac{1}{4}$, 1, $1\frac{1}{2}$ and 2, and plotting leads to the full lines of Fig. 3.2 in which the fractional conversion is plotted as a function of the dimensionless time ratio $t/t_{0\cdot 9}$. These lines are called reference curves and it is, of course, possible to plot alternative reference curves based on fractional conversions other than 0·9.

The first two columns of the following table contain experimental concentration–time data and the remaining two columns show the processed results needed for use of the reference curves of Fig. 3.2.

t (min)	C_t g litre^{-1}	$(t/t_{0\cdot 9})$	f
0	5·19	0	0
0·1	4·39	0·035	0·155
0·2	3·77	0·069	0·272
0·4	2·89	0·139	0·442
0·6	2·30	0·208	0·556
0·8	1·88	0·277	0·638
1·0	1·57	0·347	0·697
1·5	1·07	0·520	0·794
2·0	0·80	0·693	0·848
5·0	0·25	1·733	0·952
10·0	0·094	3·466	0·982

These results have been plotted on Fig. 3.2 from which it can be seen that the reaction order lies between 1·5 and 2·0 and linear interpolation would suggest an order of about 1·62–1·63.

3.7 Variable Volume Reactions

While, as already suggested, reactions normally take place at essentially constant volume, it is the purpose of this section to show how equation (3.9) may be applied to situations in which volume changes cannot be neglected.

Consider the reaction:

$$2NO + O_2 \rightarrow 2NO_2$$

converting a stoichiometric mixture of reactants at a fixed pressure and constant temperature with observations of total volume against time. At temperatures below 290°C the reaction is known to be forward only and to have the kinetics:

$$\frac{d}{dt}(NO_2) = k_3 V[NO]^2[O_2]. \tag{3.51}$$

Let V_0 be the initial volume of reaction mixture at pressure P and absolute temperature T. Then using the ideal gas law, the moles of each component and the total moles at zero time and at time 't' when the fractional conversion of either reactant is f are as follows:

	NO	O_2	NO_2	Total
Moles at $t = 0$	$\frac{2}{3}\frac{PV_0}{RT}$	$\frac{1}{3}\frac{PV_0}{RT}$	0	$\frac{PV_0}{RT}$
Moles at t	$\frac{2}{3}\frac{PV_0}{RT}(1-f)$	$\frac{1}{3}\frac{PV_0}{RT}(1-f)$	$\frac{1}{3}\frac{PV_0}{RT}2f$	$\frac{PV_0}{RT}(1-\frac{1}{3}f)$

Again using the ideal gas law and the total moles at time 't', the volume at time t is:

$$V_0(1 - \tfrac{1}{3}f).$$

Hence the concentrations of NO and O_2 are respectively:

$$\frac{2P(1-f)}{3RT(1-\frac{1}{3}f)} \quad \text{and} \quad \frac{P(1-f)}{3RT(1-\frac{1}{3}f)}.$$

The various terms required by the kinetic equation have now been developed and their substitution gives:

$$\frac{d}{dt}\left(\frac{PV_0}{3RT}2f\right) = k_3 V_0 (1-\tfrac{1}{3}f)\left[\frac{2P(1-f)}{3RT(1-\frac{1}{3}f)}\right]^2\left[\frac{P(1-f)}{3RT(1-\frac{1}{3}f)}\right] \tag{3.52}$$

which on rearrangement leads to:

$$\frac{2k_3P^2}{(3RT)^2}\int_0^t dt = \int_0^f \frac{(1-\frac{1}{3}f)^2}{(1-f)^3}\,df \tag{3.53}$$

and on integration to:

$$\frac{2k_3P^2}{(3RT)^2} = \frac{1}{9}\left[\frac{2(4f-3f^2)}{(1-f)^2} + \text{Ln}\frac{1}{1-f}\right]. \tag{3.54}$$

Using equation (3.13) an alternative form of the kinetic equation could be developed as:

$$[NO_2]\frac{dV}{dt} + V\frac{d}{dt}[NO_2] = k_3 V[NO]^2[O_2]. \quad (3.55)$$

Using the above general values of V and nitric oxide concentration the following further terms may be developed:

$$\frac{dV}{dt} = -\frac{V_0}{3}\frac{df}{dt} \quad (3.56)$$

and

$$\frac{d}{dt}[NO_2] = \frac{2P}{3RT}\frac{d}{dt}\left[\frac{f}{1-\frac{1}{3}f}\right]. \quad (3.57)$$

On substitution into equation (3.55) the identical integral to equation (3.54) again follows.

3.8 The Effect of Temperature on the Specific Rate Constant

It has been stated earlier that the specific reaction rate constant and the order of reaction must be determined experimentally. However, attempts have been made to predict rate constants from a knowledge of the properties of the molecules taking part in the reaction, and the resultant expressions reveal the manner in which the rate constant may be expected to depend on temperature.

The first criterion to be satisfied if reaction is to occur is that the reacting molecules must collide and hence an estimate of reaction rate in gases might be expected from the collision rate as predicted by the simple kinetic theory. However, the further condition of sufficient energy to effect reaction must also be satisfied and it is found that in only a small fraction of collisions is there the necessary energy to lead to fruitful reaction. This situation may be summarised by the expression:

$$k = Zq \quad (3.58)$$

in which Z is the collision rate and q is the fraction of collisions in which there is sufficient energy for reaction to occur.

The kinetic theory of gases shows Z to be proportional to the square root of the absolute temperature and that:

$$q = \exp(-E/RT) \quad (3.59)$$

where E is the molecular translation energy.

Application of the resultant expressions to the dissociation of hydrogen iodide has led to the classical result of $k = 5{\cdot}2 \times 10^{-7}$ mole litre^{-1} sec^{-1} in comparison with the experimental value of $3{\cdot}5 \times 10^{-7}$ mole litre^{-1} sec^{-1}.

In general, agreement is by no means as close as this and over-estimates of reaction rate by as much as a factor of 10^9 sometimes occur. This discrepancy is partly accounted for by an orientation factor P and partly by entropy of activation. In the case of two simple hydrogen iodide molecules colliding prior to decomposition, the result would not be expected to depend very much on their relative orientation. However, if two organic molecules are to react, the success may be expected to depend on the coming together of the functional groups, and thus while there may be sufficient energy for reaction there may not be any reaction because of poor relative orientation of the functional groups. Equation (3.58) is thus modified to:

$$k = PZq. \qquad (3.60)$$

The mathematical form of Z and q enables the dependence of rate constant on temperature to be expressed as:

$$k = A'\sqrt{T}.\exp(-E/RT). \qquad (3.61)$$

More advanced theories of reaction rate have also predicted the temperature dependence of k to be of the form:

$$k = \phi(T)\exp(-E/RT). \qquad (3.62)$$

No further discussion of other theories is given in this text but the interested reader should consult Glasstone, Laidler and Eyring (1941).

The exponential term of equation (3.61) is much more sensitive to changes in temperature than the remaining factor and hence for relatively small ranges of temperature the expression may be simplified to:

$$k = A\exp(-E/RT) \qquad (3.63a)$$

or

$$\mathrm{Ln}\,k = \mathrm{Ln}\,A - E/RT. \qquad (3.63b)$$

Equation (3.63a) is known as the 'Arrhenius Equation', the constant A being referred to as the 'pre-exponential factor' or the 'frequency factor' and E the 'activation energy'. This equation is used extensively in the design and analysis of chemical reactors operating under non-isothermal conditions.

The activation energy of a reaction is usually determined by plotting either Ln k or Log k against $1/T$ and deduction of E from the slope of the resultant straight line. The development of an Arrhenius Equation from experimental data is illustrated by the following example:

Example 4. Using the following experimental results for the variation of the specific reaction rate with temperature for the decomposition of an organic acid establish an expression relating k and T.

T°C	$k_1 \times 10^4$ h^{-1}
50·0	1·08
70·1	7·34
89·4	45·4
101·0	138·0

Solution. This problem could be conveniently solved by plotting (Ln k) against $(1/T)$. However if there is a fair amount of scatter in the results it is difficult to predict the best straight line by eye. The calculation below illustrates the use of the method of least squares (Davies, 1958) to obtain a reliable equation from a set of scattered points. The slope of the best straight line is then:

$$m = \frac{N\Sigma\left(\frac{1}{T}\right)(\log k) - \Sigma\left(\frac{1}{T}\right)\Sigma(\log k)}{N\Sigma\left(\frac{1}{T}\right)^2 - \left[\Sigma\left(\frac{1}{T}\right)\right]^2} \quad \text{(I)}$$

and the intercept is:

$$C = \frac{\Sigma\left(\frac{1}{T}\right)^2\Sigma(\log k) - \Sigma\left(\frac{1}{T}\right)(\log k)\,\Sigma\left(\frac{1}{T}\right)}{N\Sigma\left(\frac{1}{T}\right)^2 - \left[\Sigma\left(\frac{1}{T}\right)\right]^2}. \quad \text{(II)}$$

Then

T°C	T°K	$\frac{1}{T}$	$k \times 10^4$	$\log(k \times 10^4)$	$\left(\frac{1}{T}\right)^2$	$\left(\frac{1}{T}\right)(\log k \times 10^4)$
50·0	323·0	$3{\cdot}09 \times 10^{-3}$	1·08	0·0334	$9{\cdot}55 \times 10^{-6}$	$1{\cdot}032 \times 10^{-4}$
70·1	343·1	$2{\cdot}91 \times 10^{-3}$	7·34	0·8657	$8{\cdot}46 \times 10^{-6}$	$25{\cdot}192 \times 10^{-4}$
89·4	362·4	$2{\cdot}76 \times 10^{-3}$	45·4	1·6571	$7{\cdot}62 \times 10^{-6}$	$45{\cdot}736 \times 10^{-4}$
101·0	374·0	$2{\cdot}67 \times 10^{-3}$	138·0	2·1399	$7{\cdot}13 \times 10^{-6}$	$57{\cdot}135 \times 10^{-4}$
		$11{\cdot}43 \times 10^{-3}$		4·6961	$32{\cdot}76 \times 10^{-6}$	$129{\cdot}095 \times 10^{-4}$

Then

$$m = \frac{(4 \times 1{\cdot}291 \times 10^{-2}) - (1{\cdot}143 \times 10^{-2} \times 4{\cdot}6961)}{(4 \times 3{\cdot}276 \times 10^{-5}) - (1{\cdot}143 \times 10^{-2})^2}$$

$$= \frac{5{\cdot}164 \times 10^{-2} - 5{\cdot}368 \times 10^{-2}}{0{\cdot}44 \times 10^{-6}}$$

$$m = -4636$$

and

$$C = \frac{(3{\cdot}276 \times 10^{-5} \times 4{\cdot}6961) - (129{\cdot}095 \times 10^{-4} \times 1{\cdot}143 \times 10^{-2})}{0{\cdot}44 \times 10^{-6}}$$

$$C = 14{\cdot}43.$$

Therefore

$$\log(k \times 10^4) = 14{\cdot}43 - \frac{4636}{T} \qquad \text{(III)}$$

Then

$$\text{Ln}(k \times 10^4) = 33{\cdot}24 - \frac{10680}{T} \qquad \text{(IV)}$$

but $10680 = E/R$ and so $E = 21\,200$ cal (g mole)$^{-1}$.
The Arrhenius equation may be written as

$$k \times 10^4 = 2{\cdot}69 \times 10^{14} \exp\left(\frac{-21\ 200}{RT}\right) \qquad \text{(V)}$$

or as

$$k \times 10^4 = \exp\left(33{\cdot}24 - \frac{21\,200}{RT}\right). \qquad \text{(VI)}$$

Substitution of the experimental temperatures into any of equations (III), (IV), (V) or (VI) leads to the following comparison between the experimental values of k and those implied by the fitted straight line:

T°K	323·0	343·1	362·4	374
$k \times 10^4$ Experimental	1·08	7·34	45·4	138
$k \times 10^4$ Predicted	1·19	8·27	43·2	108

Equation (3.63b) may be evaluated between limits of temperature T_1 and T_2 and corresponding rate constants k_{T_1} and k_{T_2} to give:

$$\text{Ln}\frac{k_{T_1}}{k_{T_2}} = -\frac{E}{R}\left[\frac{1}{T_1} - \frac{1}{T_2}\right] = -\frac{E}{R}\left[\frac{T_2 - T_1}{T_1 T_2}\right]. \tag{3.64}$$

Since E is always positive, it will be seen that the rate constant always increases with temperature. The magnitude of E will determine the sensitivity of the rate constant to the changes in temperature. A rough working guide that the rate constant is doubled for every increase of 10 degC has often been applied. The above equation will now be used to examine the implication of this statement on the activation energy in the region of 100°C.

Putting $T_1 = 383$ and $T_2 = 373$ with $k_{T_1} = 2k_{T_2}$:

$$\text{Ln}\,2 = -\frac{E}{1{\cdot}986}\left[\frac{-10}{383 \times 373}\right]$$

or

$$E = 19650 \text{ cal (g mole)}^{-1}.$$

In practice E may range from 5000 to 50000 cal (g mole)$^{-1}$ or even wider and so a 10 degC change in temperature may lead to a smaller or greater factor than two in the rate constant. A secondary effect is the level of temperature; in the above case an activation energy of 19650 cal (g mole)$^{-1}$ gives a factor of 2·51 between 50 and 60°C and a factor of only 1·18 between 500 and 510°C.

A connection exists between the heat of reaction and the activation energies of the forward and reverse rate constants in any particular reaction.

From equations (2.66) and (3.69) of the following section:

$$\frac{d}{dT}\text{Ln}\frac{k_1}{k_2} = \frac{\Delta H}{RT^2}$$

and differentiation of equation (3.63b) leads to:

$$\frac{d}{dT}\operatorname{Ln} k = \frac{E}{RT^2} \tag{3.65}$$

which if applied to the forward and reverse rate constants k_1 and k_2 with respective activation energies E_1 and E_2 and substituted into equation (2.66) gives:

$$E_1 - E_2 = \Delta H. \tag{3.66}$$

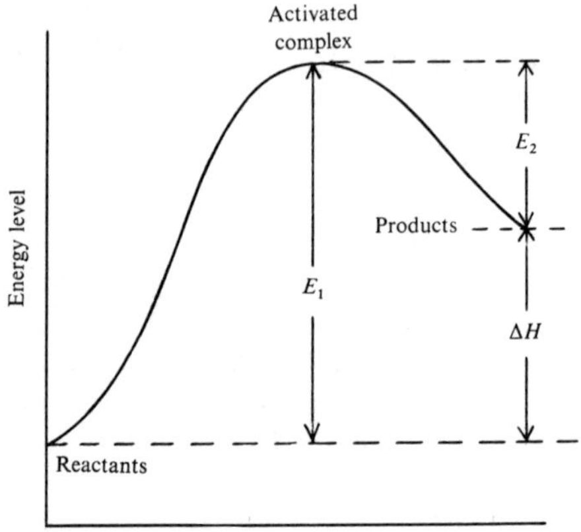

Endothermic reaction

($E_1 > E_2$; ΔH positive)

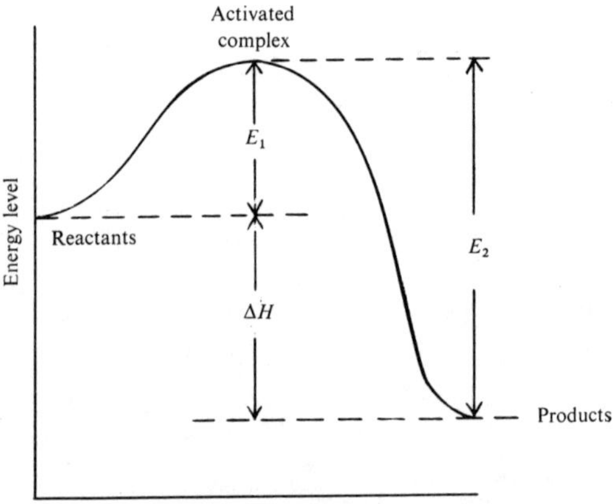

Exothermic reaction

($E_1 < E_2$; ΔH negative)

FIG. 3.3. Activation energies and heat of reaction

This connection is illustrated in Fig. 3.3 for both endothermic and exothermic reactions. The figure shows the energy level which must be attained on collision for reaction to take place, and the highest energy level is associated with the formation of an intermediate activated compound of the reactants which subsequently decomposes to the products. This intermediate activated compound is called the activated complex and it is now agreed that all chemical reactions proceed through such complexes.

3.9 The Reverse Reaction

The above introduction to kinetics has assumed forward reactions only. The same principles apply equally to reverse reactions and in practice the forward and reverse reactions are taking place simultaneously. In the previous chapter equilibrium conversions, at which the forward and reverse rates just balance, were calculated. The kinetic considerations which now follow are concerned with the time taken to approach equilibrium and are illustrated by first and second order situations.

3.9.1 First order reversible reaction

For the reaction:

$$A \underset{k_2}{\overset{k_1}{\rightleftharpoons}} B \tag{3.67}$$

k_1 is the rate constant for the conversion of A to B and k_2 for the back reaction of B to A. Hence the overall rate of conversion of A is:

$$-\frac{dc_A}{dt} = k_1 c_A - k_2 c_B. \tag{3.68}$$

Equilibrium is defined when dc_A/dt is zero and then the equilibrium values of concentration c'_A and c'_B are given by:

$$\frac{k_1}{k_2} = \frac{c'_B}{c'_A} = K \tag{3.69}$$

which is seen to be consistent with equation (2.54) of the preceding chapter.

Whatever the kinetic form of a reaction it is always possible to express the equilibrium constants as the ratio of the two velocity constants by setting the net rate of reaction at zero as above.

Let the initial reaction mixture contain A at a concentration c_{A0} and B at c_{B0} and let f be the fractional conversion of A at time t. Then from mass balances

$$c_A = c_{A0}(1-f) \quad \text{and} \quad c_B = c_{B0} + fc_{A0} \tag{3.70}$$

which on substitution into the kinetic equation give:

$$c_{A0}\frac{df}{dt} = k_1 c_{A0}(1-f) - k_2(c_{B0} + fc_{A0}) \tag{3.71}$$

i.e.

$$\frac{df}{dt} = \left(k_1 - k_2\frac{c_{B0}}{c_{A0}}\right) - (k_1 + k_2)f. \tag{3.72}$$

The forward rate constant is often factorised from such expressions giving in this instance:

$$\frac{df}{dt} = k_1\left[\left(1 - \frac{1}{K}\frac{c_{B0}}{c_{A0}}\right) - \left(1 + \frac{1}{K}\right)f\right] \tag{3.73}$$

in terms of the equilibrium constant, K. Hence

$$\int_0^t dt = \frac{1}{k_1}\int_0^f \frac{df}{\left(1 - \frac{1}{K}\frac{c_{B0}}{c_{A0}}\right) - \left(1 + \frac{1}{K}\right)f} \tag{3.74}$$

i.e.

$$t = \frac{1}{\left(1 + \frac{1}{K}\right)k_1}\,\mathrm{Ln}\,\frac{\left(1 - \frac{1}{K}\frac{c_{B0}}{c_{A0}}\right)}{\left(1 - \frac{1}{K}\frac{c_{B0}}{c_{A0}}\right) - \left(1 + \frac{1}{K}\right)f} \tag{3.75}$$

or

$$t = \frac{1}{k_1 + k_2}\,\mathrm{Ln}\,\frac{\left(1 - \frac{1}{K}\frac{c_{B0}}{c_{A0}}\right)}{\left(1 - \frac{1}{K}\frac{c_{B0}}{c_{A0}}\right) - \left(1 + \frac{1}{K}\right)f}. \tag{3.76}$$

A less cumbersome procedure is the re-arrangement of the equation before integration by introduction of the equilibrium fractional f_e as follows:

$$k_1 c_{A0}(1 - f_e) = k_2(c_{B0} + f_e c_{A0}) \tag{3.77}$$

from which

$$k_2 c_{B0} = k_1 c_{A0} - f_e c_{A0}(k_1 + k_2) \tag{3.78}$$

and on substitution of c_{B0} from this expression into equation (3.72):

$$\frac{df}{dt} = k_1(1 - f) - k_1 + f_e(k_1 + k_2) - k_2 f \tag{3.79}$$

i.e.

$$\frac{df}{dt} = (k_1 + k_2)(f_e - f) \tag{3.80}$$

which on integration between limits gives the simpler equivalent of equation (3.76):

$$t = \frac{1}{k_1 + k_2} \text{Ln} \frac{f_e}{f_e - f} = \frac{1}{k_1 + k_2} \text{Ln} \frac{c_{A0} - c_{Ae}}{c_A - c_{Ae}} \tag{3.81}$$

where c_{Ae} is the equilibrium concentration of A.

Thermodynamic considerations enable K, f_e or c_{Ae} to be evaluated and then the insertion of experimental results of f or c_A and time into equations (3·76) or (3.81) enable $(k_1 + k_2)$ to be determined either numerically or graphically. This result is then solved simultaneously with the thermodynamic ratio of $K = k_1/k_2$ to yield the individual forward and backward rate constants.

3.9.2 Second order reversible reaction

Second order reversible reactions of the types:

$$A + B \underset{k_2}{\overset{k_1}{\rightleftarrows}} C + D \tag{3.82}$$

with

$$-\frac{dc_A}{dt} = k_1 c_A c_B - k_2 c_C c_D \tag{3.83}$$

or

$$A + B \underset{k_2}{\overset{k_1}{\rightleftarrows}} 2C \tag{3.84}$$

with

$$-\frac{dc_A}{dt} = k_1 c_A c_B - k_2 c_C^2 \tag{3.85}$$

are common in organic synthesis, esterification and saponification reactions.

Example 5. Integration of equation (3.83) is illustrated below in order to analyse the following experimental data for the reaction in a solution of nitrobenzene between dimethyl-p-toluidine and methyl iodide to give a quaternary ammonium salt. The reaction is:

$$CH_3I + CH_3 . C_6H_5 . N(CH_3)_2 \rightleftharpoons CH_3 . C_6H_5 . N(CH_3)_3^+ I^-$$

Time (mins)	Fractional conversion
10·2	0·175
26·5	0·343
36·0	0·402
78·0	0·523

The initial mixture contained the reactants at equal concentrations of 0·05 g mole litre^{-1} and the equilibrium constant is 1·43. The data is to be tested to establish whether it can be represented adequately by forward kinetics only or whether the reverse reaction also needs to be considered.

For the irreversible hypothesis, equation (3.24) was used and the deduced values of rate constant are shown in the final column of the table below. The drift in these values is sufficiently large for this hypothesis to be rejected.

Using equation (3.83) to allow for the reverse reaction gives:

$$c_{A0}\frac{df}{dt} = k_2\left[c_{A0}^2(1-f)^2 - \frac{c_{A0}^2}{K}f^2\right] \tag{I}$$

or

$$t = \frac{1}{k_2 c_{A0}}\int_0^f \frac{df}{(1-f)^2 - (\alpha f)^2}$$

$$= \frac{1}{k_2 c_{A0}}\int_0^f \frac{df}{[1-f(\alpha+1)][1+f(\alpha-1)]} \tag{II}$$

where

$$\alpha = \frac{1}{\sqrt{K}}. \tag{III}$$

On resolution into partial fractions, integration and substitution of limits:

$$t = \frac{1}{2\alpha k_2 c_{A0}} \operatorname{Ln} \frac{1 + f(\alpha - 1)}{1 - f(\alpha + 1)}. \qquad \text{(IV)}$$

Inserting the experimental values of t and f into (IV) gives the results for k_2 shown in the table below.

Time (min)	f	k_2 (Rev) litre (g mole)$^{-1}$ min^{-1}	k_2 (Irrev) litre (g mole)$^{-1}$ min^{-1}
10·2	0·175	0·425	0·415
26·5	0·343	0·426	0·394
36·0	0·402	0·426	0·375
78·0	0·523	0·475	0·280

The specific reaction rates for the second order reversible reaction give reasonably constant values of k_2, and so this hypothesis for explanation of the overall reaction rate is accepted.

3.10 Simultaneous and Consecutive Reactions

The above introduction to kinetics has treated single irreversible reactions for the most part. However, in section (3.9) the concept of simultaneous forward and backward rate processes in a single reaction was introduced, and it is the purpose of the present section to consider more complicated simultaneous and consecutive reactions.

In a chemical process, conversion of A to B may be the objective, but the following two possibilities might serve to reduce the yield of B:

(*i*)

$$\begin{array}{l} A \xrightarrow{k_1} B \\ k_2 \downarrow \\ C \end{array} \qquad (3.86)$$

(*ii*)

$$A \xrightarrow{k_1} B \xrightarrow{k_2} C \qquad (3.87)$$

In (*i*) the yield of B is reduced by the simultaneous side reaction of reactant A being converted to unwanted C, while in (*ii*) B undergoes a consecutive reaction to produce unwanted C. In either of these examples any, or all of the reactions could be reversible, and the various possibilities are elaborated later in this section. It will be shown that A, B and C concentrations may be obtained as functions of time using calculus techniques although analogue computation will provide solutions more conveniently.

In practice many more complicated situations arise; an example is the chlorination of benzene (MacMillin, 1948; Jenson and Jeffreys, 1965) by the following set of second order reactions:

$$C_6H_6 + Cl_2 \xrightarrow{k_1} C_6H_5Cl + HCl \tag{3.88}$$

$$C_6H_5Cl + Cl_2 \xrightarrow{k_2} C_6H_4Cl_2 + HCl \tag{3.89}$$

$$C_6H_5Cl_2 + Cl_2 \xrightarrow{k_3} C_6H_3Cl_3 + HCl \tag{3.90}$$

The above reference explains how time may be eliminated from the initial rate equations to give other equations that enable the relative concentration of benzene and the three chlorobenzenes to be obtained at corresponding times. This procedure is adopted when it is not possible to solve for the concentrations as functions of time by the usual calculus techniques. In order to retain time in the analysis, analogue computation is employed and examples of the application of three computer techniques are given later in this chapter.

3.10.1 Simultaneous irreversible side reaction

Consider the chemical reactions:

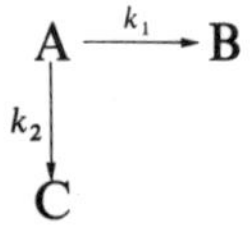

(3.86)

in which the reactions are first order and the initial concentration of A is 'a' and B and C are both zero.

The rate equations representing the reactions are:

$$-\frac{dc_A}{dt} = k_1c_A + k_2c_A = (k_1 + k_2)c_A \tag{3.91}$$

$$\frac{dc_B}{dt} = k_1c_A \tag{3.92}$$

and

$$\frac{dc_C}{dt} = k_2 c_A. \tag{3.93}$$

Solution of equation (3.91) gives:

$$c_A = a \exp\{-(k_1 + k_2)t\}. \tag{3.94}$$

Substituting this expression into equations (3.92) and (3.93) give:

$$c_B = \frac{k_1 a}{k_1 + k_2}[1 - \exp\{-(k_1 + k_2)t\}] \tag{3.95}$$

and

$$c_C = \frac{k_2 a}{k_1 + k_2}[1 - \exp\{-(k_1 + k_2)t\}]. \tag{3.96}$$

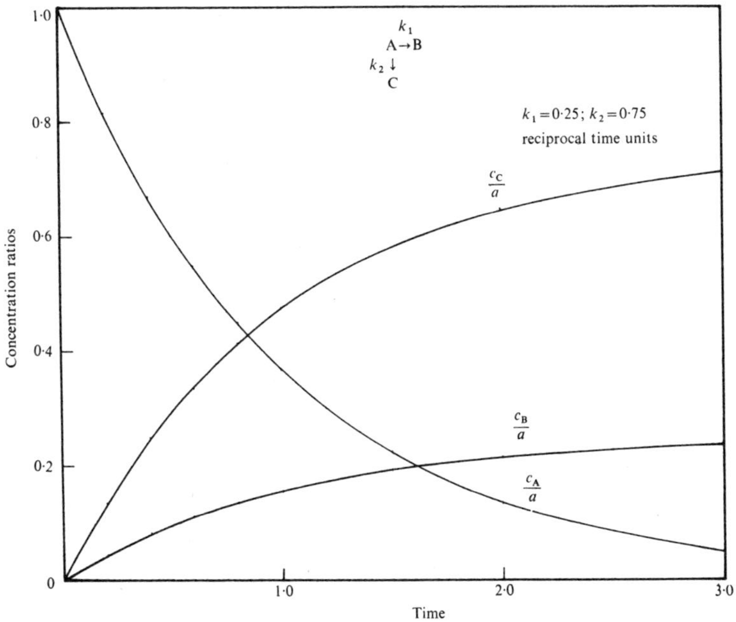

FIG. 3.4. Simultaneous conversion of A into B and C by first order mechanisms

Thus A decreases exponentially with time while B and C increase asymptotically to their ultimate concentrations:

$$\frac{k_1 a}{k_1 + k_2}$$

and

$$\frac{k_2 a}{k_1 + k_2}$$

Furthermore at any time the relative concentrations of B and C are dependent on the fixed ratio of the rate constants. A typical plot is shown in Fig. 3.4.

3.10.2 Consecutive irreversible reaction

The differential equations which represent the reaction scheme:

$$A \xrightarrow{k_1} B \xrightarrow{k_2} C \qquad (3.97)$$

are:

$$\frac{dc_A}{dt} = -k_1 c_A \qquad (3.98)$$

$$\frac{dc_B}{dt} = k_1 c_A - k_2 c_B \qquad (3.99)$$

$$\frac{dc_C}{dt} = k_2 c_B. \qquad (3.100)$$

The Laplace transforms of these equations for the initial conditions: $c_A = a$, $c_B = 0$ and $c_C = 0$ are:

$$s\bar{c}_A - a = -k_1 \bar{c}_A \qquad (3.101)$$

$$s\bar{c}_B - 0 = k_1 \bar{c}_A - k_2 \bar{c}_B \qquad (3.102)$$

and

$$s\bar{c}_C - 0 = k_2 \bar{c}_B. \qquad (3.103)$$

Hence from equation (3.101):

$$(s + k_1)\bar{c}_A = a \tag{3.104}$$

and

$$c_A = \mathscr{L}^{-1}\left[\frac{a}{s + k_1}\right] = a \exp(-k_1 t) \tag{3.105}$$

and by elimination of $\bar{c}_A$:

$$(s + k_2)\bar{c}_B = \frac{k_1 a}{s + k_1} \tag{3.106}$$

and

$$\bar{c}_B = \frac{k_1 a}{(s + k_1)(s + k_2)} = \frac{k_1 a}{k_2 - k_1}\left[\frac{1}{s + k_1} - \frac{1}{s + k_2}\right] \tag{3.107}$$

or

$$c_B = \frac{k_1 a}{k_2 - k_1}[\exp(-k_1 t) - \exp(-k_2 t)] \tag{3.108}$$

similarly elimination of $\bar{c}_B$ gives:

$$c_C = \mathscr{L}^{-1}\left[\frac{k_1 k_2 a \; 1}{s(s + k_1)(s + k_2)}\right] \tag{3.109}$$

$$= k_1 k a \, \mathscr{L}^{-1}\left[\frac{1}{k_1 k_2 s} + \frac{1}{k_2 - k_1}\left\{\frac{1}{k_2(s + k_2)} - \frac{1}{k_1(s+k_1)}\right\}\right]. \tag{3.110}$$

On inversion and re-arrangement:

$$c_C = a\left[1 + \frac{1}{k_2 - k_1}\{k_1 \exp(-k_2 t) - k_2 \exp(-k_1 t)\}\right]. \tag{3.111}$$

A plot of these results in Fig. 3.5 shows A decreasing exponentially, B passing through a maximum and C increasing monotonously via a point of inflexion. Analysis of the equations shows that the maximum concentration of B is:

$$a\left(\frac{k_2}{k_1}\right)^{k_2/(k_1 - k_2)}$$

occurring at a time:

$$\frac{\mathrm{Ln}\left(\frac{k_1}{k_2}\right)}{k_1 - k_2}$$

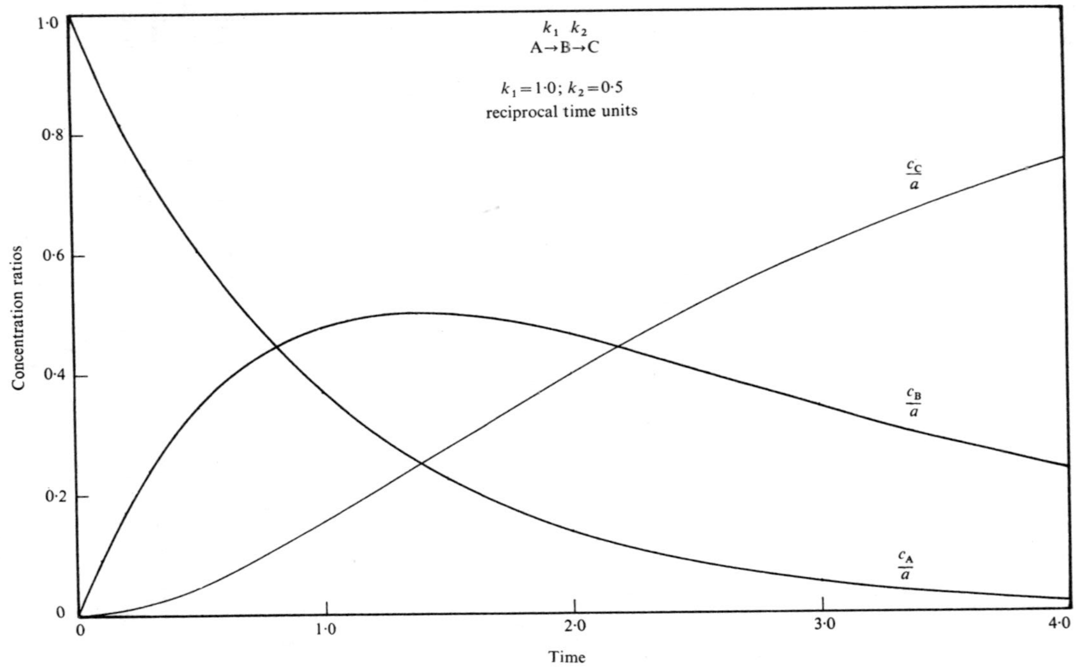

FIG. 3.5. Consecutive reactions

and that the time of the C curve inflexion coincides with the time of maximum B.

3.10.3 Consecutive reversible reactions

The analysis of the preceding section is now extended to cover the case in which both reactions are reversible:

$$A \underset{k_2}{\overset{k_1}{\rightleftharpoons}} B \underset{k_4}{\overset{k_3}{\rightleftharpoons}} C. \tag{3.112}$$

For first order kinetics the three differential equations are:

$$\frac{dc_A}{dt} = k_2 c_B - k_1 c_A \tag{3.113}$$

$$\frac{dc_B}{dt} = k_1 c_A + k_4 c_C - (k_2 + k_3) c_B \tag{3.114}$$

$$\frac{dc_C}{dt} = k_3 c_B - k_4 c_C. \tag{3.115}$$

From (3.113)

$$c_A = \frac{k_2 c_B}{D + k_1} \tag{3.116}$$

and from (3.115)

$$c_C = \frac{k_3 c_B}{D + k_4} \tag{3.117}$$

where D is the differential operator d/dt.

Eliminating c_A and c_C from equation (3.114) and re-arranging gives:

$$D^3 c_B + (k_1 + k_2 + k_3 + k_4) D^2 c_B + (k_1 k_4 + k_1 k_3 + k_4 k_2) D c_B = 0. \tag{3.118}$$

Let the initial values of c_B, Dc_B and $D^2 c_B$ be denoted:

$$c_B(0), \quad \dot{c}_B(0) \quad \text{and} \quad \ddot{c}_B(0).$$

and consider, as an example, the case where the initial concentration of A = a and of both B and C is zero:

i.e.

$$c_A(0) = a \tag{3.119}$$

$$c_B(0) = 0 \tag{3.120}$$

$$c_C(0) = 0. \tag{3.121}$$

Then applying zero time to equations (3.113), (3.114) and (3.115) gives:

$$\dot{c}_A(0) = k_1 a \tag{3.122}$$

$$\dot{c}_B(0) = k_1 c_A(0) + k_4 c_C(0) - (k_2 + k_3) c_B(0) = k_1 a \tag{3.123}$$

$$\dot{c}_C(0) = 0. \tag{3.124}$$

Differentiation of (3.114) and applying to zero time gives:

$$\ddot{c}_B(0) = k_1 \dot{c}_A(0) + k_4 \dot{c}_C(0) - (k_2 + k_3)\, \dot{c}_B(0) \tag{3.125}$$

Hence using the results of (3.122) and (3.124):

$$\ddot{c}_B(0) = k_1(-k_1 a) + 0 - (k_2 + k_3) k_1 a = -k_1 a(k_1 + k_2 + k_3). \tag{3.126}$$

Having established the initial values $c_B(0)$, $\dot{c}_B(0)$ and $\ddot{c}_B(0)$ the transformations of the various terms of equation (3.118) are:

$$D^3 c_B : s^3 \bar{c}_B - [s^2 \times 0 + s k_1 a - k_1 a(k_1 + k_2 + k_3)]$$

$$D^2 c_B : s^2 \bar{c}_B - [0 + k_1 a]$$

and

$$D c_B : s \bar{c}_B - [0].$$

Hence writing

$$L \equiv k_1 + k_2 + k_3 + k_4 \tag{3.127}$$

and

$$M \equiv k_1 k_4 + k_1 k_3 + k_4 k_2 \tag{3.128}$$

the transform of the differential equation is:

$$[s^3 \bar{c}_B - s k_1 a + k_1 a(k_1 + k_2 + k_3)] + L\,[s^2 \bar{c}_B - k_1 a] + M\,[s \bar{c}_B] = 0 \tag{3.129}$$

or

$$[s^3 + Ls^2 + Ms]\bar{c}_B = sk_1a + k_1k_4a = k_1a(s + k_4) \quad (3.130)$$

or

$$c_B = k_1a\,\mathscr{L}^{-1}\left[\frac{s + k_4}{s(s^2 + Ls + M)}\right]. \quad (3.131)$$

Hence

$$c_B = k_1a\,\mathscr{L}^{-1}\left[\frac{s + k_4}{s(s - \alpha)(s - \beta)}\right] \quad (3.132)$$

where α and β are the roots of the quadratic term, and are related to L and M by:

$$\alpha\beta = M \quad (3.133)$$

and

$$\alpha + \beta = -L. \quad (3.134)$$

Resolution into partial fractions and inversion yields the following result in terms of α and β:

$$c_B = a\left[\frac{k_1k_4}{\alpha\beta} + \frac{k_1}{(\alpha - \beta)}\left\{\frac{(\alpha + k_4)}{\alpha}\exp(\alpha t) - \frac{(\beta + k_4)}{\beta}\exp(\beta t)\right\}\right]. \quad (3.135)$$

The same procedure may be applied in turn to obtain third order differential equations for c_A and c_C, or alternatively the above result for c_B may be back substituted into equations (3.113) and (3.115) for c_A and c_C respectively. In either case it may be shown that:

$$c_A = a\left[\frac{k_4k_2}{\alpha\beta} + \frac{k_1k_2}{\alpha - \beta}\left\{\frac{\alpha + k_4}{\alpha(\alpha + k_1)}\exp(\alpha t) - \frac{\beta + k_4}{\beta(\beta + k_1)}\exp(\beta t)\right\}\right] \quad (3.136)$$

and

$$c_C = a\left[\frac{k_1k_3}{\alpha\beta} + \frac{k_1k_3}{\alpha - \beta}\left\{\frac{\exp(\alpha t)}{\alpha} - \frac{\exp(\beta t)}{\beta}\right\}\right]. \quad (3.137)$$

Addition of these three results for c_A, c_B and c_C leads, for all values of time, to the total concentration 'a' of the three components in the mixture. The constant sum of c_A, c_B and c_C is special to the reaction system chosen since each reaction yields one mole of product for one mole of reactant converted. This fact could have been utilised in conjunction with equations (3.116) and (3.117) to develop the following second order differential equation for c_B:

$$D^2 c_B + Dc_B [k_1 + k_2 + k_3 + k_4]$$
$$+ [k_2 k_4 + k_1 k_3 + k_1 k_4] c_B = k_1 k_4 a. \qquad (3.138)$$

However, transformation of this differential equation yields the result of equation (3.132) again and hence no simplification of the inversion procedure is achieved. It is likewise possible to obtain second order differential equations for both c_A and c_C which differ from equation (3.138) only in the constant on the right-hand side.

The equilibrium concentrations may be checked by substitution of $t = \infty$ into equations (3.136), (3.135) and (3.137) to give:

$$c_A(\infty) = \frac{k_2 k_4 a}{\alpha\beta} \qquad (3.139)$$

$$c_B(\infty) = \frac{k_1 k_4 a}{\alpha\beta} \qquad (3.140)$$

$$c_C(\infty) = \frac{k_1 k_3 a}{\alpha\beta}. \qquad (3.141)$$

Example 6. These results will be illustrated by a numerical example.

Let:

$$k_1 = 1 \qquad k_2 = 2 \qquad k_3 = 3 \qquad k_4 = 4.$$

Then

$$L = 10; \qquad M = 15; \qquad \alpha = -8{\cdot}162; \qquad \beta = -1{\cdot}838.$$

Hence:

$$c_A = a[\tfrac{8}{15} + 0{\cdot}0225 \exp(-8{\cdot}162t) + 0{\cdot}444 \exp(-1{\cdot}838t)] \qquad \text{(I)}$$

$$c_B = a[\tfrac{4}{15} - 0{\cdot}0806 \exp(-8{\cdot}162t) - 0{\cdot}1865 \exp(-1{\cdot}838t)] \qquad \text{(II)}$$

$$c_C = a[\tfrac{3}{15} + 0{\cdot}0581 \exp(-8{\cdot}162t) - 0{\cdot}258 \exp(-1{\cdot}838t)]. \qquad \text{(III)}$$

These equations may be checked for the initial conditions by putting $t = 0$ while the leading term of each bracket, or equations (3.139), (3.140) and (3.141), will give the equilibrium composition. Plots of these results are shown in Fig. 3.6(a). For the particular chosen values of the rate constants, the concentration of B steadily increases, and if an attempt is made to find a turning point by differentiation of (II) and equating to zero, no meaningful value of time is found. However, other combinations of the rate constants lead to a time-concentration history of B passing through a maximum. This is illustrated below in the analogue computer solution Fig. 3.6(a), (b) and (c).

While, as already stated above, the analogue computer is not essential for solution of these reversible consecutive rate equations, this is a convenient example to illustrate the convenience of analogue computation and demonstrates the versatility and ease with which the computer is able to solve this kind of problem. The reader is referred to (Jenson and Jeffreys, 1965) for an explanation of the computer symbols used in Fig. 3.7, which is a circuit capable of implementing equations (3.113), (3.114) and (3.115). The values of the rate constants are set on the potentiometers and the initial values of c_A, c_B and c_C if non-zero are introduced at the integrators.

3.11 Chain Reactions

Instances have arisen in which the principles already discussed have been unable to account for the concentration–time observations made on an apparently straightforward reaction. A classical example is the rate of formation of hydrogen bromide from the combination of bromine and hydrogen molecules. The rate of reaction has been found to be the following function of reactant concentrations:

$$\frac{d(\mathrm{HBr})}{dt} = \frac{A(\mathrm{H_2})(\mathrm{Br_2})^{\frac{1}{2}}}{1 + \dfrac{B(\mathrm{HBr})}{(\mathrm{Br_2})}} \tag{3.142}$$

in which A and B are constants.

The following mechanism has been postulated to explain this rate equation:

$$\mathrm{Br_2} \xrightarrow{k_1} 2\mathrm{Br} \tag{3.143}$$

$$\mathrm{Br} + \mathrm{H_2} \xrightarrow{k_2} \mathrm{HBr} + \mathrm{H} \tag{3.144}$$

$$\mathrm{H} + \mathrm{Br_2} \xrightarrow{k_3} \mathrm{HBr} + \mathrm{Br} \tag{3.145}$$

$$\mathrm{H} + \mathrm{HBr} \xrightarrow{k_4} \mathrm{H_2} + \mathrm{Br} \tag{3.146}$$

$$\mathrm{Br} + \mathrm{Br} \xrightarrow{k_5} \mathrm{Br_2} \tag{3.147}$$

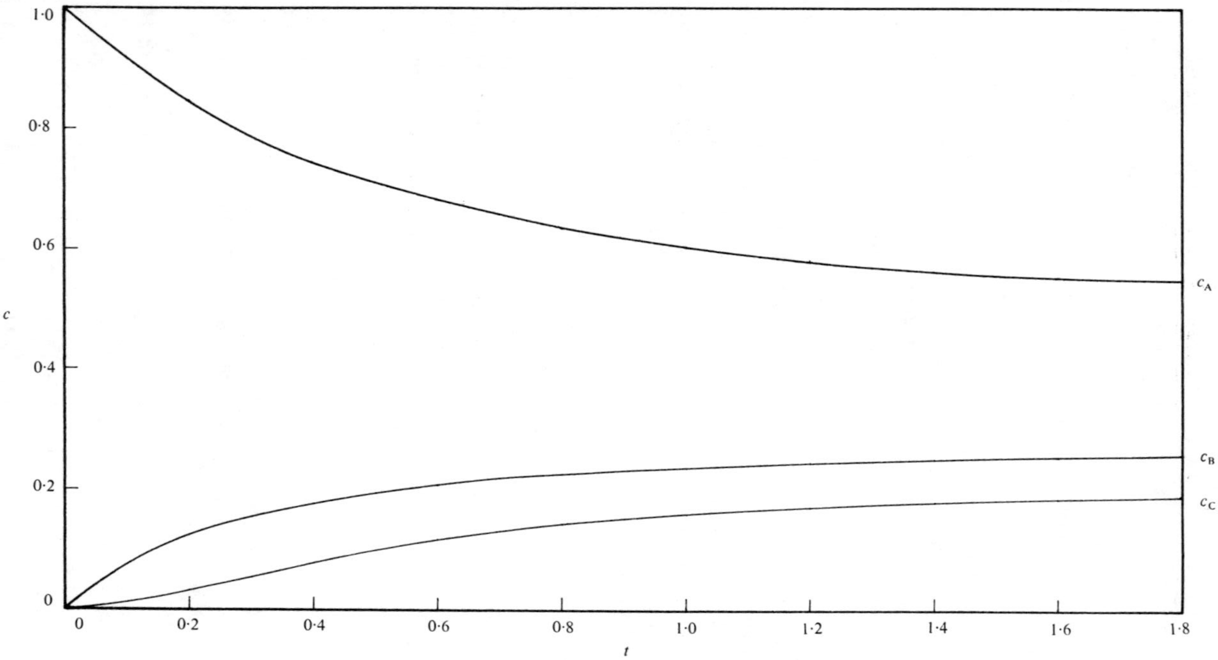

FIG.3.6(a). Product distribution in a consecutive reversible reaction scheme: $A \underset{k_2}{\overset{k_1}{\rightleftharpoons}} B \underset{k_4}{\overset{k_3}{\rightleftharpoons}} C$ $k_2 = 2; k_3 = 3; k_4 = 4; k_1$ varied; all first order

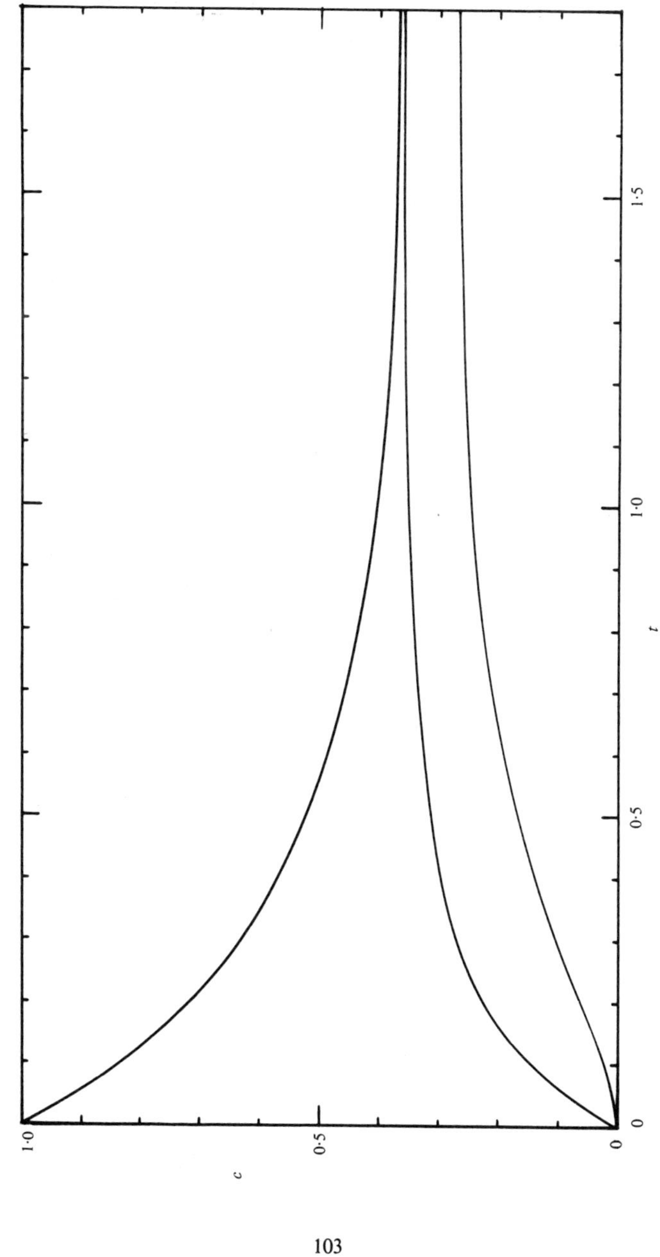

FIG. 3.6(b)

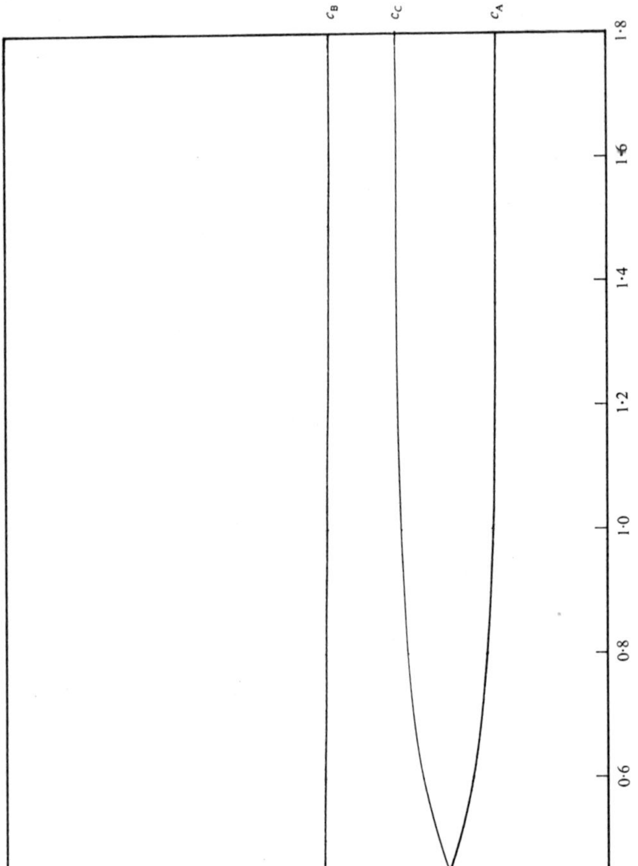

FIG. 3.6(c)

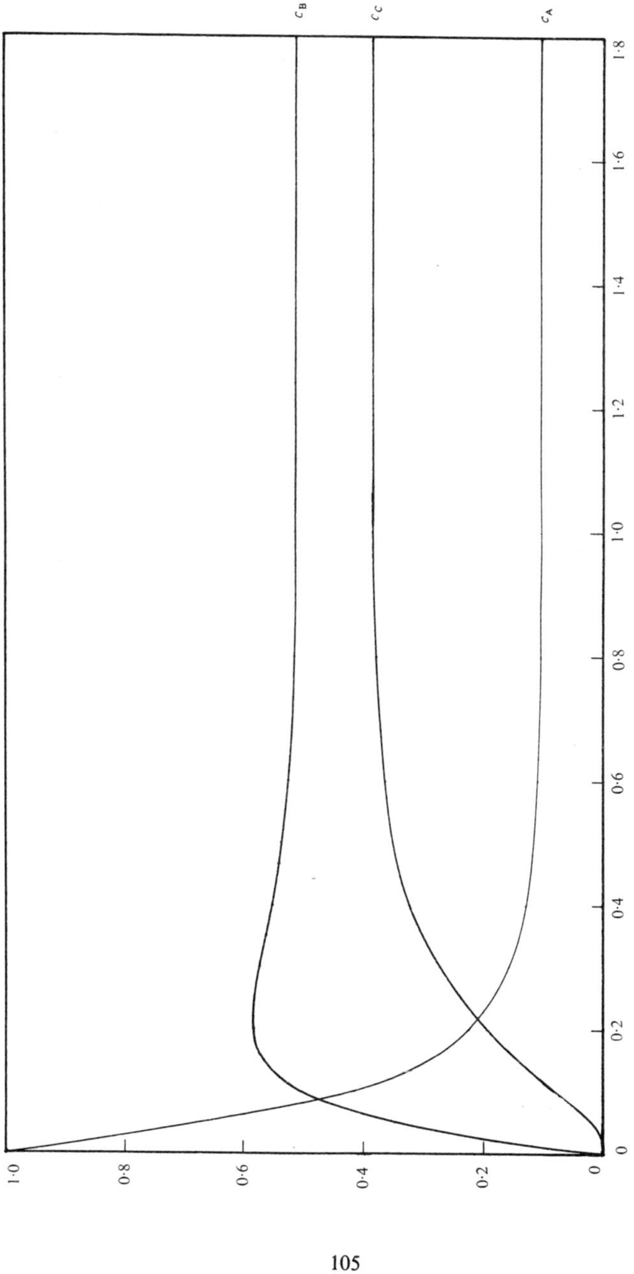

FIG. 3.6(d)

In this process, the chain of events is initiated by the formation of bromine atoms, which are then able to propagate the chain sequence shown.

The events may be seen in more detail by considering a simple material balance. Let one lb atom of hydrogen be formed in reaction (3.144) from a lb mole of reactant hydrogen and let a fraction 'f' of this react by (3.145) and let the remaining fraction $(1 - f)$ be reconverted to hydrogen molecules and bromine atoms by (3.146). The balance of atoms and molecules in the propagation sequence is then:

$$\begin{aligned} \mathrm{HBr}&: +1 + f - (1 - f) = 2f \\ \mathrm{Br}&: -1 + f + (1 - f) = 0 \\ \mathrm{H_2}&: -1 + (1 - f) = -f \\ \mathrm{Br_2}&: -f \\ \mathrm{H}&: +1 - f - (1 - f) = 0 \end{aligned}$$

The net result is seen to be the reaction of equal quantities of hydrogen and bromine molecules to form HBr as expected from the stoichiometric equation:

$$H_2 + Br_2 \rightarrow 2HBr \tag{3.148}$$

without change in the amount of either hydrogen or bromine atoms. The kinetic equation is in fact formulated from a mathematical statement of the zero net consumption of the hydrogen and bromine atoms. Thus for hydrogen atoms:

$$k_2(\mathrm{Br})(\mathrm{H_2}) - k_3(\mathrm{H})(\mathrm{Br_2}) - k_4(\mathrm{H})(\mathrm{HBr}) = 0 \tag{3.149}$$

i.e.

$$(\mathrm{H}) = \frac{k_2(\mathrm{Br})(\mathrm{H_2})}{k_3(\mathrm{Br_2}) + k_4(\mathrm{HBr})}. \tag{3.150}$$

While for bromine atoms:

$$k_1(\mathrm{Br_2}) - k_5(\mathrm{Br})^2 + k_3(\mathrm{H})(\mathrm{Br_2}) + k_4(\mathrm{H})(\mathrm{HBr}) - k_2(\mathrm{Br})(\mathrm{H_2}) = 0. \tag{3.151}$$

Since the three terms of equation (3.149) are necessarily present in equation (3.151), the latter simply reduces to:

$$k_1(\mathrm{Br_2}) = k_5(\mathrm{Br})^2 \tag{3.152}$$

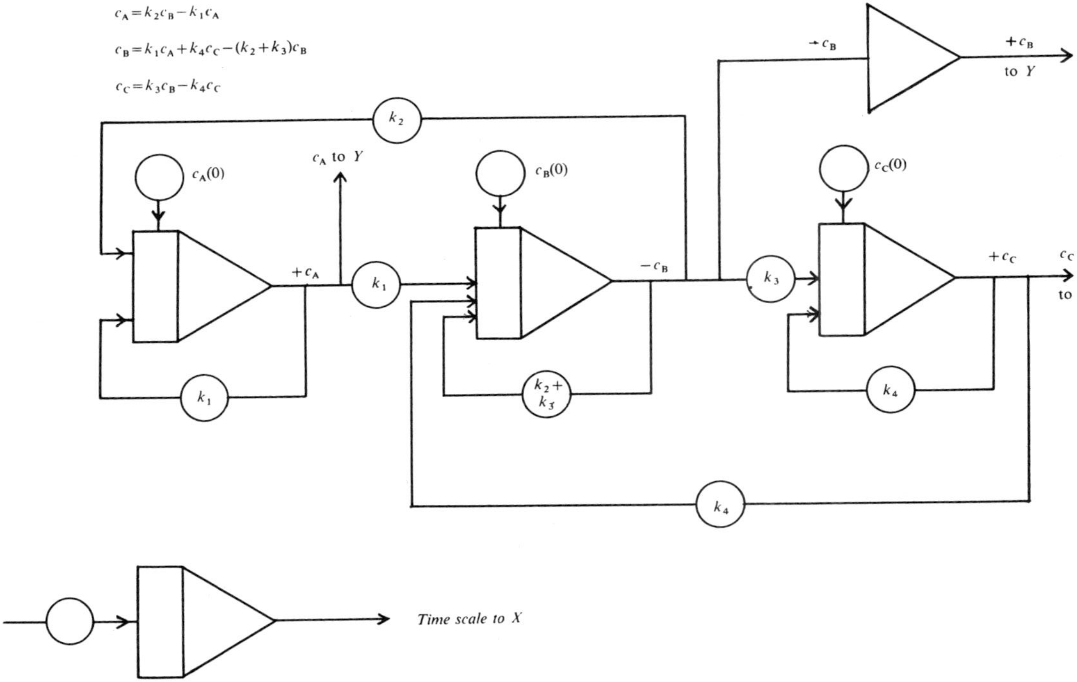

FIG. 3.7. Analogue computer implementation of the reaction scheme A ⇌ B ⇌ C

and this equation may be written directly as the equilibrium between the initiation and termination processes for bromine atoms. The ratio of k_1 to k_5 is an equilibrium constant for the initiation–termination process and the concentration of bromine atoms may now be expressed as:

$$(\mathrm{Br}) = \sqrt{\frac{k_1}{k_5}}(\mathrm{Br_2})^{\frac{1}{2}} \quad \text{or} \quad \sqrt{K}(\mathrm{Br_2})^{\frac{1}{2}}. \tag{3.153}$$

The kinetic equation is finally developed by formulation of a rate equation for net formation of hydrogen bromide with the elimination of (H) and (Br) terms.

i.e.
$$\frac{d(\mathrm{HBr})}{dt} = k_2(\mathrm{Br})(\mathrm{H_2}) + k_3(\mathrm{H})(\mathrm{Br_2}) - k_4(\mathrm{H})(\mathrm{HBr}) \tag{3.154}$$

$$= k_2\sqrt{K}(\mathrm{Br_2})^{\frac{1}{2}}(\mathrm{H_2}) + (\mathrm{H})[k_3(\mathrm{Br_2}) - k_4(\mathrm{HBr})]$$

$$= k_2\sqrt{K}(\mathrm{Br_2})^{\frac{1}{2}}(\mathrm{H_2}) + \frac{k_2(\mathrm{H_2})\sqrt{K}(\mathrm{Br_2})^{\frac{1}{2}}}{k_3(\mathrm{Br_2}) + k_4(\mathrm{HBr})} \times [k_3(\mathrm{Br_2}) - k_4(\mathrm{HBr})]$$

$$= \frac{2k_2\sqrt{K}(\mathrm{Br_2})^{\frac{1}{2}}(\mathrm{H_2})k_3(\mathrm{Br_2})}{k_3(\mathrm{Br_2}) + k_4(\mathrm{HBr})}$$

and finally
$$\frac{d(\mathrm{HBr})}{dt} = \frac{2k_2\sqrt{K}(\mathrm{Br_2})^{\frac{1}{2}}(\mathrm{H_2})}{1 + \dfrac{k_4}{k_3}\dfrac{(\mathrm{HBr})}{(\mathrm{Br_2})}}. \tag{3.155}$$

The constants A and B of equation (3.142) may now be identified from this equation.

It must be emphasised that the successful postulates of equations (3.143) to (3.147) are not the only mechanisms that might have been considered. It is a matter of trial and error to compare the consequences of such postulates with the experimental observations and to continue until the deduced overall kinetic equation accounts for the experimental observations.

This brief introduction to chain reactions may be extended by further reading with particular reference to the following aspects. The example given represents a material chain reaction; energy chain reactions also exist in which the heat of reaction is transferred from a product molecule to a reactant molecule thereby activating the latter. Other methods of initiation must also be studied. In the

HBr example above it was assumed that the whole process was governed by thermal activation as discussed in section 3.8. It is perfectly possible, for instance, for the energy barrier to be transgressed by photochemical energy or by energy from nuclear irradiation. Both initiation and termination may be promoted by adsorption on the wall of the containing vessel. Propagation may well be via free radicals such as CH_3— or CH_3CO— in many organic reactions. It is also possible to have branched chain propagation in which more of the propagating species is produced than consumed; acceleration of the rate in this way may lead to an explosive reaction.

3.12 Photochemical Reactions

The modification to the rate equation resulting from photochemical initiation of hydrogen bromide formation will be considered. The only difference from the thermal initiation case is the rate of formation of bromine atoms:

i.e.
$$Br_2 \xrightarrow{k_1 I_A} 2Br. \tag{3.156}$$

The term $k_1 I_A$ signifies a rate proportional to the rate of light absorption I_A and the equilibrium initiation and termination processes are now governed by:

$$k_5(Br)^2 = k_1 I_A \tag{3.157}$$

with the bromine atom concentration:

$$(Br) = I_A^{\frac{1}{2}} \sqrt{\frac{k_1}{k_5}} = I_A^{\frac{1}{2}} \sqrt{K}. \tag{3.158}$$

If this difference is incorporated into the treatment of section 3.11 the resulting rate of formation of hydrogen bromide is:

$$\frac{d(HBr)}{dt} = \frac{2k_2 \sqrt{K}(H_2) I_A^{\frac{1}{2}}}{1 + \dfrac{k_4}{k_3} \dfrac{(HBr)}{(Br_2)}}. \tag{3.159}$$

3.13 Catalysis

It frequently occurs that a thermodynamically feasible reaction will not proceed at an economic rate for industrial manufacture. Fortunately it often happens in such cases that the reaction mechanism can be altered by the presence of a non-reactant with a greater resultant reaction rate. A catalyst is in fact defined as a substance accelerating a reaction process without itself being consumed; the catalyst may well take part in the changed reaction mechanism but if so it is regenerated in a subsequent step. However, a catalyst does not alter the position of equilibrium; it simply reduces the time to attain or approach this equilibrium. Thus at a given temperature a catalyst changes the forward and reverse rate constants in proportion to one another leaving the value of the equilibrium constant at that temperature unchanged.

The following paragraphs outline some of the important properties of catalysts, but a more detailed treatment of the very important category of gas–solid reactions is deferred to section 3.13.4.

3.13.1 Homogeneous and heterogeneous

There are some catalytic processes that proceed entirely in the gas phase and others entirely in the liquid phase. Such situations are described by the term homogeneous catalysis and the chamber oxidation of sulphur dioxide catalysed by NO is a gas phase example:

$$NO + \tfrac{1}{2}O_2 \rightarrow NO_2 \tag{3.160}$$

$$NO_2 + SO_2 \rightarrow SO_3 + NO \tag{3.161}$$

The same overall result may be achieved without the NO catalyst by the adsorption of SO_2 and O_2 on an active solid surface as in the more recently developed "contact" process. This latter process is described as heterogeneous catalysis and may also occur between a liquid phase and a solid catalyst. An example of such liquid–solid heterogeneous catalysis arises in the desulphurisation of petroleum hydrocarbons.

3.13.2 Specificity

Two catalysts may be capable of yielding different reaction products from a given reactant

e.g.

$$C_2H_5OH \begin{cases} \xrightarrow{Cu} CH_3CHO + H_2 \\ \xrightarrow{Al_2O_3} C_2H_4 + H_2O \end{cases} \tag{3.162}$$

Alumina is specific to the conversion of alcohol into ethylene while copper is specific to the conversion to acetaldehyde.

Specificity is much more pronounced in fermentation reactions and other biological processes. The catalysts are called enzymes in this field of study and each step in a biological process is promoted by its own enzyme. The enzyme is often named after the reactant which it is capable of converting. Thus in the conversion of maltose to alcohol by the enzymes maltase and zymase, the one enzyme, maltase, is so named.

$$\text{Maltose} \xrightarrow{\text{Maltase}} \text{Glucose} \xrightarrow{\text{Zymase}} C_2H_5OH \tag{3.163}$$

3.13.3 Homogeneous catalysis

It has already been mentioned that a catalyst may modify the reaction mechanism by participation in the reaction, and that the final step will regenerate the catalyst which is then free to promote further reaction. The steps making up the overall reaction are each likely to have an activation energy barrier lower than that of the direct uncatalysed reaction.

In homogeneous catalysis the rate of reaction is likely to depend on the catalyst concentration 'c' and on the reactant (or substrate) concentration 's'.

i.e.
$$-\frac{ds}{dt} = kc^n s^m. \tag{3.164}$$

An example of homogeneous gas phase catalysis is the decomposition of acetaldehyde using iodine as the catalyst. The overall reaction:

$$CH_3CHO \xrightarrow{I_2} CO + CH_4 \tag{3.165}$$

proceeds by the two steps:

$$CH_3CHO + I_2 \rightarrow CH_3I + HI + CO \tag{3.166}$$

$$CH_3I + HI \rightarrow CH_4 + I_2 \tag{3.167}$$

The first of these steps is the rate determining one and the kinetics have been found to be first order with respect to the iodine and acetaldehyde concentrations.

i.e.
$$-\frac{d}{dt}(c_{CH_3CHO}) = k\, c_{CH_3CHO}\, c_{I_2}. \tag{3.168}$$

Homogeneous liquid phase catalytic reactions are frequently found to be catalysed by acids or bases. In some instances the reaction is catalysed only by hydrogen ions while in others, all acidic components (e.g. H_2O, H_3O^+, H^+) are effective; these two situations are described respectively as specific hydrogen ion catalysis and generalised acid catalysis. Similarly there exists specific hydroxyl ion catalysis and generalised base catalysis. In addition there is the possibility of simultaneous catalysis by all acidic and basic components, referred to as generalised acid/base catalysis. The specific hydrogen ion catalysis of the hydrolysis of an ester E may take place as follows:

$$E + H^+ \underset{k_2}{\overset{k_1}{\rightleftharpoons}} EH^+ \tag{3.169}$$

$$EH^+ + H_2O \xrightarrow{k_3} \text{Acid} + \text{alcohol} + H^+ \tag{3.170}$$

The first step is rapid and equilibrium is established while the second step is slow and rate determining

i.e.
$$\frac{d}{dt}(E) = k_3(EH^+)(H_2O) \tag{3.171}$$

and using the equilibrium relationship of equation (3.169), (EH^+) may be eliminated to give:

$$\frac{d}{dt}(E) = \frac{k_3 k_1}{k_2}(E)(H^+)(H_2O). \tag{3.172}$$

Furthermore if the reaction takes place in dilute solution with an essentially constant water concentration:

$$\frac{d}{dt}(E) = k(E)(H^+). \tag{3.173}$$

Sometimes a reaction is catalysed by one of its products and is then described as autocatalytic. For example in the reaction:

$$A \rightarrow B + C \tag{3.174}$$

the kinetics may be represented by:

$$-\frac{dc_A}{dt} = kc_A c_B. \tag{3.175}$$

If the initial reaction mixture contains no B then the reaction will not be able to start and so some B would have to be introduced to 'seed' the reaction. If the initial concentrations are 'a' and 'b' and the change in concentration is 'x' at time 't':

$$-\frac{d}{dt}(a-x) = k(a-x)(b+x) \tag{3.176}$$

and

$$t = \frac{1}{k(a+b)} \mathrm{Ln} \frac{a(b+x)}{b(a-x)}. \tag{3.177}$$

A plot of the reaction progress will be sigmoidal in shape (Fig. 3.8). The slow initial rate results from the low concentration of B and the low final rate from the lack of A while a maximum rate exists when the concentrations of both A and B are significant.

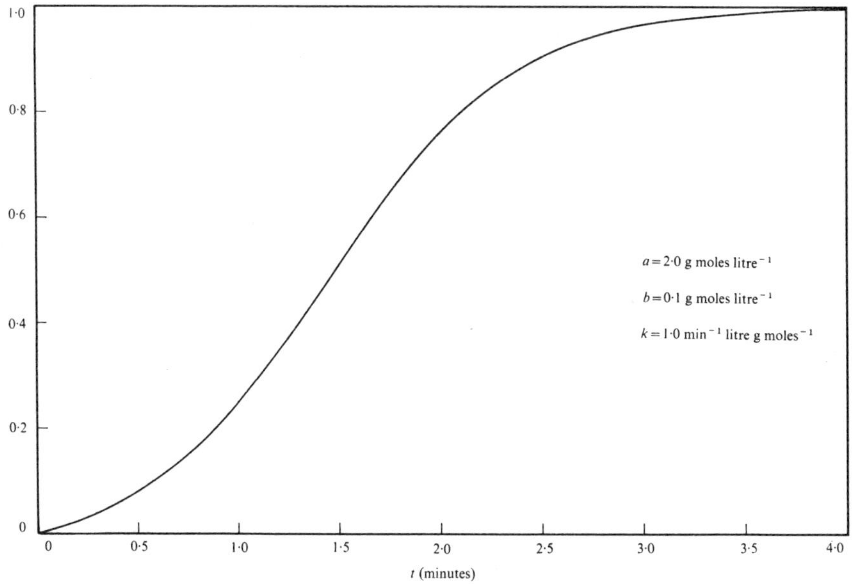

FIG. 3.8. Autocatalytic kinetics

3.13.4 HETEROGENEOUS CATALYSIS

In heterogeneous catalysis, the reactants are adsorbed onto the surface of a solid and after reaction are desorbed and diffuse back into the bulk gas phase. The surface of a solid catalyst is believed to possess many active sites on which the reactants are adsorbed. The activity of a catalyst then depends on the number of these sites per unit mass of catalyst. The art of catalyst development depends partly upon the production of a solid with extended internal surface since the number of active sites can be expected to be proportional to the surface area of the solid. Thus an effective solid catalyst is likely to be manufactured from a highly porous material, although the pore diameter must not be so small that the entry of reactants or discharge of the products is impeded. It is possible to increase the effectiveness of a porous material by the incorporation of relatively small amounts of 'promoters' in the structure. The effect of a promoter is first to increase the actual surface area per unit mass and secondly to increase the density of active sites per unit area; both of these factors lead to an increase in the number of active sites per unit mass of catalyst.

The overall rate of a heterogeneous gas-solid reaction is made up of a series of physical steps as well as the actual chemical reaction. The possible steps are as follows:

(*i*) Diffusion of reactant from the bulk gas phase to the external solid surface.

(*ii*) Diffusion from the solid surface to the internal active sites.

(*iii*) Adsorption on solid surface.

(*iv*) Activation of the adsorbed reactants.

(*v*) Chemical reaction.

(*vi*) Desorption of products.

(*vii*) Internal diffusion of products to the external solid surface.

(*viii*) Diffusion to the bulk gas phase.

All of these steps are rate processes and are temperature dependent. While it is normally assumed at the present state of knowledge that the overall reaction process takes place isothermally, it is important to realise that very large temperature gradients may exist between active sites and the bulk gas phase, and that the detailed mechanism may be far from an isothermal process.

Usually one step is slower than the others and it is this rate controlling step which is studied and leads to the formulation of the kinetic rate. In an extreme case all the steps could have similar rates and the overall rate equation would depend on each step. The formulation of a satisfactory rate equation follows from trial and error attempts to fit experimental results to mathematical models.

representing individual steps or combinations of the various possible steps. However, many reactions of commerical importance are satisfactorily represented by considering the chemical reaction as rate determining and therefore the corresponding mathematical model is now considered in detail, while some of the other rate equations are developed in Chapter 8.

Since all of the steps except reaction are rapid, the adsorption and desorption processes in particular are rapid, and a model is constructed by considering the equilibrium of the adsorption–desorption rate processes. From the kinetic theory of gases the rate of collision between a gas and unit area of a solid surface is given by:

$$\frac{p_A}{(2\pi mkT)^{\frac{1}{2}}}$$

where p_A is the partial pressure of the adsorbent gas A.

The rate of adsorption depends not only on the rate of collision but also on the fraction of the surface which is free from adsorbed molecules. Therefore the rate of adsorption of A may be expressed as:

$$\frac{k'_{1A}p_A(1-\theta)}{(2\pi mkT)^{\frac{1}{2}}} \quad \text{or} \quad k_{1A}p_A(1-\theta)$$

where θ is the fraction of surface covered by all adsorbed gases,
$(1-\theta)$ is the fraction of surface still free,
k'_{1A} is a constant of proportionality,
and k_{1A} is an adsorption rate constant for a particular temperature T.

The rate of desorption of a particular gas A is proportional to the fraction of the surface θ_A which that gas occupies. Thus the rate of desorption may be expressed as:

$$k_{2A}\theta_A.$$

Hence if three gases A, B and C are adsorbed, the equilibrium adsorption–desorption equations are:

$$k_{1A}p_A(1-\theta) = k_{2A}\theta_A \tag{3.178}$$

$$k_{1B}p_B(1-\theta) = k_{2B}\theta_B \tag{3.179}$$

$$k_{1C}p_C(1-\theta) = k_{2C}\theta_C \tag{3.180}$$

where

$$\theta = \theta_A + \theta_B + \theta_C. \tag{3.181}$$

Equations (3.178), (3·179) and (3.180) are rewritten with the ratios:

$$\frac{k_{1A}}{k_{2A}}, \frac{k_{1B}}{k_{2B}}, \text{and} \frac{k_{1C}}{k_{2C}}$$

replaced by adsorption–desorption equilibrium constants K_A, K_B and K_C respectively.

Thus:

$$K_A p_A(1 - \theta) = \theta_A \qquad (3.182)$$

$$K_B p_B(1 - \theta) = \theta_B \qquad (3.183)$$

$$K_C p_C(1 - \theta) = \theta_C. \qquad (3.184)$$

This set of equations is then solved to give the fractions of the surface occupied by the individual gases as:

$$\theta_A = \frac{K_A p_A}{1 + K_A p_A + K_B p_B + K_C p_C} \qquad (3.185)$$

$$\theta_B = \frac{K_B p_B}{1 + K_A p_A + K_B p_B + K_C p_C} \qquad (3.186)$$

$$\theta_C = \frac{K_C p_C}{1 + K_A p_A + K_B p_B + K_C p_C}. \qquad (3.187)$$

This analysis could be extended to the case of 'n' adsorbed gases to give for the ith component:

$$\theta_i = \frac{K_i p_i}{1 + \sum_1^n K_n p_n} \qquad (3.188)$$

in which the summation contains a term for each of the gases adsorbed regardless of whether the gas will actually take part in a chemical reaction.

The final step in the formulation of the model is to express the rate of reaction of a molecule as proportional to the fraction of surface it occupies. Examples are:

$A \rightarrow$ Products	Rate $= k_1\theta_A$	(3.189)
$A \rightleftharpoons B$	Rate $= k_1\theta_A - k_2\theta_B$	(3.190)
$A + B \rightarrow$ Products	Rate $= k_1\theta_A\theta_B$	(3.191)
$A + B \rightleftharpoons C + D$	Rate $= k_1\theta_A\theta_B - k_2\theta_C\theta_D$	(3.192)

The integrations involved will now be considered for a number of important cases.

(*i*) A → B with A adsorbed but B not held

$$-\frac{dp_A}{dt} = k\theta_A = \frac{k\,K_A p_A}{1 + K_A p_A}. \tag{3.193}$$

If the initial pressure P_A has fallen to p_A at time t:

$$\int_0^t dt = \frac{1}{kK_A}\int_{P_A}^{p_A}\left(K_A + \frac{1}{p_A}\right)dp_A \tag{3.194}$$

or

$$t = \frac{1}{kK_A}\left[K_A(P_A - p_A) + \mathrm{Ln}\frac{P_A}{p_A}\right]. \tag{3.195}$$

Two special cases may arise in which A is either weakly adsorbed or strongly adsorbed. In the former case $K_A p_A$ is small compared with unity and so the denominator in equation (3.193) reduces to unity:

i.e.

$$-\frac{dp_A}{dt} = kK_A p_A \tag{3.196}$$

and this simplified form is referred to as a 'first order heterogeneous reaction'.

In the case of strong adsorption $K_A p_A$ is large compared with unity, and equation (3.193) reduces to the zero order heterogeneous rate equation:

$$-\frac{dp_A}{dt} = \mathrm{k}. \tag{3.197}$$

(*ii*) A → B with both A and B adsorbed

$$-\frac{dp_A}{dt} = k\theta_A = \frac{kK_A p_A}{1 + K_A p_A + K_B p_B}. \tag{3.198}$$

Before this equation can be integrated p_B must be expressed as a function of p_A and if for example the initial reactant is pure A at pressure P_A, then when A's pressure has fallen to p_A, B will have a pressure $(P_A - p_A)$ and the reaction time is:

$$t = -\frac{1}{kK_A}\int_{p_A^0}^{p_A}\left[(K_A - K_B) + \frac{1 + K_B P_A}{p_A}\right] dp_A. \qquad (3.199)$$

(*iii*) A ⇌ B with A, B and an inert gas I all adsorbed

$$-\frac{dp_A}{dt} = k_1\theta_A - k_2\theta_B \qquad (3.200)$$

$$= \frac{k_1 K_A p_A - k_2 K_B p_B}{(1 + K_A p_A + K_B p_B + K_I p_I)} \qquad (3.201)$$

and with the same starting conditions as in the previous case:

$$t = -\frac{1}{k_1 K_A}\int_{p_A^0}^{p_A}\frac{(1 + K_B p_A + K_I p_I) + (K_A - K_B)p_A}{p_A\left(1 + \dfrac{k_2 K_B}{k_1 K_A}\right) - \dfrac{k_2 K_B}{k_1 K_A} P_A} dp_A. \qquad (3.202)$$

(*iv*) A + B ⇌ C + D All adsorbed

$$-\frac{dp_A}{dt} = k_1\theta_A\theta_B - k_2\theta_C\theta_D \qquad (3.203)$$

$$= \frac{k_1 K_A K_B p_A p_B - k_2 K_C K_D p_C p_D}{(1 + K_A p_A + K_B p_B + K_C p_C + K_D p_D)^2}. \qquad (3.204)$$

3.13.5 Poisoning of solid catalysts

The effectiveness of a catalyst may be diminished or even completely lost by the retention on the active sites of adsorbed components. In this way, sulphur, silicon, and arsenic are able to form chemical compounds with many catalyst materials and since the resultant compounds are stable it is difficult to remove them and restore the active sites. The catalyst is said to be permanently poisoned and to prevent this situation arising it may be necessary to purify the reactor feedstock to remove the potential poison

In contrast a catalyst may suffer temporary poisoning by simple physical obstruction of the catalyst structure to reactant molecules. An example is the deposition of carbon on the catalyst used for petroleum hydrocarbon cracking. The catalyst becomes coated after a very short time of usage and has to be transferred to a catalytic cracker regenerator unit in which the carbon is burnt off by air. The catalyst is thus cycled alternately between the cracking and regeneration units.

Problems

1. The decomposition of H_2O_2 in the presence of a silver catalyst is first order. 50 ml of a solution of H_2O_2 gave 12·8 ml oxygen in 6·0 minutes and the maximum amount of oxygen obtained from the solution was 40 ml. What is the rate constant and what additional volume of oxygen would be evolved in the next 9·0 minutes?

ANS. 0·0642 min^{-1} and 12 ml additional

2. The following temperature-time-conversion data was obtained for the batchwise conversion of pure gas A to product gas B where the molecularity was given by A → B. The equilibrium constant for the reaction is large over the temperature range concerned:

t (sec)	0	100	500	1000	2000	5000
f at 100°C	0	0·048	0·20	0·33	0·50	0·71
f at 125°C	0	0·092	0·34	0·50	0·67	0·84
f at 150°C	0	0·167	0·50	0·667	0·80	0·91

Show that the reaction is first order with respect to A and deduce the activation energy for the forward reaction.

ANS. Vel-csts. $4{\cdot}92 \times 10^{-4}$; $1{\cdot}025 \times 10^{-3}$; $2{\cdot}0 \times 10^{-3}$ sec^{-1}

$$\Delta E = 8900\text{–}9000 \text{ cal(g mole)}^{-1}$$

3. Butadiene is dimerised at 350°C homogeneously in the gas phase according to second order kinetics. Evaluate the velocity constant from the following data obtained at constant volume:

t (min)	0	6	12	26	38	60
AbsP (mmHg)	500	467	442	401	378	350

ANS. 1·9–2·0 litre min^{-1}(g mole)$^{-1}$

4. The following results were obtained for reaction between trimethylamine and n-propyl bromide each having an initial concentration of 0·1 g mole $litre^{-1}$.

Time (min)	13·0	34·0	59·0	120
% Conversion	11·2	25·7	36·7	55·2

Using the integration method and assuming no reverse reaction determine the order of the reaction and its specific reaction rate.

ANS. $k = 0{\cdot}1\ \text{min}^{-1}$ (concentration dependence to be decided)

5. For a constant volume reaction of the form:

$$n\text{A} \rightarrow m\text{B}$$

containing initially A at a partial pressure p_A, B at a partial pressure p_B, and inerts at a partial pressure p_I, show that at a subsequent time when the total pressure has changed to P, the partial pressure of A will be:

$$\frac{mp_A + n(p_B + p_I - P)}{m - n}.$$

Derive the corresponding expression for the partial pressure of B at this time. Use the results to account for the changes which have occurred when a total pressure of 250 mm has developed from an initial mixture of $p_A = 100$ mm, $p_B = 20$ mm and $p_I = 80$ mm in the reaction:

$$2\text{A} \rightarrow 3\text{B}.$$

6. Show that the following observations conform to kinetics of order 1.5 and evaluate the velocity constant:

Conc. (g litre^{-1})	16·0	13·2	11·1	8·8	7·1
Time (min)	0	10	20	35	50

ANS. $k = 0{\cdot}005\ \text{min}^{-1}(\text{g litre}^{-1})^{-\frac{1}{2}}$

7. The irreversible decomposition of a reagent A in solution is catalysed by a material B. The time of half-completion of the batch reaction was determined by varying the initial concentrations of A and B. From the results estimate:

(*a*) the order of reaction with respect to A,
(*b*) the order of reaction with respect to B, and
(*c*) the velocity constant.

A g mole litre^{-1} at $t = 0$	B g mole litre^{-1} at $t = 0$	$t_{\frac{1}{2}}$ min
1·0	0·001	17·0
1·0	0·002	12·0
2·0	0·003	9·8
2·0	0·004	8·5

Vel. cst. 1·29 litre$^{\frac{1}{2}}$(g mole)$^{-\frac{1}{2}}$ min^{-1}

8. The half-life of a first order reaction at 60°C was found to be 2·76 minutes. The corresponding values at 20°C, 30°C and 50°C were respectively 426, 102 and 10·1 minutes. Express the rate equation in the form:

$$k = A \exp(-\Delta E/RT).$$

ANS. $k_{20} = 0{\cdot}0016$; $k_{30} = 0{\cdot}0068$; $k_{50} = 0{\cdot}069$; $k_{60} = 0{\cdot}25$ min^{-1}
$A = 1{\cdot}31 \times 10^{15}$, $\Delta E = 24000$

9. The following first order specific reaction rates were obtained for the transformation of nitrogen pentoxide at different temperatures.

k minutes $\times 10^5$	4·72	203	2990	9000	29200
Temperature °C	0	25	45	55	65

Express the Arrhenius Equation for this reaction in the form:

$$\ln k = A - (B/R\,T)$$

and check that the tabulated values can be deduced from the expression.

10. In the hydrolysis of methyl acetate, catalysed by hydrochloric acid at 25°C, it was found that an initial solution of the ester of 1·15 g mole litre^{-1} was 40·5% hydrolysed in 1·0 h. What is the specific reaction rate for the forward reaction if the equilibrium constant is 0·22?

ANS. $1{\cdot}5 \times 10^{-4}$ litre(g mole)$^{-1}$ min^{-1}

11. The rate of decomposition of an optically active compound was followed by a polarimeter when the following results were obtained:

Time (min)	0	180	360	540	900	∞
Polarimeter reading	115·6°	102·0°	90·5°	80·6°	65·1°	23·5°

Show that the integrated form of the rate equation is:

$$(k_1 + k_2)t = -\ln\left(1 - \frac{f_A}{f_e}\right)$$

where f_A is the fraction converted at time t,

f_e is the fraction of the original compound remaining at equilibrium,

k_1 is the first order rate constant for the forward reaction,

k_2 is the first order rate constant for the reverse reaction.

Estimate k_1 and k_2, given that the equilibrium constant is 3·89.

ANS. $k_1 = 0{\cdot}7 \times 10^{-4}\ \text{min}^{-1}$ and $k_2 = 0{\cdot}18 \times 10^{-4}\ \text{min}^{-1}$

12. The decomposition of hydrogen iodide is a reversible second order reaction in both directions:

i.e. $$HI \underset{k_2}{\overset{k_1}{\rightleftharpoons}} \tfrac{1}{2}H_2 + \tfrac{1}{2}I_2$$

A flask of volume 2·00 litres was filled with pure HI at 1·24 atm pressure and 683°K and the decomposition was followed by measuring the absorption of light by the iodine produced in the reaction. The optical density quoted is proportional to the iodine concentration existing at any time.

Time (min)	42	118	230	397	680	770	940
Optical density	0·81	2·13	3·66	5·04	6·00	6·12	6·21

Immediately after the last reading the flask was chilled and an analysis for iodine showed 1·17 g. Show that this experimental data agrees with the reaction scheme and evaluate k_1 and k_2.

ANS. $k_1 = 0{\cdot}0303$; $k_2 = 0{\cdot}1{\cdot}75$

13. For the first order reaction sequence:

$$A \xrightarrow{k_1} B \xrightarrow{k_2} C \xrightarrow{k_3} D$$

find the variation of the concentration with time of the various components:

(*a*) when the initial solution contains A only at concentration '*a*',
(*b*) when the initial concentrations of the various components are '*a*', '*b*', '*c*' and '*d*'.

ANS. (*a*) Put $b = c = d = 0$ in answers to part (*b*).

(*b*) $c_A = a \exp(-k_1 t)$

$$c_B = \frac{k_1 a}{k_2 - k_1}[\exp(-k_1 t) - \exp(-k_2 t)] + b \exp(-k_2 t)$$

$$c_C = \frac{k_1 k_2 a \exp(-k_3 t)}{k_1 - k_2}\left[\frac{1 - \exp[-(k_2 - k_3)t]}{k_2 - k_3} - \frac{1 - \exp[-(k_1 - k_3)t]}{k_1 - k_3}\right] + \frac{k_2 b}{k_3 - k_2}[\exp(-k_2 t) - \exp(-k_3 t)] + c \exp(-k_3 t)$$

$$c_D = a\left[1 - \frac{k_2 k_3 \exp(-k_1 t)}{(k_2 - k_1)(k_3 - k_1)} - \frac{k_1 k_3 \exp(-k_2 t)}{(k_1 - k_2)(k_3 - k_2)} - \frac{k_1 k_2 \exp(-k_3 t)}{(k_1 - k_3)(k_2 - k_3)}\right] + b\left[1 + \frac{k_3 \exp(-k_2 t) - k_2 \exp(-k_3 t)}{k_2 - k_3}\right] + c[1 - \exp(-k_3 t)] + d$$

14. In the constant-volume reaction sequence:

$$A + B \xrightarrow{k_1} C \xrightarrow{k_2} D$$

the reverse reactions proceed at negligible rates and the forward rates are first order with respect to each reactant. If the initial reaction mixture contains '*a*' and '*b*' g mole litre^{-1} of A and B respectively but no C or D and the velocity constants are in consistent units, show that the concentration of A at time '*t*' is:

$$\frac{a(a - b)}{a - b \exp[-k_1 t(a - b)]} \text{ g mole litre}^{-1}$$

Examine this answer for the ultimate concentration of A for the cases $a > b$ and $a < b$. Make a sketch of the expected time-concentrations of the four compounds.

15. The rate of the autocatalytic reaction at constant volume

$$A + B \rightarrow 2C$$

in the presence of a large excess of B is given by the expression:

$$\frac{d(\mathrm{A})}{dt} = -k(\mathrm{A})(\mathrm{C}).$$

For an initial concentration a of A and c of C, develop an integral for the fractional conversion of A with time.

Evaluate the integral and hence obtain an expression for the time taken for the concentration of A to fall by a factor 2 when the value of 'a' is n times the value of 'c'.

$$\text{ANS. } ktc = \int_0^f \frac{df}{(1-f)\left(1 + \frac{2af}{c}\right)}$$

$$t = \frac{n \,\mathrm{Ln}\, 2(1+n)}{ak(1+2n)}$$

16. The heterogeneous catalytic decomposition of ammonia was observed to give the following pressure variation with time:

Time (min)	0	6·5	17·0	35·0
Pressure (mmHg)	300	339	402	510

Assuming adsorption–desorption equilibrium find the order of the reaction.

17. For the non-reversible reaction of A on a catalyst to form 2B in the presence of inerts I with all three components adsorbed find an expression for the fraction of the surface covered by A in terms of the partial pressures of each component and defined constants.

Hence assuming reaction rate to be proportional to the fraction of surface covered by A obtain an expression for the time taken for the partial pressure of A to fall by a factor of two at constant volume starting from a batch mixture of 50% A, 10% B and 40% I at P atm total pressure.

$$\text{ANS. } t = \frac{1}{k}\left[1 + \frac{11K_B P}{10} + \frac{2K_I P}{5}\,\frac{\mathrm{Ln}\, 2}{K_A} + \left(1 - \frac{2K_B}{K_A}\right)\frac{P}{4}\right].$$

18. The consecutive second order non-reversible reactions:

$$C_6H_6 + Cl_2 \xrightarrow{k_1} C_6H_5Cl + HCl$$

and

$$C_6H_5Cl + Cl_2 \xrightarrow{k_2} C_6H_4Cl_2 + HCl$$

are effected in a batch reactor with the chlorine gas bubbled through the liquid reaction mixture. The initial charge contains benzene at concentration B_o and monochlorobenzene at concentration M_o.

Show by elimination of time from the differential equations that the corresponding concentrations of benzene (B) and monochlorobenzene M at any given time are connected by:

$$M = \frac{B}{1 - R}\left\{\left(\frac{B_o}{B}\right)^{1-R}\left[1 + \frac{M_o}{B_o}(1 - R)\right] - 1\right\}$$

where

$$R = \frac{k_2}{k_1}$$

Indicate how you would use an analogue computer to find actual values of M and B as functions of time for given values of the constants k_1 and k_2.

REFERENCES

DAVIES, O. L. 1958. *Statistical methods in research and production.* 3rd edn. Oliver and Boyd, Edinburgh.

GLASSTONE, S., LAIDLER, K. J., and EYRING, H. 1941. *Theory of rate processes.* McGraw-Hill, New York.

JENSON, V. G., and JEFFREYS, G. V. 1965. *Mathematical methods in chemical engineering.* Academic Press, New York.

MACMULLIN, R. B. 1948. Distribution reaction products in benzene chlorination. Batch versus continuous process procedures. *Chem. Engng Prog.*, **44**, 183.

4

BATCH REACTORS

4.1 Introduction

A batch reactor is simply a vessel that accommodates a charge of the reactants and causes them to react at an acceptable rate to produce the required products. The vessel may have a loose fitting cover to exclude dust, when it is called a 'reaction kettle'. Alternatively, the cover may be securely bolted to the reactor body so that the contents may be processed at a pressure either greater or less than atmospheric, and the vessel is then called an 'autoclave'.

Most batch reactors are equipped with an agitator of a type depending on the viscosity of the mixture. An anchor type paddle would be expected for thick slurries and a high speed propellor or turbine for low viscosity liquids. The choice of the agitator is important and has received considerable attention in the literature. While the selection of an agitator is outside the scope of this text, a particularly good account of the factors affecting the choice of an agitator is to be found in *Mixing in the Chemical Industry* (Sterbacek and Tausk, 1965).

All chemical reactions proceed with the absorption or evolution of heat and therefore the kettle or autoclave will generally be equipped with some means of heating or cooling in order to control the temperature of the reaction mixture at the desired level. If the reaction is endothermic, heat may be supplied electrically from heating pads or immersion heaters or by means of a steam jacket. Steam jackets are preferred to steam coils because the steam jacket has the greater holding capacity, may readily have a greater heat transfer area and condensate removal is easier. On the other hand cooling coils are preferable to a cooling jacket for heat removal from batch reactors carrying out exothermic reactions. The reason for this is that, by increasing the flow of coolant through the coils, it is possible to obtain a high heat transfer coefficient which will result in a far greater heat transfer rate, even with a coil area smaller than that of the jacket.

Since there is a limit to the size of a pressure vessel and since any significant mass of gas requires considerable space, gaseous reactions are seldom affected batchwise on a commercial scale. Batch reactors are thus employed almost exclusively for liquid phase reactions but may involve reaction between a liquid and a solid or between a liquid and a gas bubbled into the liquid continuously. A batch reactor has all the inherent difficulties of any batch operated process. These difficulties include high labour costs, controllability and loss of production while changing from one batch to the next. Batch reactors are nowadays confined, therefore, to the production of expensive fine chemicals and pharmaceuticals which do not warrant large scale manufacture.

Finally the batch reactor may be operated under isothermal or non-isothermal conditions. Generally when the reaction rate is low and the reactant concentration is small, the major part of the reaction process may be conducted isothermally if this is desired. At the other extreme, a high initial reaction rate at high reactant concentration will have very little diluent solvent to act as a thermal buffer and it may consequently be impossible to maintain constant temperature conditions in the early stages of the conversion. The energy balances describing these situations are combined with the material balances in the following sections.

4.2 Batch Reactor Analysis

Consider a batch reactor of effective volume V_R charged with a reaction mixture of total mass M_T and containing initially M_A moles of reactant. At time 't' and fractional conversion 'f', let the change in conversion of the initially charged reactant be 'df' in a further time 'dt'. Then since:

$$f = \frac{\text{mass reacted}}{\text{initial mass of reactant}} \tag{4.1}$$

it follows that:

$$d(\text{mass reacted}) = M_A df. \tag{4.2}$$

If 'r' is the reaction rate expressed in units consistent with the units of M_A, V_R and t, the material balance over the time interval 'dt' is:

$$M_A df = rV_R dt \tag{4.3}$$

i.e.
$$t = M_{\mathrm{A}} \int \frac{df}{rV_{\mathrm{R}}} \tag{4.4}$$

which for the normal case of constant volume of reactor contents reduces to:

$$t = \frac{M_{\mathrm{A}}}{V_{\mathrm{R}}} \int \frac{df}{r} = c_{\mathrm{A0}} \int \frac{df}{r} \tag{4.5}$$

where c_{A0} is the initial reactant concentration. Before equation (4.5) can be integrated, 'r' must be expressed as a function of fractional conversion; in addition to this, integration of equation (4.4) would also require V_{R} to be so expressed.

4.3 Isothermal Operation

The earlier chapter on introductory kinetics dealt with batch kinetics at a single temperature and therefore many of the results are readily adaptable to the present considerations. The consistency between application of equation (4.5) and some of the results of Chapter 3 is now examined.

For an nth order irreversible isothermal reaction:

$$r = kc^n = kc_{\mathrm{A0}}^n(1 - f)^n \tag{4.6}$$

and hence using equation (4.5):

$$t = \frac{c_{\mathrm{A0}}}{kc_{\mathrm{A0}}^n} \int_0^f \frac{df}{(1 - f)^n} = \frac{1}{kc_{\mathrm{A0}}^{n-1}} \int_0^f \frac{df}{(1 - f)^n}. \tag{4.7}$$

As before this integral can be evaluated for the general value 'n' and for the special case of $n = 1$ giving the results of equations (3.32) and (3.18). It will again be seen from this analysis that the batch time for a first order reaction is not dependent on the initial concentration.

In a like manner, the reversible reaction results of Chapter 3 may be applied. However, in many commercial reaction processes the reactions are very complex and because of the uncertainty and complexity of the resultant rate equation it may be more rational to use an empirical representation of the rate 'r'.

It will readily be seen from equation (4.5) that the time of batch operation is independent of actual reactor volume for constant volume systems. Furthermore even if the volume of the system varies it would be expressed as:

$$V_0 \, \phi(f)$$

where V_0 is the initial reactor volume and then equation (4.4) would become:

$$t = c_{A0} \int \frac{df}{r\phi(f)} \tag{4.8}$$

and again the batch time does not depend on the amount of material being processed.

The size of a batch reactor would be decided in the following way. Let the average hourly rate required to meet the market be Q volume units and let the time taken to discharge one batch and load the next batch be T. Then the total batch hold-up volume V_R is simply:

$$V_R = Q(t + T) \tag{4.9}$$

and if this V_R is inconveniently large for a single reaction vessel, then it is subdivided into parallel units of acceptable size.

An illustration of average batch production rate is considered in section 5.3 where a comparison is made between batch and continuous well-stirred reactor operation.

4.4 Non-isothermal Batch Reactor Analysis

Industrial scale chemical reactions are frequently complicated by variations occurring in the temperature of the reactor during operation, and these fluctuations can seriously affect the production rate of a reaction process. Consequently the effects of variation in temperature must be considered in the analysis of a design of a new reactor or the prediction of the production rate from an existing reactor.

Temperature changes always affect the physical properties of the materials being processed and these changes must be taken into account when assessing the agitator power requirements and the rate of heat transfer to or from the contents of the kettle. Frequently, sufficient correction can be made to these quantities for estimating the power and heating by assuming that the physical properties are constant at some suitable mean temperature. On the other hand, changes in temperature bring about such large changes in the rate of chemical reaction that the application of a mean rate would result in large errors in the reactor design or analysis. Consequently changes in the rate of chemical reaction with temperature must be introduced into the rate equation before solution of the basic equation. Usually this is done by expressing the specific reaction rate as a function of temperature by means of the Arrhenius equation, equation (3.63a):

$$k = A \exp(-E/RT) \tag{4.10}$$

and expressing the temperature in this as a function of f by a differential energy balance. It is the variation of temperature throughout the course of reaction that ultimately affects the capacity of the reactor, and the temperature-conversion history is governed by the relative magnitudes of the heat of reaction, the total mass of the reactor contents and the possible heat transfer rate. The temperature affects the batch time through the variation of specific rate constant and the possible conversion by variation of the equilibrium constant. If the temperature history under adiabatic operation adversely affects the rate or yield of the reaction, then means of heat transfer will have to be provided. Thus the temperature in an adiabatic endothermic process could fall to such an extent that the batch time could be unacceptably large. Conversely in an adiabatic exothermic process, such as a nitration, the temperature can rise alarmingly and lead to an explosion.

There are alternative methods to direct heat transfer for controlling the rate of a chemical reaction. Thus the reaction could be made to take place in a semi-batch reactor (section 4.5) or in a two phase liquid system in which the rate of chemical reaction is controlled by the rate of mass transfer of one of the reactants to the reaction phase. Alternatives of this kind will be discussed later in the text. Finally there are a number of reaction processes where the temperature changes assist the reaction process and it would be a disadvantage to introduce heaters or coolers into the reaction that would tend to make the process isothermal.

Whatever the reasons for the non-isothermal operation of a batch reactor, the design or analysis is carried out by employing the basic material balance equation in conjunction with an energy balance and an Arrhenius type expression for the reaction rate.

The energy balance is an extension of equation (2.17) which was developed for unit mass:

$$q = (\Delta U_f^0)_P - (\Delta U_f^0)_R + \int_{T_0}^{T_2} (C_v)_P \, dT - \int_{T_0}^{T_1} (C_v)_R \, dT$$

$$= \Delta U_{T_0}^0 + \int_{T_0}^{T_2} (C_v)_P \, dT - \int_{T_0}^{T_1} (C_v)_R \, dT. \qquad (2.17)$$

As discussed in section 2.2.2 there is no significant numerical difference between ΔU and ΔH in a liquid phase; likewise C_v and C_P values will be identical. The notation ΔH and C_P will therefore be used in the ensuing development.

The two integrals in (2.17) represent the energy accumulation in the reactor and the physical significance may be seen as:

$$\begin{bmatrix}\text{HEAT TRANSFER FROM}\\ \text{SURROUNDINGS}\end{bmatrix} = \begin{bmatrix}\text{HEAT OF}\\ \text{REACTION}\end{bmatrix} + \begin{bmatrix}\text{ENERGY ACCUMULATION}\\ \text{IN REACTOR}\end{bmatrix} \qquad (4.11)$$

The use of signs in equation (4.11) has caused considerable confusion, and errors in conclusions about direction of reactor temperature changes are best avoided by a thorough understanding of the other two terms in the equation. Thus:

(i) The heat generated by reaction may be positive or negative, and ΔH should be substituted together with its algebraic sign as defined in equations (2.13) and (2.18).

For a reactor volume V_R, the total heat generated in time dt is:

$$rV_R \, dt \,.\, \Delta H.$$

(ii) The heat exchanged with the surroundings in time dt is given by:

$$UA(T_s - T)\, dt$$

where U is the overall heat transfer coefficient,
A is the heat transfer area,
T_s is the temperature of the heating or cooling medium,
and T is the temperature of the reaction mixture.

The heat exchanged may be positive ($T_s > T$), negative ($T_s < T$) or in the case of adiabatic operation, zero.

(iii) The direction of temperature change in the reactor (i.e. the accumulation term) will take its sign from the net algebraic result of terms (*i*) and (*ii*) which will themselves generally turn out to have opposite signs.

Before substitution into equation (2.17), the terms:

$$\int_{T_0}^{T_2} (C_P)_P \, dT - \int_{T_0}^{T_1} (C_P)_R \, dT$$

may be simplified in many instances by assuming the total reactor charge M_T to be constant and the specific heat to be independent of composition giving:

$$\int_{T_0}^{T_1} M_T C_P \, dT + \int_{T_1}^{T_2} M_T C_P \, dT - \int_{T_0}^{T_1} M_T C_P \, dT$$

or simply:

$$M_T \int_{T_1}^{T_2} C_P \, dT$$

and for an incremental change in temperature dT with a mean specific heat $\bar{C}_P$, the term becomes:

$$M_T \bar{C}_P \, dT.$$

In terms of reaction rate, the energy balance for a time dt is then:

$$UA(T_s - T)\, dt = rV_R \, \Delta H \, dt + M_T \bar{C}_P \, dT \tag{4.12}$$

which may be expressed in terms of conversion using (4.3) as:

$$UA(T_s - T)\, dt = M_A \Delta H \, df + M_T \bar{C}_P \, dT. \tag{4.13}$$

4.4.1 Adiabatic operation

For adiabatic operation, equation (4.13) is employed with the heat transfer term set at zero,

i.e.

$$-M_A \Delta H \, df = M_T \bar{C}_P \, dT. \tag{4.14}$$

On integration this equation gives the time independent connection between f and T:

$$\int_{T_0}^{T} dT = -\frac{M_A \Delta H}{M_T \bar{C}_P} \int_{0}^{f} df \tag{4.15}$$

or

$$T = T_0 - \frac{M_A \Delta H}{M_T \bar{C}_P} f \equiv T_0 - \alpha f \tag{4.16}$$

where T_0 is the initial batch temperature at $f = 0$.

In order to find the required batch time, equation (4.5) is used with rate expressed as:

$$r = k\phi(f) = A\phi(f) \exp\left(-\frac{E}{RT}\right) \tag{4.17}$$

and eliminating temperature by equation (4.16):

$$r = A\phi(f)\exp\left[-\frac{E}{R(T_0 - \alpha f)}\right] \tag{4.18}$$

and hence t from equation (4.5) is:

$$t = \frac{c_{A0}}{A}\int_0^f \frac{df}{\phi(f)\exp\left[-\dfrac{E}{R(T_0 - \alpha f)}\right]}. \tag{4.19}$$

Such an integral is unlikely to be solved analytically but may readily be solved either graphically or by numerical computer methods.

By substituting values of f separately into (4.16) and (4.19), the time-temperature-conversion history can be established. It should be noted that the batch time result is again independent of the reactor volume since any change in M_A will entail a proportional change in M_T.

The details of a calculation for operation in this way are shown in the next example.

Example 1. A batch reactor has a 500 lb charge of a solution of acetic anhydride at 60°F containing acetic anhydride at a concentration of 0·0135 lb mole ft^{-3}. The solution density is 65·5 lb ft^{-3} and its specific heat 0·9 Btu lb^{-1} degF^{-1}. The first order reaction has $\Delta H = -90000$ Btu lb mole^{-1} of anhydride hydrolysed and the variation of specific rate constant with temperature is given in the following table. The time for 80% conversion under adiabatic conditions is required.

Temperature (°F)	40	50	60	70	80	90
Rate Constant (min^{-1})	0·035	0·057	0·084	0·123	0·174	0·245

For adiabatic operation, equation (4.14) applies,

i.e.
$$-M_A \Delta H\, df = M_T \bar{C}_P\, dT. \tag{I}$$

Hence:
$$-\frac{500 \times 0{\cdot}0135 \times (-90000)}{65{\cdot}5}\, df = 500 \times 0{\cdot}9\, dT \tag{II}$$

or $$dT = 20{\cdot}6\, df \tag{III}$$

and $$T = 60 + 20{\cdot}6\, f \text{ in °F.} \tag{IV}$$

Applying equation (4.5) to a first order chemical reaction:

$$t = c_{A0}\int_0^{0\cdot8} \frac{df}{k_1 c_{A0}(1-f)} = \int_0^{0\cdot8} \frac{df}{k_1(1-f)} \tag{V}$$

The plot of k_1 versus temperature shown in Fig. 4.1 is used to interpolate values of k_1 for the integration and the resulting data for the graphical solution of this equation taking 0·1 increments for df are tabulated below. A more accurate solution would be

CONVERSION f	0·0	0·1	0·2	0·3	0·4	0·5	0·6	0·7	0·8
TEMPERATURE °F	60·0	62·0	64·1	66·1	68·3	70·3	72·2	74·4	76·4
RATE $k_1(1-f)$	0·084	0·083	0·08	0·074	0·07	0·062	0·054	0·043	0·032
$[k_1(1-f)]^{-1}$	11·9	12·1	12·5	13·4	14·3	16·1	18·5	23·2	31·4

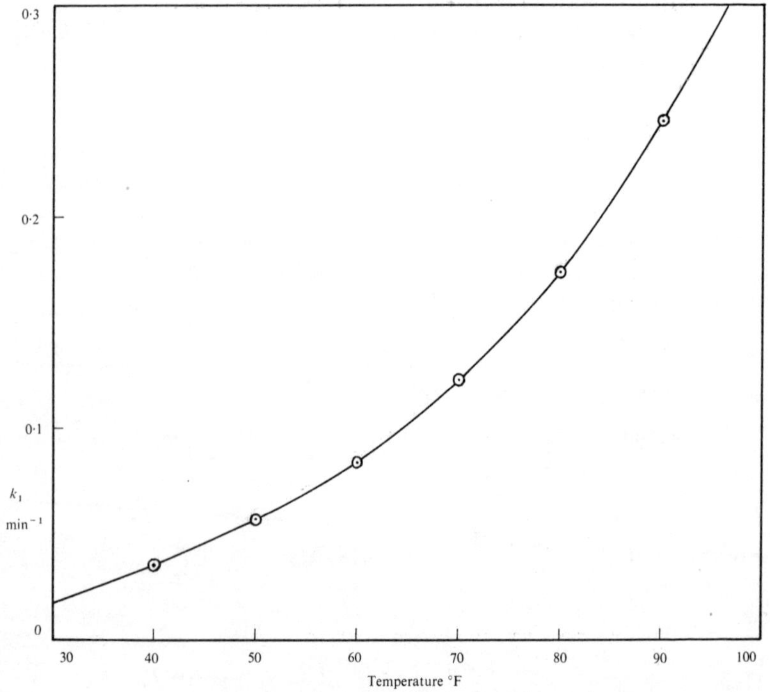

FIG. 4.1a. Variation of rate constant with temperature

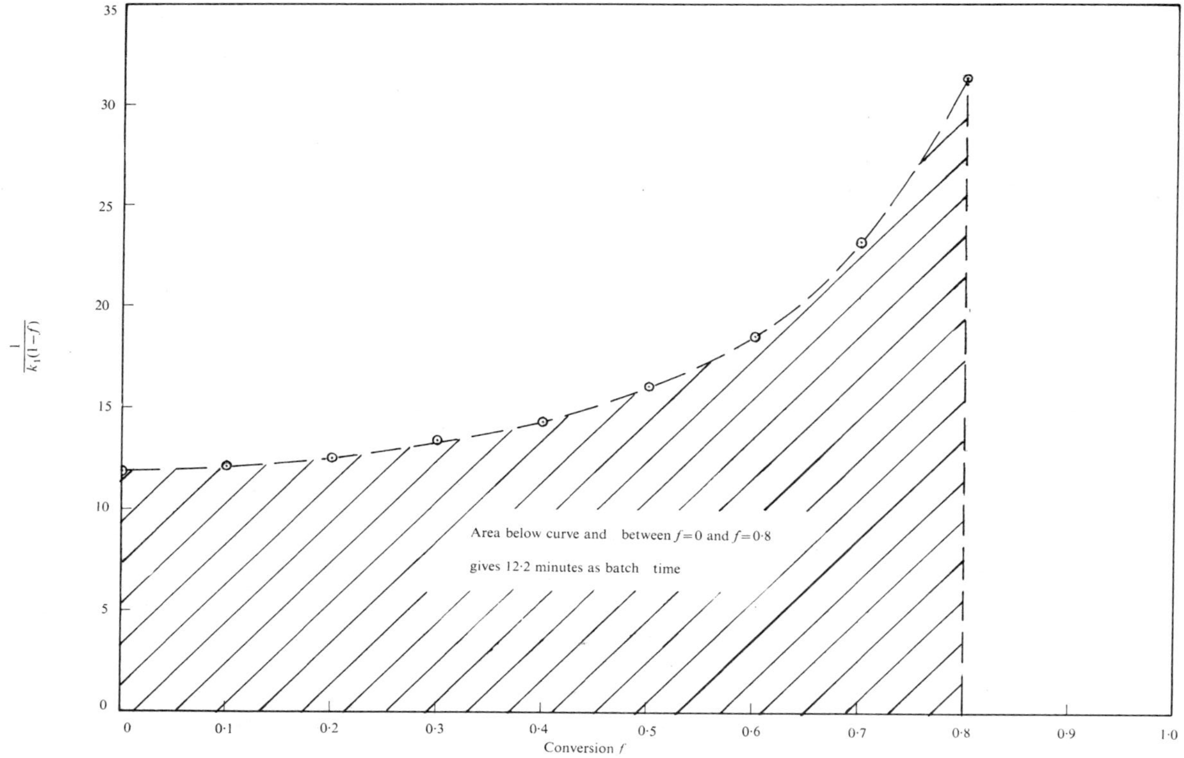

FIG. 4.1b. Graphical integration to obtain batch time

readily obtained using a smaller increment and a digital computer. A plot of $[k_1(1 - f)]^{-1}$ against f is also shown in Fig. 4.1. The area under the curve between the values of $f = 0$ and $f = 0{\cdot}8$ gives the adiabatic batch processing time as 12·2 minutes. The final temperature (76·4°F) is not dangerously excessive.

For isothermal conditions of 60°F, the constant specific reaction rate of 0·084 min^{-1} gives 22 minutes as the reaction time for the same conversion.

4.4.2 Non-adiabatic operation

The solution of equation (4.13) in conjunction with the rate equation is a trial and error procedure and the nth increment of the calculation could be solved by the following sequence of steps:

(*i*) Choose an increment size Δf.

(*ii*) From the initial temperature T_n and fractional conversion f_n, obtain a first estimate of Δt_n assuming the initial rate r_n to apply to the whole increment.

(*iii*) Deduce a first estimate of T_{n+1} from the energy balance equation.

(*iv*) Use this estimate of T_{n+1} and the current value of f_{n+1} to get a first estimate of r_{n+1}.

(*v*) Use $\frac{1}{2}(r_n + r_{n+1})$ to revise Δt_n and hence revise T_{n+1}.

(*vi*) Continue this procedure until two successive estimates of T_{n+1} are acceptably convergent.

The satisfactory answer to T_{n+1} is then used to start the calculation of the next increment and the Δt_n accumulated towards the total batch time. The procedure is illustrated in the following example and shown as a computer flow diagram in Fig. 4.2.

Example 2. Acetylated caster oil is hydrolysed for the manufacture of drying oils in kettles operated batchwise. The charges are 500 lb (227 kg) and the initial temperature is 613°K. Complete hydrolysis yields 0·156 lb acetic acid per lb of ester and 70% conversion is required.

For the reaction, Grummit and Fleming (1945) have found pseudo-first order kinetics given by:

$$r = c \exp\left(35{\cdot}2 - \frac{22450}{T}\right) \text{ g acid produced cm}^{-3}\text{ min}^{-1}$$

with T in °K and c in g cm^{-3}.

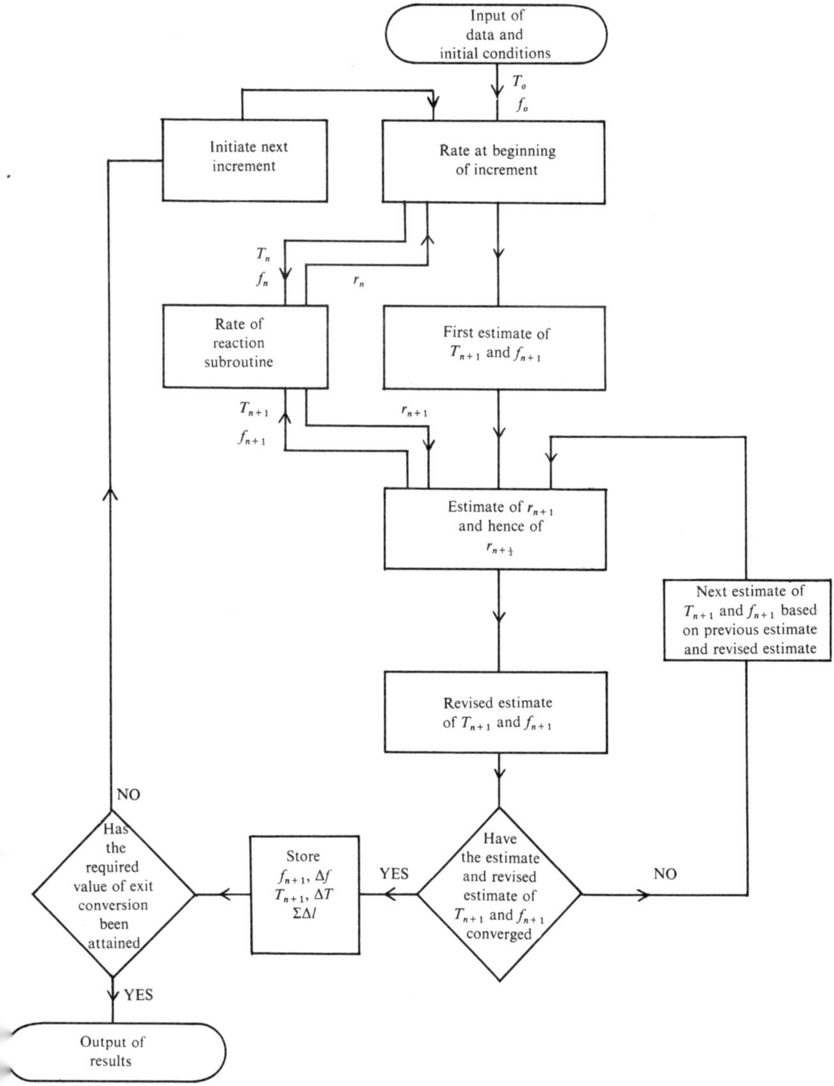

FIG. 4.2. Digital computer flow diagram for non-adiabatic batch calculation

The specific heat of the solution is constant at 0·6 cal g^{-1} $degK^{-1}$ and $\Delta H = +15000$ cal (g mole)$^{-1}$ of acid produced. Heat is generated in the system by electrical immersion heaters.

The rate of heat input will be varied in the following illustrations and the consequences on the temperature-time history and the required conversion time examined. However, within each example the rate of heat input will be constant throughout the conversion.

(*i*) Immersion heaters rated at 52·8 kwatt (or 756000 cal min^{-1}) are considered initially.

The terms of equation (4.13) then have the numerical values:

$$M_T\bar{C}_P = 227000 \times 0{\cdot}6 = 136200 \text{ cal degC}^{-1},$$

$$M_A = \left(\frac{227000 \times 0{\cdot}156}{60}\right) = 591 \text{ g moles acid,}$$

$$M_A \Delta H = 591 \times 15000 = 8{\cdot}86 \times 10^6 \text{ cal.}$$

Hence for a change in temperature ΔT and fractional conversion Δf occurring in time Δt min

$$136200\,\Delta T = 756000\,\Delta t - 8{\cdot}86 \times 10^6\,\Delta f$$

or $\Delta T = 5{\cdot}55\,\Delta t - 65\,\Delta f$.

The temperature T at time t and fractional conversion f is then given by:

$$T = 613 + 5{\cdot}55\,t - 65\,f.$$

Using the iterative scheme, shown in Fig. 4.2 for increments of $\Delta f = 0{\cdot}05$, gives the following results:

f	T°K	Time for increment (min)	Cumulative time (min)
0	613·0	—	—
0·05	611·0	0·227	0·227
0·10	609·2	0·268	0·495
0·15	607·7	0·312	0·807
0·20	606·5	0·360	1·167
0·25	605·5	0·410	1·578
0·30	604·8	0·461	2·039
0·35	604·4	0·512	2·551
0·40	604·3	0·562	3·114
0·45	604·4	0·611	3·724
0·50	604·8	0·658	4·383
0·55	605·5	0·704	5·087
0·60	606·4	0·750	5·837
0·65	607·6	0·798	6·635
0·70	609·0	0·850	7·485

This particular rate of heat input is not sufficient to balance the heat required by the endothermic reaction until approximately 40% conversion is reached. At higher levels of conversion the heat imput is able to supply the reaction heat and also increase the temperature of the vessel contents from the minimum value of 604·3°K. The fractional conversion–temperature history is shown in Fig. 4.3.

The times taken for each 0·05 increase in fractional conversion are shown in the table along with the cumulative times for each level of conversion.

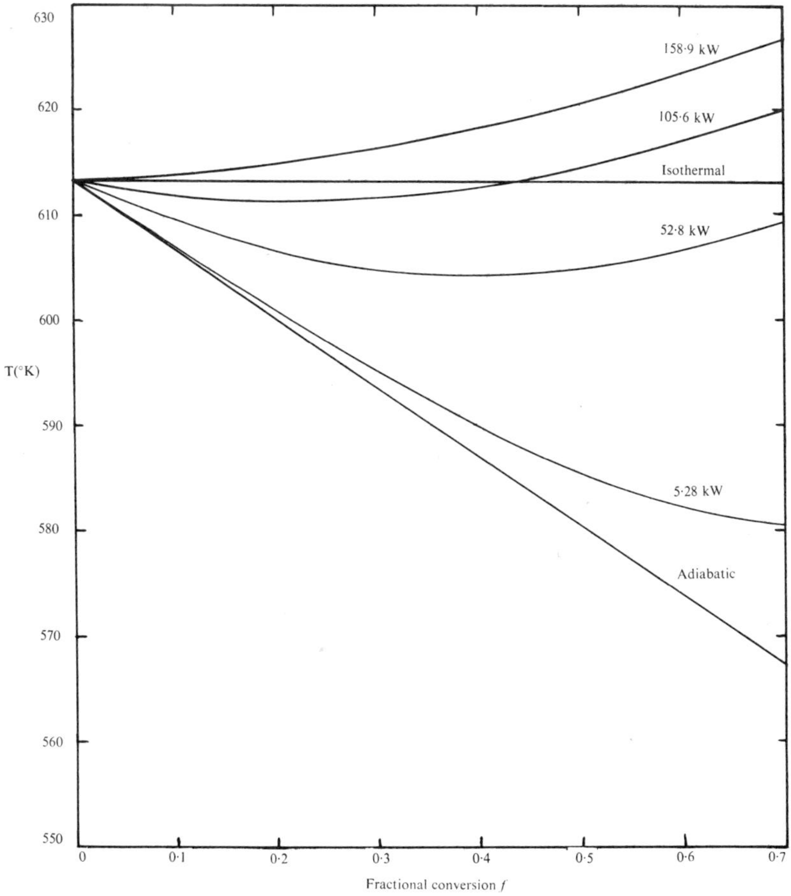

FIG. 4.3. Temperature-conversion progress for various rates of heat input

(*ii*) The influence of increasing the heat input rate from zero on the time taken to reach 70% conversion is shown in the following table and the temperature progress is plotted in Fig. 4.3. For comparison, the time for isothermal conversion at 613°K obtained by use of equation (3.32) is 4·97 min.

kwatt	kcal min^{-1}	$q/M_T\bar{C}_P$	Time (min) for 70% Conversion
0	0	0	38·25
5·28	75·6	0·555	23·64
52·8	756	5·55	7·485
105·6	1512	11·1	4·72
158·4	2268	16·65	3·55

In adiabatic operation, the temperature falls linearly with '*f*'. For the lowest heat input rate considered the temperature falls monotonously up to the conversion level considered. The highest rate of heat input is sufficient to supply the initial reaction heat and also increase the temperature of the reaction mixture. The intermediate heating rates shown lead to an initial fall in temperature followed by an increase, which in the one case exceeds the initial vessel temperature at about 0·425 conversion.

(*iii*) The specification of a heater to meet a design requirement of 20 minutes for a batch conversion is now considered. From the above table, it can be seen that the heater will have to be rated between 5·28 and 52·8 kwatt and nearer to the lower value. Adaptation of the computer program to iterate on the conversion time by alteration of the heat input term, gave the solution:

$$\frac{q}{M_T\bar{C}_P} = 0{\cdot}873 \text{ at } t = 20 \text{ min}$$

with the corresponding power rating of 8·31 kwatt.

Assuming the overall temperature difference to be limited to 150 degC, the surface area of immersion heater to be installed is now calculated from the overall heat transfer coefficient (based on the outside coil area) of Btu h^{-1} ft^{-2} $degF^{-1}$ between the heating element and the vessel charge.

Then using $q = UA\Delta T$

$$3413 \times 8{\cdot}31 = 60 \times A \times 150 \times 1{\cdot}8$$

giving $A = 1{\cdot}75\ \text{ft}^2$ and could conveniently be fabricated from 13·4 ft of $\frac{1}{2}$ in O.D. heating element.

4.5 Semi-batch Operation

The possible use of semi-batch operation to control the rate of reaction was mentioned early in section 4.4. Such operation could be effected by initially loading a reaction vessel with the whole mass of one reactant to be used in a conversion, and then feeding the other reactant at such a rate that the heat released could be adequately removed by the cooling system without elevation of reactor temperature. During operation no product is removed and the vessel contents increase.

For a first order irreversible reaction effected isothermally, the unsteady state mass balance on the continuously fed reactant is:

$$Qc_0 = kVc_A + \frac{d}{dt}Vc_A \tag{4.20}$$

with both V and c_A as time dependent variables.

If the feed rate is constant at Q,

$$\frac{dV}{dt} = Q \tag{4.21}$$

and
$$V = V_0 + Qt \tag{4.22}$$

where V_0 is the charge present in the reactor at the time the flow Q is started and the reaction commences.

Solving equation (4.20) for the combined variable Vc_A gives:

$$Vc_A = I\,\mathrm{e}^{-kt} + \frac{Qc_0}{k}. \tag{4.23}$$

If the initial volume V_0 contains none of the reactant, $Vc = 0$ at $t = 0$ and

$$I = \frac{Qc_0}{k}. \tag{4.24}$$

Hence
$$(Vc_A) = \frac{Qc_0}{k}(1 - \mathrm{e}^{-kt}) \tag{4.25}$$

and from equation (4.22):

$$\frac{c_A}{c_0} = \frac{Q(1 - \mathrm{e}^{-kt})}{k(V_0 + Qt)} = \frac{(1 - \mathrm{e}^{-kt})}{k\left(\dfrac{V_0}{Q} + t\right)}. \tag{4.26}$$

The concentration 'c_A' reaches a maximum value and thereafter decreases with time as the reactant becomes diluted by the ever increasing volume of products in the vessel.

The time of greatest concentration is found by putting equal to zero the differential coefficient dc_A/dt obtained from equation (4.26), leading to the implicit equation:

$$1 + k\left(\frac{V_0}{Q} + t\right) = e^{kt}. \tag{4.27}$$

For a reaction in which a single mole of product is formed for each mole of reactant converted, the amount of product B is:

$$Qc_0 t - \frac{Qc_0}{k}(1 - e^{-kt}) \tag{4.28}$$

and hence the concentration of B is given by:

$$\frac{c_B}{c_0} = \frac{kt - (1 - e^{-kt})}{k\left(\frac{V_0}{Q} + t\right)}. \tag{4.29}$$

The variation of concentrations c_A/c_0 and c_B/c_0 with time are shown in Fig. 4.4 for values of $k = 0{\cdot}2\ h^{-1}$ and $V_0/Q = 0{\cdot}5$ hr.

The rate of heat release at any instant is:

$$kVc_A\Delta H$$

which from equation (4.25) gives:

$$Qc_0\Delta H\,(1 - e^{-kt}).$$

Thus the rate of heat release is zero initially, and as time proceeds it tends to the value $Qc_0\Delta H$. Since for a steady feed rate the heat release is less than that corresponding to the reactant input rate, the above manner of operation affords a means of controlling the reaction.

Problems

1. Methyl acetate is to be hydrolysed by water in a batch reactor at constant temperature. The forward velocity constant with some hydrogen ion as catalyst is 0·000148 litre (g mole)$^{-1}$ min^{-1} and the equilibrium constant is 0·219. The initial concentration of the ester is

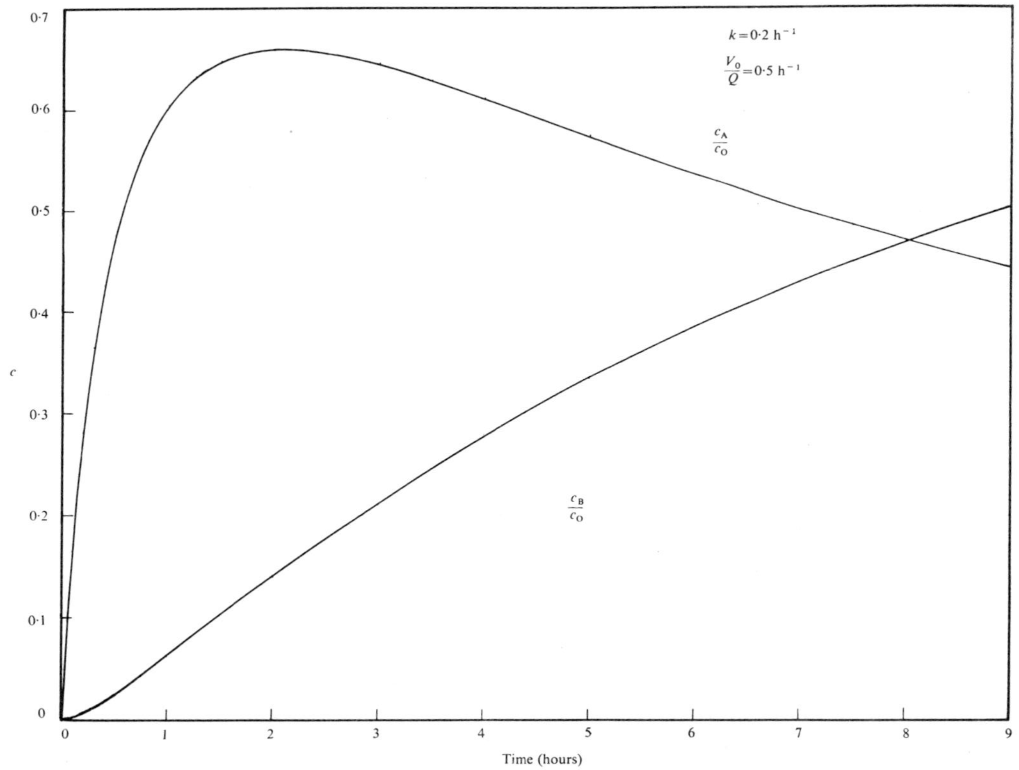

FIG. 4.4. Semi-batch first order reaction

1·151 g mole litre^{-1} and of water 48·76 g mole litre^{-1}. Find (*a*) the equilibrium conversion of ester, (*b*) the time required for 90% of this equilibrium yield to be obtained, and (*c*) the final composition of the reactor charge assuming no volume changes.

ANS. (*a*) 0·91 (*b*) 1·9 min.
(*c*) Ester 0·41 mole %; H_2O 95·7 mole %; products 1·89 mole % each

2. Butyl acetate is to be manufactured in a batch reactor operating at 100°C. If the feed streams are in the ratio of 4·97 moles of butanol per mole of acetic acid estimate the time required for 50% conversion of the acid. Determine the size of the reactor and the original mass of reactants that must be charged in order to produce ester at an average rate of 100 lb h^{-1}. The change round time is 0·75 h.
Rate equation: $r = kc_A^2$ where c_A is acetic acid concentration g moles cm^{-3}; and

$$k = 17{\cdot}4\ \text{cm}^3\ (\text{g mole})^{-1}\ \text{min}^{-1}$$

$$\rho(\text{acetic acid}) = 0{\cdot}958;\ \rho(\text{butanol}) = 0{\cdot}742;\ \rho(\text{butyl acetate}) = 0{\cdot}796$$

ANS. 20 ft^3.

3. The compound A is to be reacted batchwise by the following scheme with initial concentrations of '*a*', '*b*', '*c*' and '*e*':

$$\begin{array}{ccccc} A & \underset{k_2}{\overset{k_1}{\rightleftharpoons}} & B & \overset{k_3}{\rightarrow} & C \\ & & \downarrow k_4 & & \\ & & E & & \end{array}$$

Find general algebraic expressions for the concentrations of A, B, C and E with time, and check that their sum is:

$$a + b + c + e.$$

If the initial reaction mixture contains A at a concentration 50 g mole litre^{-1}, B at 5·0 g mole litre^{-1} and E at 3·0 g mole litre^{-1}, find the concentrations of B and E as functions of time, for the following numerical values of the velocity constants, and show that the maximum concentration of B is 20·1 g mole litre^{-1}

$$\begin{aligned} k_1 &= 1{\cdot}75\ \text{sec}^{-1} \\ k_2 &= 1{\cdot}25\ \text{sec}^{-1} \\ k_3 &= 0{\cdot}2\ \text{sec}^{-1} \\ k_4 &= 0{\cdot}8\ \text{sec}^{-1} \end{aligned}$$

ANS. $B = 31{\cdot}25 \exp(-\tfrac{1}{2}t) - 26{\cdot}25 \exp(-7t/2)$
$E = 47 - 50 \exp(-\tfrac{1}{2}t) + 6 \exp(-7t/2)$

4. In the production of 50 kg batches of molten thorium metal by the reaction

$$ThF_4 + 2Ca \rightarrow Th + 2CaF_2$$

there is an enthalpy deficit between cold solid reactants and hot molten products of 0·5 kcal for each g mole of thorium product. This deficit is to be made up by the simultaneous calcium reduction of zinc chloride having an enthalpy surplus of 50 kcal for each g mole of zinc formed. The latter reaction has also to make available heat to counteract the furnace losses for a period of 20 min. The furnace exterior has area 15 ft^2, temperature above ambient of 300 degF and the heat transfer coefficient is 4·0 Btu h^{-1} ft^{-2} $degF^{-1}$. Assuming stoichiometric ratios of reactants and complete conversion show that the separated thorium-zinc alloy would be expected to contain slightly more than 4% w/w zinc.

(A.W. thorium = 232, and zinc = 65·4.)

5. Show that the concentration of reactant at time t in a semi-batch reactor effecting the first order reversible reaction:

$$A \underset{k_2}{\overset{k_1}{\rightleftharpoons}} B$$

is:
$$c_A = \frac{Qc_0}{(k_1 + k_2)^2}\left[\frac{k_1 + k_2(k_1 + k_2)t - k_1 \exp\{-(k_1 + k_2)t\}}{V_0 + Qt}\right]$$

where V_0 is the initial volume of charge

and Q is the feedrate containing A at concentration c_0.

Develop an expression to show how the reaction heat release varies with time.

REFERENCES

GRUMMITT, O., and FLEMING, H. 1945. Acetylated castor oil. Preparation and thermal decomposition. *Ind. Engng Chem.*, **37**, 485.

STERBACEK, Z., and TAUSK, P. 1965. *Mixing in the chemical industry*. Pergamon Press, Oxford.

5

THE CONTINUOUS STIRRED TANK REACTOR

5.1 Introduction to Continuous Reactors

In continuously operated chemical reactors, the reactants are pumped at constant rate into the reaction vessel and chemical reaction takes place while the reactant mixture flows through. The reaction products are continuously discharged to the subsequent separation and purification stages. The extent of the required separation and purification processes depend on the efficiency of the reactor, and so the selection of the correct type of reactor for a given duty is most important since the economics of the whole process could hinge on this choice. Normally the efficiency of a chemical reactor is measured by its ability to convert the reactants into the desired products with the exclusion of unwanted by-products; this is measured by yield. However, additional factors such as safety, ease of control and stability of the process must also be considered. Many of these factors depend on the size and shape of the reactor. The size of reactor for a proposed feed rate depends on the reaction kinetics of the materials undergoing chemical change and on the flow conditions in the reactor. The flow conditions are determined by the cross-sectional area of the path through which the reaction mixture flows, i.e. on the shape of the reactor. Thus it is apparent that size and shape are interrelated factors which must be taken together when considering continuous reactors.

The extreme types of continuous reactor are:

(*i*) The plug flow tubular reactor.
(*ii*) The well stirred tank reactor.

In (*i*) the ability of the contents to mix is limited to radial mixing whereas in (*ii*) the contents are assumed to be completely mixed. Each of these types of continuous reactor has special fields of application and these should be considered when selecting a continuous reactor for a particular duty. The factors governing the choice between a tubular and a well stirred tank reactor have been discussed by Denbigh (1965) and are as follows:

(*i*) If the reaction rate has a large activation energy, i.e. is sensitive to temperature and it is advantageous to operate the reactor isothermally, a stirred tank reactor would be the more suitable, especially if the heat of reaction is large (either positive or negative).

(*ii*) If the reaction rate is low and a long residence time in the reactor is desirable, a stirred tank would be preferred.

(*iii*) If one of the reactants is of an explosive nature at high concentrations, a stirred tank reactor would be preferred because the reactant concentration in the feed would be reduced immediately to that of the reactor effluent.

(*iv*) If the chemical reaction is complex, the choice depends on the relative orders of the desirable and the unwanted reactions. For example consider two simultaneous reactions of different order represented by:

$$\mathrm{A + B \rightarrow X} \quad \text{with} \quad r = k_1 c_\mathrm{A} c_\mathrm{B} \tag{5.1}$$

$$\mathrm{2A + B \rightarrow Y} \quad \text{with} \quad r = k_2 c_\mathrm{A}^2 c_\mathrm{B}. \tag{5.2}$$

If X is the desired product and Y a by-product, the choice between a stirred tank and a tubular reactor can be made by expressing the ratio of the concentration of the product X to that of Y as a function of time and estimating how this ratio varies with time. Thus if at any time 't' after the start of the reaction, 'x' is the concentration of X and 'y' that of Y, then the relative rates of production of X and Y are:

$$\frac{dx}{dy} = \frac{k_1}{k_2 c_\mathrm{A}}. \tag{5.3}$$

Since the desired product X is produced by the reaction of lower order, a stirred tank reactor should be used because the concentration c_A for a fixed feed composition will be lower in a C.S.T.R. than the sequence of values in a tubular reactor. Conversely if Y is the desired product a tubular reactor should be employed.

(*v*) If the reaction mixture is gaseous, a tubular reactor is generally more suitable.

(*vi*) If the reaction process involves consecutive reactions of the type:

$$A \rightarrow B \rightarrow C \tag{5.4}$$

in which B is the desired product, a tubular reactor is generally more suitable because of the lack of spread of residence times of the fluid elements in the reactor.

(*vii*) High pressure reactions are best effected in tubular reactors.

(*viii*) Strongly endothermic reactions to be effected at high temperatures are best carried out in a tubular reactor.

The comparison presented above between a continuous stirred tank reactor and a tubular reactor enables a choice of reactor type to be made in a general manner only. However, each reaction process must be considered on its own merits because, on occasion, special circumstances arise which might overrule the above items of selection.

This chapter now continues with a detailed consideration of the well stirred reactor, and the next two chapters deal with plug flow reactors and reactors having residence-time distributions between these extremes.

5.2 Introduction to the Continuous Stirred Tank Reactor

This type of reactor consists of one or more cylindrical tanks. Normally the tanks are arranged with their axes vertical, although this is not essential. The stirring of the contents of each tank is effected by an agitator mounted on a shaft inserted through the vessel lid. In addition, the tank is fitted with the auxillary equipment necessary to maintain the desired reaction temperature and pressure conditions.

The well stirred reactor is used almost exclusively for liquid phase reactions, although instances of gaseous reactions have been reported. In normal operation a steady continuous feed of reactants is pumped into the vessel and since there is usually negligible density change on reaction, an equal volume of the reactor contents is displaced through an overflow pipe situated near the top of the vessel.

Consider the single stirred tank reactor shown in Fig. 5.1. It is assumed that the design of the agitator is such that the whole of

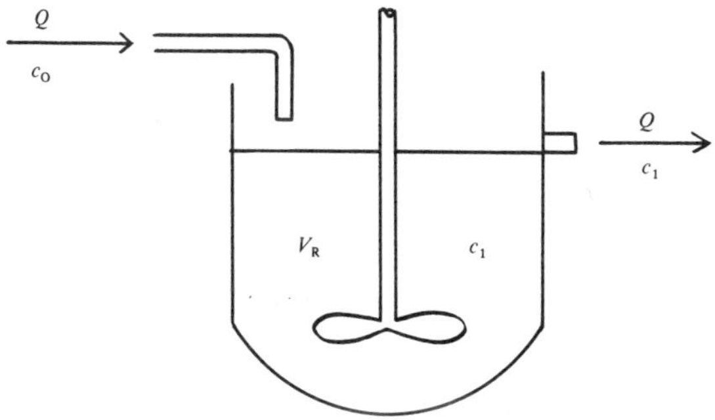

FIG. 5.1. Single stirred tank reactor

the vessel contents are completely mixed and at both a uniform temperature and composition. This premise of complete mixing then implies that the reactor outlet stream is identical in temperature and composition to the bulk reactor contents. Let the concentration of the reactant in the feed be c_0 and in the exit stream c_1. Then for an effective reactor volume V_R and volumetric input and output Q, a steady state mass balance is:

$$Qc_0 = Qc_1 + rV_R. \tag{5.5}$$

The reaction rate 'r' is constant with reactor position and also with time when steady state operation has been established. Hence under steady state operation, equation (5.5) is an algebraic equation and its application to some simple reactions is now considered.

5.3 Single Tanks with Simple Reactions

For irreversible reactions, the reaction rate term is:

$$kc_1^n.$$

For a first order reaction:

$$c_0 = c_1 + \frac{kV_Rc_1}{Q} = c_1\left(1 + \frac{kV_R}{Q}\right). \tag{5.6}$$

V_R/Q is the space time and will be given the symbol θ. Hence:

$$\frac{c_1}{c_0} = \frac{1}{1 + k\theta}. \tag{5.7}$$

Since the input and outlet volumetric flow rates are equal, the fractional conversion 'f' may be written:

$$f = 1 - \frac{c_1}{c_0} = \frac{k\theta}{1 + k\theta} \tag{5.8}$$

and as is usual with first order reactions the result is independent of the feed concentration.

For a second order reaction:

$$c_0 = c_1 + k\theta c_1^2 \tag{5.9}$$

and the positive root of this quadratic gives:

$$c_1 = \frac{-1 + (1 + 4k\theta c_0)^{\frac{1}{2}}}{2k\theta} \tag{5.10}$$

and hence

$$f = 1 + \frac{1 - (1 + 4k\theta c_0)^{\frac{1}{2}}}{2k\theta c_0}. \tag{5.11}$$

The values of c_1 and f are seen to be no longer independent of the feed concentration, while on extension to non-integral reaction orders the algebraic equations are also likely to require numerical iteration for their solution.

Having developed the equation for a single continuous well stirred reactor it is convenient to compare the throughput with that obtained from the same vessel operated batchwise. This is done in the next example.

Example 1. 90% conversion is required in a first order reaction system having a velocity constant 1·0 min^{-1}. Compare batch with continuous operation for the cases where:

(*i*) changeround from one batch to the next takes a negligible time;
(*ii*) the changeround takes 5 min;
(*iii*) the changeround takes 10 min.

Consider the situation in general terms with the production of a quantity of product R in a vessel having a hold-up V. The number of batch operations is R/V and hence the total batch conversion time is:

$$\frac{R}{V}\left[\frac{1}{k}\operatorname{Ln}\frac{c_0}{c_1}\right].$$

If each batch requires a changeround time T, then the total batch operation time is:

$$\frac{R}{V}\left[T + \frac{1}{k}\,\mathrm{Ln}\,\frac{c_0}{c}\right]$$

and the average batch production rate Q_B is:

$$Q_B = \frac{V}{T + \dfrac{1}{k}\,\mathrm{Ln}\,\dfrac{c_0}{c_1}}. \tag{I}$$

The steady rate of production Q_C for continuous operation may be obtained by rearrangement of equation (5.7) as:

$$Q_C = \frac{kV}{\dfrac{c_0}{c_1} - 1} \tag{II}$$

Hence:

$$\frac{Q_B}{Q_C} = \frac{\dfrac{c_0}{c_1} - 1}{\left[kT + \mathrm{Ln}\,\dfrac{c_0}{c_1}\right]}. \tag{III}$$

When T is negligible:

$$\frac{Q_B}{Q_C} = \frac{10 - 1}{\mathrm{Ln}\,10} = 3{\cdot}9.$$

When T is 5 min:

$$\frac{Q_B}{Q_C} = \frac{9}{5 + 2{\cdot}303} = 1{\cdot}23$$

and when $T = 10$ min:

$$\frac{Q_B}{Q_C} = \frac{9}{12{\cdot}303} = 0{\cdot}733,$$

i.e. as the changeround time increases, the continuous throughput exceeds the average batch rate of production. The critical changeround time may be established from equations (I) and (II). Thus continuous operation will be advantageous if:

$$\frac{kV}{\dfrac{c_0}{c_1} - 1} > \frac{V}{T + \dfrac{1}{k}\,\mathrm{Ln}\,\dfrac{c_0}{c_1}}$$

or

$$T > \frac{1}{k}\left[\frac{c_0}{c_1} - 1 - \mathrm{Ln}\,\frac{c_0}{c_1}\right].$$

For the above numerical values, continuous operation is advantageous if $T > 6{\cdot}7$ min.

5.4 Multiple Tank Cascade with Simple Reactions

A major shortcoming of a single stirred tank is that all of the reaction takes place at the low final reactant concentration and hence requires an unduly large reactor hold-up. If a number of smaller well stirred reactors are arranged in series, only the last one will have a reaction rate governed by the final reactant concentration and all of the others will have higher rates. Hence for a given duty the total reactor hold-up will be less than for a single tank. This saving in reactor volume increases as the required fractional conversion increases and also as the number of installed tanks increases. In fact all of the desirable features of the C.S.T.R. may be retained while the low hold-up characteristics of a plug flow tubular reactor may be approached. It will be seen in Example 2 below that the order of five to ten tanks in cascade are likely to give a close approximation to plug flow and it is a matter of economics to balance the cost of a number of tanks against their reduced size.

There are other operational advantages in carrying out reactions in a series of stirred tanks. For example if one vessel in the cascade has to be put out of commission for any reason, it may be by-passed and production continued at a slightly reduced rate whereas failure of a single C.S.T.R. would entail complete loss in production.

The analysis of a cascade of C.S.T.R.'s effecting simple reactions may be made by repeated application of equations such as (5.7) or (5.9) but allowing the possibility of different temperatures in each tank, and hence different rate constants and also of unequal tank sizes and hence varying 'space times'. With the notation of Fig. 5.2

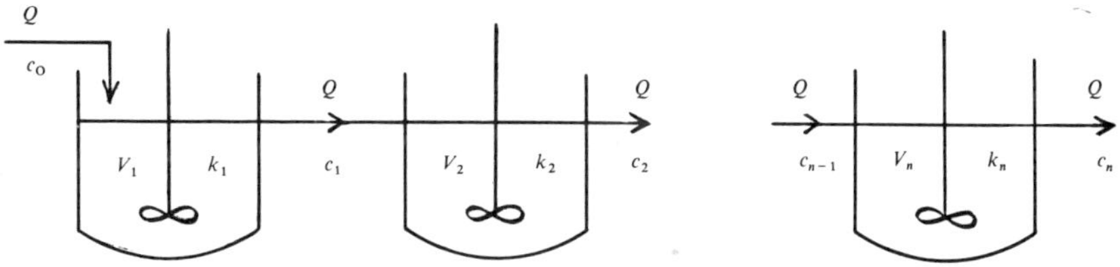

FIG. 5.2. Cascade of well stirred reactors

and using the result of equation (5.7), a first order irreversible reaction will be governed by:

$$\frac{c_1}{c_0} = \frac{1}{1 + k_1\theta_1} \tag{5.12}$$

$$\frac{c_2}{c_1} = \frac{1}{1 + k_2\theta_2} \tag{5.13}$$

$$\frac{c_n}{c_{n-1}} = \frac{1}{1 + k_n\theta_n} \tag{5.14}$$

and on multiplying these expressions together the intermediate concentrations $c_1, c_2, \ldots, c_{n-1}$ are eliminated to give:

$$\frac{c_n}{c_0} = \frac{1}{1 + k_1\theta_1} \cdot \frac{1}{1 + k_2\theta_1} \cdots \frac{1}{1 + k_n\theta_n} \equiv \frac{1}{\prod_{m=1}^{m=n} (1 + k_m\theta_m)}. \tag{5.15}$$

The overall reactor performance may therefore be evaluated without detailed calculation of the step by step concentrations. If the rate constants and space times are equal, then equation (5.15) is reduced to:

$$\frac{c_n}{c_0} = \frac{1}{(1 + k\theta)^n}. \tag{5.16}$$

The application of equation (5.16) and the effect of using multiple reactors is shown in the following example.

Example 2. Consider 100 ft^3 of total reactor volume to be available for the 90% conversion of a reactant by a first order mechanism ($k = 1$ min^{-1}). Adapting equation (5.16) to the cases of $n = 1, 2, 3,$

4, 5, 10 and 100 enables the throughput to be estimated for these cases. Thus taking $n = 5$ (i.e. $V = 20\ \text{ft}^3$):

$$\frac{c_n}{c_0} = \frac{1}{10} = \frac{1}{\left(1 + \dfrac{20}{Q}\right)^5}$$

or

$$Q = \frac{20}{10^{0\cdot2} - 1} = 34\cdot2\ \text{ft}^3\ \text{min}^{-1}.$$

This and the remaining results are tabulated below.

n	Q ft^3 min^{-1}
1	11·1
2	23·1
3	28·9
4	32·1
5	34·2
10	38·6
100	43·5

The table shows the diminishing value of subdivision beyond about two or three tanks. The throughput for an infinite number of tanks is 43·5 ft^3 min^{-1} ($n = 100$ is effectively infinite in this respect) and will be seen in section 6.3.1 to give the same result as for a plug-flow reactor operating to the same specification.

For a second order reaction, equation (5.10) applied to the first two tanks of a cascade gives:

$$\frac{c_1}{c_0} = \frac{-1 + (1 + 4k_1c_0\theta_1)^{\frac{1}{2}}}{2k_1c_0\theta_1} \tag{5.17}$$

$$\frac{c_2}{c_1} = \frac{-1 + (1 + 4k_2c_1\theta_2)^{\frac{1}{2}}}{2k_2c_1\theta_2}. \tag{5.18}$$

Multiplication of such expressions will not enable intermediate concentrations to be eliminated and it is necessary to solve the equations sequentially.

Likewise for non-integral reaction orders the numerical iteration will need to be solved for each exit concentration in turn.

5.5 Relative Sizes of Tanks in a C.S.T.R. Cascade, Effecting Simple Reactions

5.5.1 Isothermal first order reactions

The derivations in the above section have considered the general possibility of unequal reactor volumes throughout a cascade. In the interests of economy identical tanks would be fabricated and it is therefore necessary to examine the implications of such a policy on overall reactor performance. It is shown in this section that for a given total reactor volume, reactor yield is at a maximum when the tanks are in fact of equal size; it is possible to show as an alternative that the total reactor volume is at a minimum for a given yield again when the tanks are of equal size. However, equal tanks do not give the optimum condition in the case of second order reactions as shown in section 5.5.2.

The application of equation (5.15) to include a general tank 'm' for an isothermal cascade gives:

$$c_n = \frac{c_0}{(1 + k\theta_1)(1 + k\theta_2)\ldots(1 + k\theta_m)\ldots(1 + k\theta_n)}. \qquad (5.19)$$

For a maximum yield, c_n must be a minimum. The problem is now the minimisation of c_n with respect to the set of variables $\theta_1 \ldots \theta_n$ and subject to the restriction that:

$$\sum_{m=1}^{n} \theta_m = \frac{V_T}{Q} = \theta_1 + \theta_2 + \ldots + \theta_m + \ldots + \theta_n \qquad (5.20)$$

where V_T is the fixed total reactor volume.

The problem may be solved by introduction of a Lagrange multiplier 'λ' (Strain, 1961) and the function 'ϕ' where:

$$\phi = \frac{c_0}{(1 + k\theta_1)(1 + k\theta_2)\ldots(1 + k\theta_m)\ldots(1 + k\theta_n)} + \lambda\left[\theta_1 + \theta_2 + \ldots + \theta_n - \frac{V_T}{Q}\right]. \qquad (5.21)$$

The partial derivatives:

$$\frac{\partial\phi}{\partial\theta_1}, \frac{\partial\phi}{\partial\theta_2}, \ldots, \frac{\partial\phi}{\partial\theta_n}$$

are evaluated and set at zero. i.e.

$$\frac{\partial \phi}{\partial \theta_1} = -\frac{kc_0}{1 + k\theta_1} \cdot \frac{1}{(1 + k\theta_1)(1 + k\theta_2)\dots(1 + k\theta_n)} + \lambda = 0 \quad (5.22)$$

$$\frac{\partial \phi}{\partial \theta_2} = -\frac{kc_0}{1 + k\theta_2} \cdot \frac{1}{(1 + k\theta_1)(1 + k\theta_2)\dots(1 + k\theta_n)} + \lambda = 0 \quad (5.23)$$

$$\vdots$$

$$\frac{\partial \phi}{\partial \theta_n} = -\frac{kc_0}{1 + k\theta_n} \cdot \frac{1}{(1 + k\theta_1)(1 + k\theta_2)\dots(1 + k\theta_n)} + \lambda = 0. \quad (5.24)$$

Eliminating λ between equations (5.22) and (5.23) gives:

$$\frac{kc_0}{(1 + k_1\theta_1)(1 + k_2\theta_2)\dots(1 + k_n\theta_n)}\left[\frac{1}{1 + k\theta_2} - \frac{1}{1 + k\theta_1}\right] = 0 \quad (5.25)$$

i.e.

$$\theta_1 = \theta_2. \quad (5.26)$$

The corresponding elimination of λ between any pair of the above n equations will show equality between the two values of θ concerned.

i.e.

$$\theta_1 = \theta_2 = \dots = \theta_m = \dots = \theta_n \quad (5.27)$$

and since Q is constant from tank to tank

$$V_1 = V_2 = \dots = V_m = \dots = V_n = \frac{V_T}{n}. \quad (5.28)$$

5.5.2 Second order reactions in two stage systems

A general treatment allowing each of the stages to be at a different temperature will be considered initially. Thus the mass balances on the two reactors are:

$$Qc_0 = Qc_1 + k_1V_1c_1^2 \quad (5.29)$$

and

$$Qc_1 = Qc_2 + k_2V_2c_2^2. \quad (5.30)$$

The minimisation of the total system volume at a given throughput and overall fractional conversion is considered. Then Q, c_0, c_2,

k_1 and k_2 are fixed whereas V_1, V_2 and the intermediate concentration c_1 are variables.

From equations (5.29) and (5.30), the following expressions for V_1 and V_2 are obtained:

$$V_1 = \frac{Qc_0 - Qc_1}{k_1 c_1^2} = \frac{Qf_1}{k_1 c_0 (1 - f_1)^2} \tag{5.31}$$

$$V_2 = \frac{Q(f_2 - f_1)}{k_2 c_0 (1 - f_2)^2} \tag{5.32}$$

in which f_1 is the variable fractional conversion after the first reactor and f_2 is the fixed overall fractional conversion.

The total system volume V and its differential with respect to f_1 are:

$$V = \frac{Q}{c_0}\left[\frac{f_1}{k_1(1 - f_1)^2} + \frac{f_2 - f_1}{k_2(1 - f_2)^2}\right] \tag{5.33}$$

and

$$\frac{dV}{df_1} = \frac{Q}{c_0}\left[\frac{1 + f_1}{k_1(1 - f_1)^3} - \frac{1}{k_2(1 - f_2)^2}\right]. \tag{5.34}$$

For a turning point:

$$(1 - f_1)^3 = \frac{k_2(1 - f_2)^2}{k_1}(1 + f_1) \tag{5.35}$$

and on putting:

$$\frac{k_2(1 - f_2)^2}{k_1} = \beta \tag{5.36}$$

the following cubic in f_1 is obtained:

$$f_1^3 - 3f_1^2 + f_1(3 + \beta) + (\beta - 1) = 0. \tag{5.37}$$

When the vessels are operated at the same temperature, $k_1 = k_2$, so that β and hence f_1 are then functions of f_2 only. Tables 5.1 and 5.2 show how the consequent optimum total volume varies with f_2 and the velocity constant ratio.

The solutions of the above cubic equation do not always lie in the range 0 to f_2 so that in such cases there is no calculus turning

TABLE 5.1.

Minimum total volumes for two C.S.T.R.'s in series $f_2 = 0{\cdot}9$. (Range of k_1/k_2 giving a turning point is 0·01–0·19.)

k_1/k_2	f_1	$V_1 + V_2$	V_1/V_2
0·1	0·472	6·0	0·40
0·2	0·572	9·7	0·47
0·5	0·677	17·6	0·59
0·75	0·716	22·7	0·64
1·0	0·741	26·9	0·69
1·33	0·764	31·9	0·75
2·0	0·792	39·9	0·85
5·0	0·845	62·7	1·30
10·0	0·877	81·0	2·47

TABLE 5.2.

Minimum total volumes for two C.S.T.R.'s in series $f_2 = 0{\cdot}5$. (Range of k_1/k_2 giving a turning point is 0·23–3.)

k_1/k_2	f_1	$V_1 + V_2$	V_1/V_2
0·5	0·165	0·9	0·35
0·75	0·253	1·2	0·61
1·0	0·311	1·4	0·86
1·33	0·365	1·6	1·26
2·0	0·436	1·9	2·67

point in the region of physical interest. A turning point will occur between 0 and f_2 provided the ratio k_1/k_2 obeys:

$$(1 - f_2)^2 < \frac{k_1}{k_2} < \frac{1 + f_2}{1 - f_2}.$$

5.6 Extension of the Simple Reaction

The principles developed in sections 5.2 and 5.3 are now extended to other reaction mechanisms and examples are given of a side reaction, a reversible second order reaction and a set of consecutive reactions.

5.6.1 SIDE REACTION

Consider the first order system:

$$\begin{array}{l} A \xrightarrow{k_1} B \\ {\scriptstyle k_2}\downarrow \\ C \end{array} \tag{5.38}$$

in which the feed contains all three components at concentrations $c_{A.0}$, $c_{B.0}$ and $c_{C.0}$. The material balances for the three components are:

$$c_{A.0} = c_A + k_1\theta c_A + k_2\theta c_A \tag{5.39}$$

$$c_{B.0} = c_B - k_1\theta c_A \tag{5.40}$$

$$c_{C.0} = c_C - k_2\theta c_A \tag{5.41}$$

where c_A, c_B and c_C are the outlet concentrations.

Hence

$$c_A = \frac{c_{A.0}}{1 + \theta(k_1 + k_2)} \tag{5.42}$$

$$c_B = \frac{c_{B.0} + \theta[k_1(c_{A.0} + c_{B.0}) + k_2 c_{B.0}]}{1 + \theta(k_1 + k_2)} \tag{5.43}$$

and

$$c_C = \frac{c_{C.0} + \theta[k_2(c_{A.0} + c_{C.0}) + k_1 c_{C.0}]}{1 + \theta(k_1 + k_2)} \tag{5.44}$$

5.6.2 Reversible second order reaction

The following system will be considered for a feed containing A at concentration $c_{A.0}$ and B at $c_{B.0}$ but no products.

$$A + B \underset{k_2}{\overset{k_1}{\rightleftharpoons}} C + D \tag{5.45}$$

The concentrations of C and D will then be equal and so the mass balance on component A is:

$$c_{A.0} = c_A + \theta(k_1 c_A c_B - k_2 c_C^2). \tag{5.46}$$

By the stoichiometry of the reaction:

$$c_{B.0} - c_{A.0} = c_B - c_A \tag{5.47}$$

and

$$c_C = c_D = c_{A.0} - c_A \tag{5.48}$$

and so equation (5.46) may be rewritten as a function of c_A:

$$c_{A.0} = c_A + \theta[k_1 c_A(c_A + c_{B.0} - c_{A.0}) - k_2(c_{A.0} - c_A)^2] \tag{5.49}$$

and re-arrangement gives the quadratic:

$$c_A^2\theta(k_2 - k_1) - c_A[1 + k_1\theta(c_{B.0} - c_{A.0}) \\ + 2k_2\theta c_{A.0}] + c_{A.0} + k_2\theta c_{A.0}{}^2 = 0. \quad (5.50)$$

The analysis leading to equation (5.50) may be extended to the general reactor 'm' of a cascade. Let the concentration of A at inlet to vessel m be c_{m-1} and at outlet c_m. Then

$$c_{m-1} = c_m + \theta_m r_m. \quad (5.51)$$

The concentrations of B and C leaving this vessel are:

$$c_m + (c_{B.0} - c_{A.0})$$

and $$c_{A.0} - c_m.$$

Hence on denoting the difference $c_{B.0} - c_{A.0} \equiv E$, equation (5.51) gives:

$$\frac{c_{m-1}}{c_m} = 1 + \theta_m\left[k_1(c_m + E) - \frac{k_2(c_{A.0} - c_m)^2}{c_m}\right]. \quad (5.52)$$

If the space time θ_m is different for each reactor or if the reaction rate constants k_1 and k_2 vary between reactors because of non-isothermal operation, equation (5.52) must be applied sequentially along the cascade. However for equally sized vessels operating at a single temperature so that θ is constant, equation (5.52) is a non-linear first order finite difference equation in concentration. This is conveniently solved graphically by arbitrarily selecting values of c_m enclosing $c_{A.0}$ and c_N and solving equation (5.52) for c_{m-1}. Thereafter plotting $c_m \,.\, v \,.\, c_{m-1}$ gives the operating line for the stirred tank reactor battery as shown in Fig. 5.3. On the same diagram the line $c_m = c_{m-1}$ is plotted as shown. Then, starting at the concentration $c_{A.0}$ the number of reactors required for the desired conversion is 'stepped off' as shown until c_N is enclosed.

The graphical method described above can be applied to reversible and irreversible reactions of any order including fractional order, but fails when the procedure is applied to simultaneous or consecutive reactions. The reason for this is that other reactants exist in the vessels and their concentrations vary from vessel to vessel, and the manner in which they vary cannot be described by one equation of the type of (5.52). Hence the analysis of C.S.T.R. systems processing complex reactions must be carried out by stage to stage calculations.

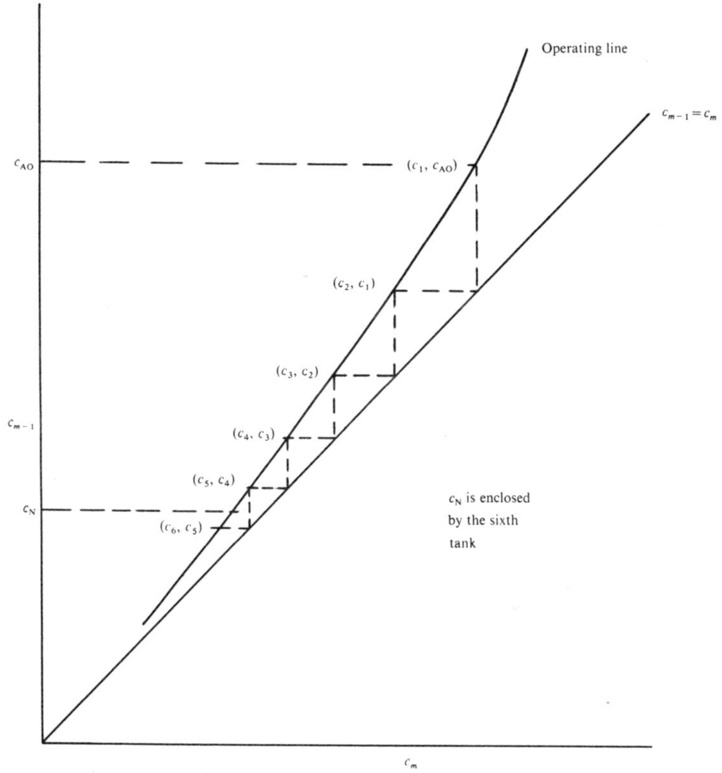

FIG. 5.3. Graphical solution of non-linear finite difference equation

An example of the graphical procedure is given in Jenson and Jeffreys (1965) and the sequential solution of equation (5.52) for tanks of different sizes now follows.

Example 3. Consider the esterification of ethyl alcohol (A) by acetic acid (B) in two tanks with $\theta_1 = 52{\cdot}6$ min^{-1} and $\theta_2 = 26{\cdot}3$ min^{-1}. The feed concentration of A is 6·63 g mole litre^{-1} and of B, 10·18 g mole litre^{-1}. The velocity constants are $k_1 = 4{\cdot}76 \times 10^{-4}$ and $k_2 = 1{\cdot}63 \times 10^{-4}$ min^{-1} (g mole)$^{-1}$ litre.

Substitution of these values into (5.52) gives:

$$165 \times 10^{-4}\, c_1^2 + 1{\cdot}20\, c_1 - 7{\cdot}01 = 0$$

or

$$c_1 = 5{\cdot}36 \text{ g mole litre}^{-1}$$

(C and D have outlet concentrations of 1·27 and that of B is 8·91 g mole litre^{-1}).

Similarly the concentration of A at the outlet from the second vessel is given by:

$$82{\cdot}5 \times 10^{-4}\, c_2^2 + 1{\cdot}09\, c_2 - 5{\cdot}48 = 0$$

or

$$c_2 = 4{\cdot}85 \text{ g mole litre}^{-1}.$$

5.6.3 Series of Consecutive Reactions

This situation is illustrated by the chlorination of benzene. The chlorine gas is sparged into the base of the tank and all of the hydrogen chloride formed is withdrawn. Consider the following second order reaction sequence with a liquid feed of pure benzene and a chlorine to benzene feed ratio of 'α':

$$C_6H_6 + Cl_2 \xrightarrow{k_1} C_6H_5Cl + HCl \tag{5.53}$$

$$C_6H_5Cl + Cl_2 \xrightarrow{k_2} C_6H_4Cl_2 + HCl \tag{5.54}$$

$$C_6H_4Cl_2 + Cl_2 \xrightarrow{k_3} C_6H_3Cl_3 + HCl \tag{5.55}$$

Let the concentrations in the reaction mixture be:

c_B for C_6H_6
c_M for C_6H_5Cl
c_D for $C_6H_4Cl_2$
c_T for $C_6H_3Cl_3$
c for Cl_2

and let the benzene feed concentration be c_{B0}. Then the kinetic equations and mass balances on the organic components give:

$$Qc_{B0} = Qc_B + Vk_1c_Bc \tag{5.56}$$

$$0 = Qc_M - Vk_1c_Bc + Vk_2c_Mc \tag{5.57}$$

$$0 = Qc_D - Vk_2c_Mc + Vk_3c_Dc \tag{5.58}$$

$$0 = Qc_T - Vk_3c_Dc \tag{5.59}$$

and a mass balance on the chlorine gives:

$$\alpha c_{B0} = c + c_M + 2c_D + 3c_T. \tag{5.60}$$

On successive elimination of c_T, c_D, c_M and c_B between equations

(5.56) to (5.59) the following quartic equation in chlorine concentration is obtained:

$$c^4 + c^3\left[(3-\alpha)\,c_{BO} + \frac{1}{\theta}\left(\frac{1}{k_1} + \frac{1}{k_2} + \frac{1}{k_3}\right)\right]$$
$$+\; c^2\left[\frac{c_{BO}}{\theta}\left(\frac{1}{k_2} + \frac{2}{k_3}\right) + \frac{1}{\theta^2}\left(\frac{1}{k_1k_2} + \frac{1}{k_1k_3} + \frac{1}{k_2k_3}\right)\right.$$
$$\left. -\frac{\alpha c_{BO}}{\theta}\left(\frac{1}{k_1} + \frac{1}{k_2} + \frac{1}{k_3}\right)\right] + c\left[\frac{1}{k_1k_2k_3\theta^3} + \frac{c_{BO}}{k_2k_3\theta^2} - \frac{\alpha c_{BO}}{\theta^2}\right.$$
$$\left.\times\left(\frac{1}{k_1k_2} + \frac{1}{k_1k_3} + \frac{1}{k_2k_3}\right)\right] - \frac{\alpha c_{BO}}{k_1k_2k_3\theta^3} = 0. \qquad (5.61)$$

A numerical solution for this reaction system is presented in the following example.

Example 4. At 55°C it is known that:

$$\frac{k_1}{k_2} = 8 \qquad \text{(I)}$$

and

$$\frac{k_2}{k_3} = 30. \qquad \text{(II)}$$

Substituting this data into equation (5.61) with the elimination of k_1 and k_3 gives:

$$c^4 + c^3\left[(3-\alpha)\,c_{BO} + \frac{249}{8\;k_2\theta}\right]$$
$$+\; c^2\left[\frac{61c_{BO}}{k_2\theta} + \frac{271}{8\;k_2^2\theta^2} - \frac{249\;\alpha c_{BO}}{8\;k_2\theta}\right]$$
$$+\; c\left[\frac{30\;c_{BO}}{k_2^2\;\theta^2} + \frac{30}{8\cdot k_2^3\theta^3} - \frac{271\;\alpha c_{BO}}{8\;k_2^2\theta^2}\right] - \frac{30}{8k_2^3}\frac{\alpha c_{BO}}{\theta^3} = 0. \qquad \text{(III)}$$

This equation may be solved for particular values of the parameters α, c_{BO} and $k_2\theta$. For pure benzene feed $c_{BO} = 11{\cdot}2$ g mole litre^{-1}; taking $\alpha = 1{\cdot}4$ and $k_2\theta = 1$, the equation to be solved is:

$$c^4 + 49{\cdot}045\,c^3 + 229{\cdot}035\,c^2 - 191{\cdot}41\,c - 58{\cdot}8 = 0 \qquad \text{(IV)}$$

from which

$$c = 0{\cdot}926 \text{ g mole litre}^{-1} \qquad \text{(V)}$$

Equations (5.56) to (5.59) are now solved successively for the particularly velocity constant ratios of (I) and (II) to give:

$$\frac{c_B}{c_{BO}} = \frac{1}{1 + 8(k_2\theta)c} \tag{VI}$$

$$\frac{c_M}{c_{BO}} = \frac{8(k_2\theta)c}{1 + (k_2\theta)c} \cdot \frac{c_B}{c_{BO}} \tag{VII}$$

$$\frac{c_D}{c_{BO}} = \frac{30(k_2\theta)c}{30 + (k_2\theta)c} \cdot \frac{c_M}{c_{BO}} \tag{VIII}$$

$$\frac{c_T}{c_{BO}} = \frac{(k_2\theta)c}{30} \cdot \frac{c_D}{c_{BO}}. \tag{IX}$$

Hence for a chosen value of $k_2\theta = 1$ and the result of $c = 0{\cdot}926$ the benzene and its derivatives have concentrations:

$$\frac{c_B}{c_{BO}} = 0{\cdot}1188,$$

$$\frac{c_M}{c_{BO}} = 0{\cdot}456,$$

$$\frac{c_D}{c_{BO}} = 0{\cdot}4095,$$

$$\frac{c_T}{c_{BO}} = 0{\cdot}01266.$$

5.6.4 Graphical analysis of complex reactions

The graphical methods of estimating the number of stirred tanks to achieve a given conversion are all based on the assumption that a material balance on a single chemical component is sufficient for the complete determination of the outlet stream composition. This is so only when the reaction process can be represented by a single stoichiometric equation; it is not true when the reaction process is complex involving consecutive or parallel steps.

When complex reactions are to be carried out in C.S.T.R. systems it is instructive and desirable to follow the progress of the reaction by 'space representation' in much the same manner as that employed in the analysis of solvent extraction processes. This type of representation is illustrated initially for simple reactions. For a simple reaction it will be shown that the product obtained from a C.S.T.R. can be identical with that obtained from a batch reaction process, but that generally for complex reactions the batch product cannot be duplicated in a C.S.T.R. operation with the same feed composition and an equivalent holding time. For example, it is possible to characterise a simple reaction of the type

$$pA + qB \rightarrow wC \tag{5.62}$$

proceeding in a batch reactor by the degree of conversion $f(t)$ as a function of time t. Thus starting with a feed containing c_{A0}, c_{B0} and c_{C0} of A, B and C respectively, the degree of conversion $f(t)$ at any time is expressed as:

$$f(t) = \frac{c_{A0} - c_A}{p} = \frac{c_{B0} - c_B}{q} = \frac{c_{C0} - c_C}{-w}. \tag{5.63}$$

These are equal because of the stoichiometry of the reaction and differ from the fractional conversion as used elsewhere in this text. The criterion that determines whether or not a single reaction such as (5.63) is able to completely describe the evolution of all the components of a batch reaction depends on whether the rates of reaction of all the components are proportional to one another during the whole of the reaction. For the reaction represented by equation (5.62) the rates are proportional as already shown in section 3.2.4. For complex reactions of the type

$$2A \rightarrow B \rightarrow C + D \tag{5.64}$$

the proportionality of the reaction rates is not generally satisfied. However, if for the reaction represented by equation (5.62), the concentrations present at different times during the batch process are plotted in a concentration space, a reaction path is obtained as shown in Fig. 5.4. This path starts with the charge or feed condition and progresses to the product composition when the reaction is complete; it can be graduated in terms of reaction time t for the batch process. Different reaction paths will be obtained for different feed conditions.

A similar path, which may not coincide with the batch path, may be obtained by plotting, for different values of the space time θ, the product concentrations from a single C.S.T.R. which receives a feed of identical composition with the batch charge. For the C.S.T.R.

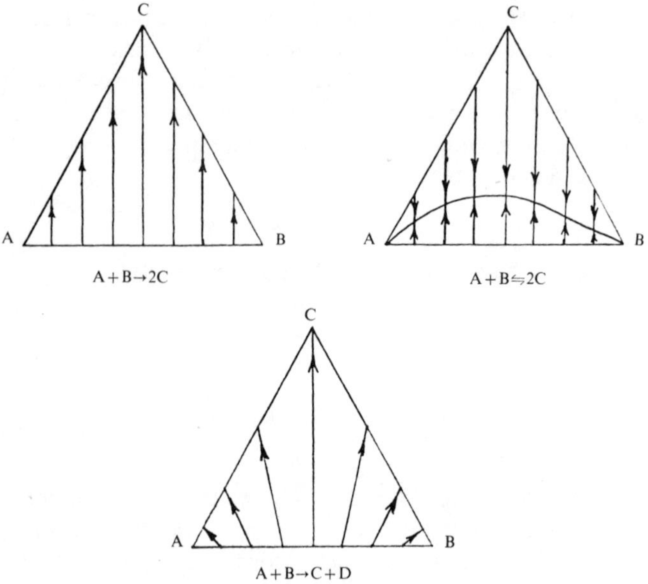

(D free basis)

FIG. 5.4. Reaction paths for simple reactions

process, the concentration change across the reactor is, for component A

$$\frac{c_{A0} - c_A}{\theta} = -\left[\frac{dc_A}{dt}\right]_{\text{batch}}. \tag{5.65}$$

Also for the batch process

$$\left[\frac{dc_A}{dt}\right]_{\text{batch}} = -p\left[\frac{df}{dt}\right]_{\text{batch}}. \tag{5.66}$$

Then combining equations (5.65) and (5.66) gives:

$$\frac{c_{A0} - c_A}{p} = \theta\left[\frac{df}{dt}\right]. \tag{5.67}$$

Since equations (5.65) to (5.67) can be written for any component in the C.S.T.R. process, then

$$\frac{c_{A0} - c_A}{p} = \frac{c_{B0} - c_B}{q} = \frac{c_{C0} - c_C}{-w} \tag{5.68}$$

which is the same as equation (5.63) for the batch process. That is, reaction path on the concentration space diagram for a C.S.T.R.

process carrying out a simple reaction will be the same as that for a batch process—a straight line as shown in Fig. 5.4, but as shown above the batch residence time will be different from the space time.

When complex reactions are involved the reaction paths for the batch and the C.S.T.R. will be different because the concentrations of the reacting species will be different. The reaction paths will not be straight lines for either the batch reactor or the C.S.T.R. and will be different for each type of reactor. That is, the products of reaction from a C.S.T.R. will be different from those obtained from a batch reactor processing the same feed. This should be realised and taken into consideration when making an analysis of a C.S.T.R. system using laboratory batch kinetic data.

5.6.5 Space representation of reaction paths

Space representation is straightforward when there are only three components involved in the reaction process, and the reaction is simple. When there are four or more components involved, such as, for example, in the reaction (A + B → C + D) it is customary to plot only the three most important and to plot the reaction path on, say 'a D-free basis'. The mole fractions of each of the pertinent components are then recalculated on this basis.

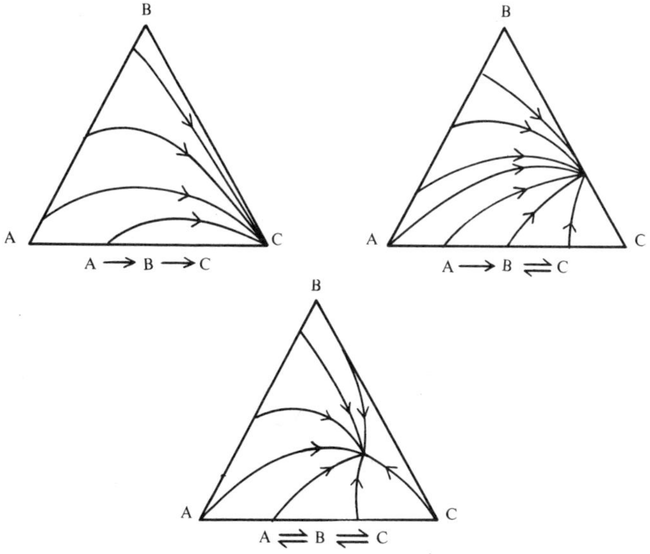

Fig. 5.5. Reaction paths for complex reactions

The reaction paths are straight for simple reactions and some examples have already been shown in Fig. 5.4. The reaction paths are not straight for complex reactions, and are different for batch and for C.S.T.R. operation. The curvature depends on the ratio of the rate constants for the different reactions. The batch reaction paths for different types of complex reactions are illustrated in Fig. 5.5. The utilisation of these curves is accomplished as follows for a C.S.T.R. having a space time θ. The mass balance for the different species taking part in the reaction can be written

$$c_{A0} - c_A = -\theta \left[\frac{dc_A}{dt}\right]_{\text{batch}} \tag{5.69}$$

and

$$c_{B0} - c_B = -\theta \left[\frac{dc_B}{dt}\right]_{\text{batch}} \tag{5.70}$$

$$c_{C0} - c_C = -\theta \left[\frac{dc_C}{dt}\right]_{\text{batch}} \tag{5.71}$$

Elimination of time from these equations gives:

$$\frac{dc_A}{c_{A0} - c_A} = \frac{dc_B}{c_{B0} - c_B} = \frac{dc_C}{c_{C0} - c_C}. \tag{5.72}$$

The geometrical interpretation of these relations is that (dc_A), (dc_B) and (dc_C) determine the tangent to the batch reaction path at the point composition c_A, c_B and c_C; and equation (5.72) implies that the vector, having the projections $(c_{A0} - c_A)$, $(c_{B0} - c_B)$ and $(c_{C0} - c_C)$, lies along the tangent. Therefore the feed composition point c_{A0}, c_{B0} and c_{C0} lies on the tangent of the batch reaction path going through the desired product composition point as shown in Fig. 5.6. This construction holds as long as the mass balances given by equations (5.69) to (5.71) are satisfied, and is not affected by the method of plotting.

In space representation the emphasis has been placed on following the change in concentration of the different species and not on reaction rate or time. Time dependence of the reaction may be indicated by constructing contours on the plots as shown in Fig. 5.7.

The above method of representing reaction paths in space can be used to predict the performance of a continuous stirred tank reactor system from batch rate data as follows.

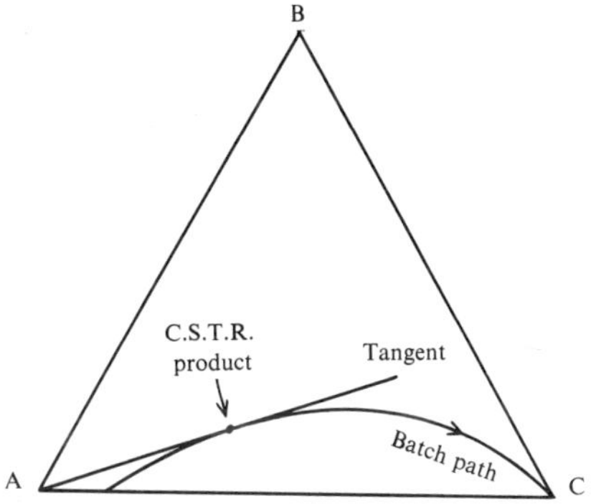

FIG. 5.6. Graphical determination of C.S.T.R. performance

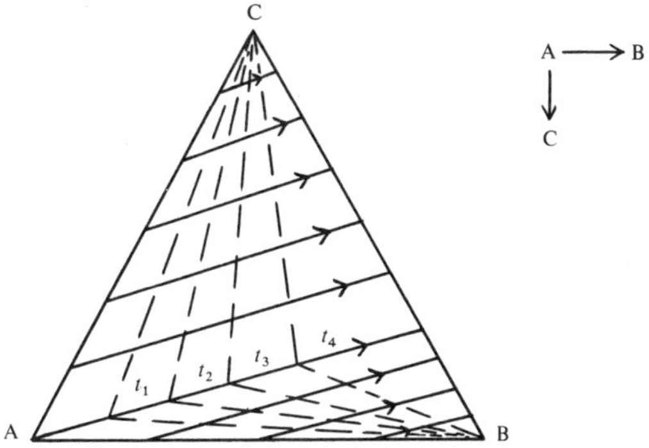

FIG. 5.7. Batch paths for simultaneous reactions showing time

(*i*) Prepare several batch reaction paths, starting with various feed combinations, on a space diagram.

(*ii*) Beginning with the feed composition, straight line tangents are drawn to the batch reaction paths and the one passing through the feed point of the C.S.T.R. is the mass balance line for the C.S.T.R. and the point of contact gives the product stream composition. The space time and thus the volume of the tank reactor for a given rate of production is calculated from the batch data at the product composition,

i.e. $$[\theta = \Delta c/r]. \tag{5.73}$$

The procedure is then repeated for the next reactor using the product composition from the first reactor as the starting point for the second.

5.7 Optimum Arrangement of C.S.T.R.'s for Reaction Processes

Most chemical engineering optimisation analyses are concerned with economic balances. However, on occasion, the primary criterion may be the yield of a desired reaction product and the number of reactors or size of reactors may have to be specified to maximise this yield. Thus, in certain organic syntheses, the reaction processes are of the types:

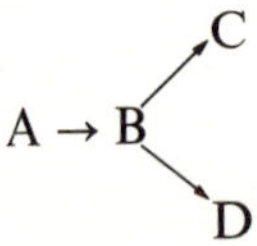

(5.74)

or

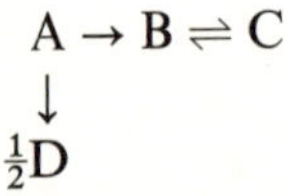

(5.75)

where B is the desired product in each case. The reactor arrangement to maximise the concentration of B is to be determined and thereby reduce the subsequent purification. In Fig. 5.8 the number of reactors corresponding to point N would be the optimum.

The arrangement of C.S.T.R.'s for optimum B at a given throughput may be resolved into the following possibilities:

(*i*) For a given size of available tank, the specification of the number of such equal tanks to be installed.

(*ii*) For a given number of tanks, the specification of the effective reaction volume required for each.

(*iii*) When neither of these constraints is applied, the specification of a combined optimum number of stages and their volume.

This will be illustrated by development of the reaction scheme given by equation (5.74) for reactors of equal size:

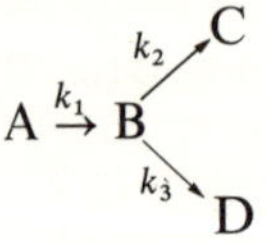

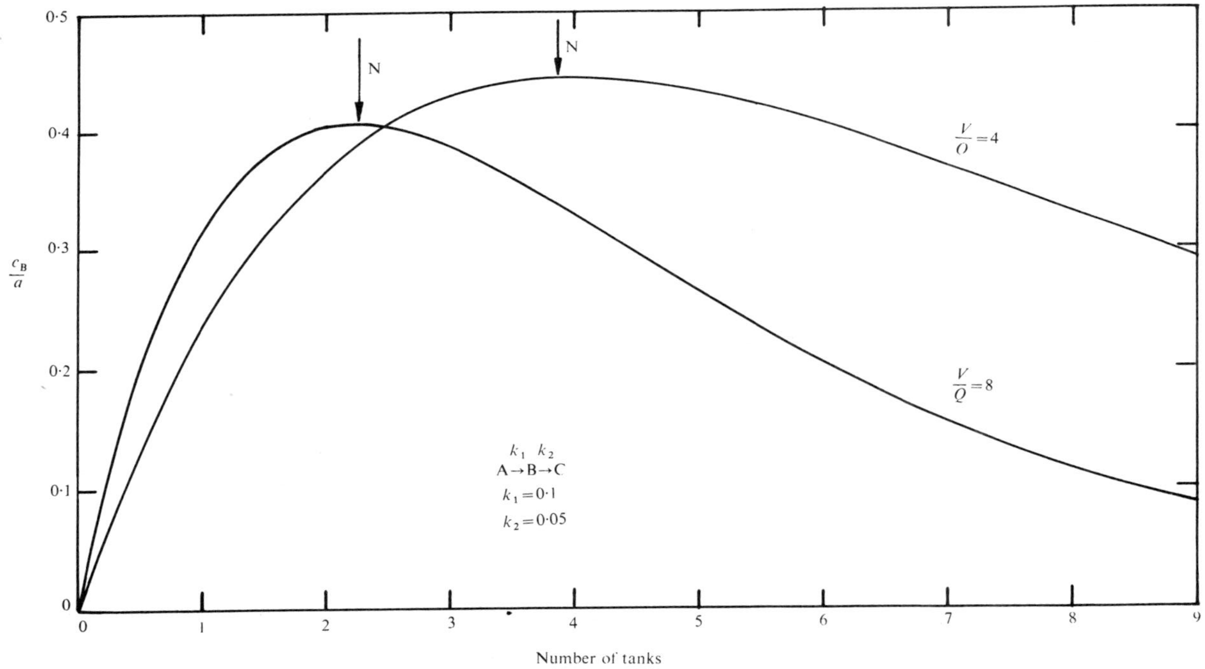

FIG. 5.8. $A \xrightarrow{k_1} B \xrightarrow{k_2} C \quad k_1 = 0{\cdot}1 \quad k_2 = 0{\cdot}05$

Let the reactions all be first order and let pure A at concentration a_0 be fed to the first reactor. If density changes are negligible, the nominal holding time in each tank is constant at θ.

Then the material balances on the nth tank are:

For A: $$a_{n-1} = a_n + k_1 a_n \theta \tag{5.76}$$

or $$a_{n-1} = a_n(1 + \alpha) \tag{5.77}$$

where $$\alpha = k_1\theta. \tag{5.78}$$

For B: $$b_{n-1} = b_n + k_2 b_n \theta + k_3 b_n \theta - k_1 a_n \theta \tag{5.79}$$

or $$b_{n-1} = b_n(1 + \beta + \gamma) - \alpha a_n \tag{5.80}$$

where $$\beta = k_2\theta \tag{5.81}$$

and $$\gamma = k_3\theta. \tag{5.82}$$

For C: $$c_{n-1} = c_n - \beta b_n \tag{5.83}$$

and for D: $$d_{n-1} = d_n - \gamma b_n. \tag{5.84}$$

Equations (5.77), (5.80), (5.83) and (5.84) are simultaneous finite difference equations and using the finite difference operator E, equation (5.77) becomes:

$$[(1 + \alpha)E - 1]\, a_{n-1} = 0 \tag{5.85}$$

from which

$$a_n = K_a\left[\frac{1}{1+\alpha}\right]^n = K_a(\rho_2)^n \tag{5.86}$$

where $$(1 + \alpha)^{-1} = \rho_2. \tag{5.87}$$

The constant K_a is evaluated from the boundary condition that:

$$a_n = a_0 \text{ when } n = 0$$

giving $$K_a = a_0 \tag{5.88}$$

and $$a_n = a_0\rho_2^n. \tag{5.89}$$

This result on substitution into (5.80) gives:

$$[(1 + \beta + \gamma)E - 1]\, b_{n-1} = \alpha a_0 \rho_2^n. \tag{5.90}$$

The complementary function part of the solution is:

$$b_n = K_b\left[\frac{1}{1+\beta+\gamma}\right]^n \equiv K_b\rho_1^n. \tag{5.91}$$

The particular solution is:

$$b_{n-1} = \frac{\alpha a_0\rho_2^n}{(1+\beta+\gamma)\rho_2 - 1} = \frac{\alpha a_0\rho_2^n(1+\alpha)}{\beta+\gamma-\alpha}$$

$$= \frac{\alpha a_0\rho_2^{n-1}}{\beta+\gamma-\alpha} \tag{5.92}$$

provided $\beta + \gamma \neq \alpha$

or

$$b_n = \frac{\alpha a_0\rho_2^n}{\beta+\gamma-\alpha}. \tag{5.93}$$

Hence the complete solution is:

$$b_n = K_b\rho_1^n + \frac{\alpha a_0\rho_2^n}{\beta+\gamma-\alpha} \tag{5.94}$$

when $n = 0, b_n = 0$.

Hence

$$K_b = -\frac{\alpha a_0}{\beta+\gamma-\alpha} \tag{5.95}$$

and

$$b_n = \frac{\alpha a_0}{\beta+\gamma-\alpha}[\rho_2^n - \rho_1^n]. \tag{5.96}$$

For C:

$$(E-1)c_{n-1} = \frac{\beta\alpha a_0[\rho_2^n - \rho_1^n]}{\beta+\gamma-\alpha} \tag{5.97}$$

and on solution as above:

$$c_n = \frac{\alpha\beta a_0}{\beta+\gamma-\alpha}\left[\frac{\rho_2(\rho_2^n - 1)}{\rho_2 - 1} - \frac{\rho_1(\rho_1^n - 1)}{\rho_1 - 1}\right]. \tag{5.98}$$

Similarly for D:

$$d_n = \frac{\alpha\gamma a_0}{\beta+\gamma-\alpha}\left[\frac{\rho_2(\rho_2^n - 1)}{\rho_2 - 1} - \frac{\rho_1(\rho_1^n - 1)}{\rho_1 - 1}\right]. \tag{5.99}$$

In the special case when $(\beta + \gamma) = \alpha$, i.e. $\rho_2 = \rho_1 = \rho$ (say):

$$a_n = a_0\rho^n \text{ as before} \tag{5.89}$$

$$b_n = \frac{\alpha a_0}{1+\alpha} n\rho^n = \alpha a_0 n\rho^{n+1} \tag{5.100}$$

$$c_n = \frac{\beta a_0}{\alpha}\left[1 - \rho^n\left(1 + \frac{n\alpha}{1+\alpha}\right)\right] \tag{5.101}$$

and

$$d_n = \frac{\gamma a_0}{a}\left[1 - \rho^n\left(1 + \frac{n\alpha}{1+\alpha}\right)\right]. \tag{5.102}$$

The optimisation is considered using equation (5.96) which is rewritten below in terms of the basic reactor variables:

$$b_n = \frac{k_1 a_0}{k_2 + k_3 - k_1}\left[\frac{1}{(1+k_1\theta)^n} - \frac{1}{(1+\{k_2+k_3\}\theta)^n}\right]. \tag{5.103}$$

The concentration of B is thus a function of the two independent variables θ and n; θ may represent variable throughput at fixed size and vice versa.

If n is fixed and θ can vary:

$$\left(\frac{\partial b_n}{\partial \theta}\right)_n = -\frac{k_1 a_0 n}{k_2 + k_3 - k_1} \times\left[\frac{k_1}{(1+k_1\theta)^{n+1}} - \frac{k_2+k_3}{(1+\{k_2+k_3\}\theta)^{n+1}}\right]. \tag{5.104}$$

Hence the optimum θ for a given number of tanks is:

$$\theta = \frac{\left(\dfrac{k_1}{k_2+k_3}\right)^{1/(n+1)} - 1}{k_1 - (k_2+k_3)\left(\dfrac{k_1}{k_2+k_3}\right)^{1/(n+1)}}. \tag{5.105}$$

If θ is fixed but n varies:

$$\left(\frac{\partial b_n}{\partial n}\right)_\theta = \frac{k_1 a_0}{k_2 + k_3 - k_1}\left[\frac{1}{(1+k_1\theta)^n}\,\mathrm{Ln}\,\frac{1}{1+k_1\theta} - \frac{1}{(1+\{k_2+k_3\}\theta)^n}\,\mathrm{Ln}\,\frac{1}{1+\{k_2+k_3\}\theta}\right] \tag{5.106}$$

and hence the optimum number of tanks is:

$$n = \frac{\mathrm{Ln}\left[\dfrac{\mathrm{Ln}\,(1 + \{k_2 + k_3\}\theta)}{\mathrm{Ln}\,(1 + k_1\theta)}\right]}{\mathrm{Ln}\left[\dfrac{1 + \{k_2 + k_3\}\theta}{1 + k_1\theta}\right]}. \tag{5.107}$$

Example 5. Consider the optimum concentrations of B for the following situations and the numerical values of velocity constants:

$$k_1 = 0{\cdot}1\ \mathrm{min}^{-1}; \quad k_2 = 0{\cdot}02\ \mathrm{min}^{-1}; \quad k_3 = 0{\cdot}03\ \mathrm{min}^{-1}$$

(*i*) For a two-tank system, equation (5.105) gives:

$$\theta = \frac{(2)^{\frac{1}{3}} - 1}{0{\cdot}1\text{–}0{\cdot}05(2)^{\frac{1}{3}}} = 7{\cdot}024,$$

i.e. B's concentration is at a maximum when

$$V = 7{\cdot}024Q$$

and equation (5.103) then gives the maximum concentration of B as $0{\cdot}41\ a_0$.

(*ii*) If the tank size is fixed and the required throughput is such that $V/Q = 4$, the number of tanks needed to maximise B's concentration is found by application of equation (5.107) giving:

$$n = \frac{\mathrm{Ln}\left[\dfrac{\mathrm{Ln}\,1 + 0{\cdot}05 \times 4}{\mathrm{Ln}\,1 + 0{\cdot}1 \times 4}\right]}{\mathrm{Ln}\left[\dfrac{1 + 0{\cdot}2}{1 + 0{\cdot}4}\right]} = 3{\cdot}975$$

and the maximum concentration of B from equation (5.103) using $n = 4$ is $0{\cdot}44\ a_0$.

5.8 Energy Balance and Stability of Operating Temperature

The steady state operating characteristics of a continuous stirred tank reactor of large hold-up suggest that the thermal conditions of the process should be easy to maintain, and isothermal operation of a series of such tanks should be a practical reality. This is, in fact, true and C.S.T.R. systems may be operated isothermally if desired. However, on occasion it is advantageous to impose non-isothermal operation on a C.S.T.R. system. For example, the output of a reaction process may be increased by imposing a temperature profile over the series of stirred tanks in order to suppress unwanted side reactions. Thus, in the manufacture of hydrazine by the Rashig process there are advantages in employing a two stage reactor. The first reactor operates at low temperature with the reaction of sodium hypochlorite and ammonia to give the maximum yield of chloramine, and the second at high temperature to produce the maximum output of hydrazine. A detailed study of this reaction process has been carried out by Ellis, Jeffreys and Wharton (1964); Jeffreys and Wharton, (1964, 1965). The imposition of a controlled temperature profile is desirable when the reaction process involves exothermic reversible reactions. Thus, fresh reactants are normally fed to the first reactor, so that in the initial reactors the concentrations of the products are low and considerably below equilibrium concentration. Therefore, it is advantageous to maintain the temperatures as high as possible in the initial reactors in order to maintain a high reaction rate. However, in the later reactors of the battery where concentrations approach the high temperature equilibrium concentrations, it is desirable to reduce the temperature and thereby increase the potential equilibrium yield from the process. Consequently for such processes it is desirable to introduce a stepwise reduction in temperature along the C.S.T.R. battery.

The analysis of non-isothermal C.S.T.R. operation must include an energy balance together with the basic C.S.T.R. mass balance equation. When this is done, the graphical procedures presented for isothermal operation are not valid and a stage by stage analysis is usually necessary. Therefore, consider a single stirred tank reactor operating under conditions where the temperature T_0 of the feed is different from that of the reactor contents T_1. Then if Q_{R} and $(C_P)_{\mathrm{R}}$ are the flow and specific heat of the individual components of the feed stream and Q_P and $(C_P)_{\mathrm{P}}$ likewise for the outlet stream, the energy balance equation (2.12) becomes:

$$UA(T_{\mathrm{S}} - T_1) = rV_{\mathrm{R}}\Delta H^0 + (T_1 - T_{\mathrm{D}})\,\Sigma Q_P(C_P)_{\mathrm{P}} - (T_0 - T_{\mathrm{D}})\,\Sigma Q_{\mathrm{R}}(C_P)_{\mathrm{R}} \qquad (5.108)$$

where T_{D} is the chosen datum temperature,
T_{S} is the temperature of the heat transfer medium,
A is the area of the heat transfer coil or jacket,
and U is the overall heat transfer coefficient.

T_S may be the temperature of condensing steam in a jacket or the mean temperature of coolant flowing through the internal coils.

It is convenient to simplify each of the summation terms of equation (5.108) to FC_P where F is the total mass throughput and C_P is the mass specific heat of both the feed and product mixtures, giving:

$$-rV_R\Delta H^0 = FC_P(T_1 - T_0) - UA(T_S - T_1). \qquad (5.109)$$

The left-hand side of equation (5.109) represents the rate of heat generation (or absorption) by the reaction, and will vary with temperature in a manner dependent upon the type of reaction. For an irreversible reaction this heat generation rate rises with temperature and at high temperatures attains values which are almost independent of further rise in temperature. For a reversible reaction, the variation passes through a maximum, and both situations are illustrated in Fig. 5.9a.

The two terms on the right-hand side of equation (5.109) represent the total heat removal as the sum of the change in enthalpy of the fluid flowing through the reactor and the heat transferred. Each of these terms may be positive or negative and may be re-arranged for steady state conditions of the feed and heat transfer medium to the following linear function of T_1, the reactor operating temperature:

$$T_1(FC_p + UA) - (FC_pT_0 + UAT_S).$$

Hence equation (5.109) may be written as:

$$-\frac{rV_R\Delta H^0}{F} = T_1\left(C_P + \frac{UA}{F}\right) - \left(C_PT_0 + \frac{UAT_S}{F}\right) \qquad (5.110)$$

per unit mass of throughput.

The slope and intercept of the heat removal line are thus functions of the reactor throughput and the intercept is in addition a function of both T_0 and T_S; a range of heat removal lines is shown in Fig. 5.9b.

The solution of equation (5.110) is accomplished by trial and error, or graphically by plotting each side of the equation against T_1 to locate the intersection of the heat generation curve with the heat removal line. The heat generation curve, Q_G, may be expressed as an analytical function of T_1 for simple forms of rate equation. Thus for an nth order forward reaction,

$$Q_G = -\frac{kV_Rc_0^n}{F}(1 - f)^n\,\Delta H^0. \qquad (5.111)$$

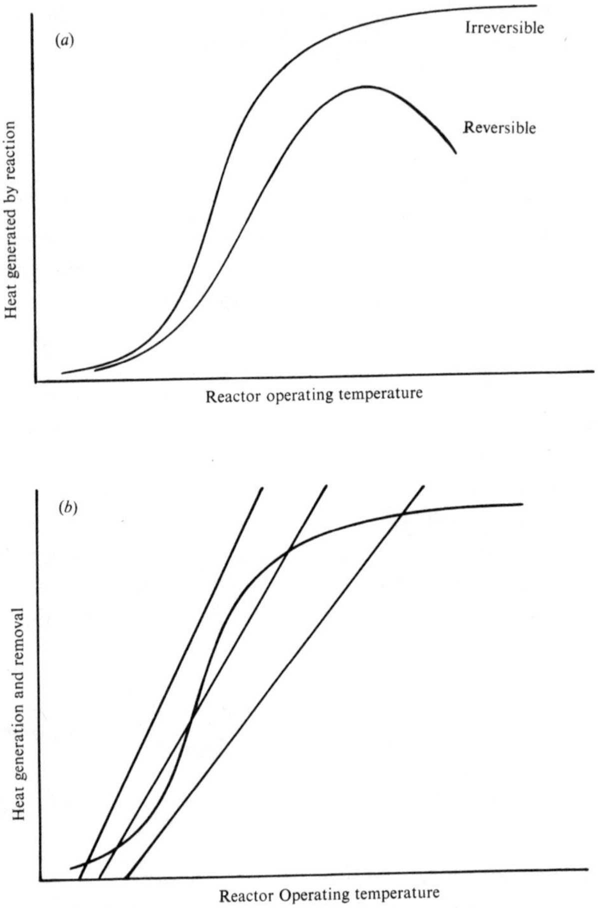

FIG. 5.9. Heat generation and heat removal plots
(a) Heat generation for reversible and irreversible reactions
(b) Solution of heat generation and heat removal equations for irreversible reactions

Expressions for $(1 - f)$ in terms of temperature for first and second order reactions may be substituted from equations (5.8) and (5.11). However, for non-integral orders, a series of values of $(1 - f)$ obtained for each nominated temperature by iteration of an equation of the form:

$$k\theta c_1^n + c_1 - c_0 = 0. \tag{5.112}$$

will have to be substituted into equation (5.111).

The heat generation expression for a first order reaction is then:

$$Q_G = \frac{-kV_R c_0 \, \Delta H^0}{F\left(1 + \dfrac{kV_R \rho}{F}\right)} \tag{5.113}$$

and on putting

$$k = a \exp\left(-\frac{E}{RT_1}\right) \tag{5.114}$$

$$Q_G = \frac{-ac_0 V_R \, \Delta H^0 \exp\left(-\dfrac{E}{RT_1}\right)}{F + aV_R \rho \exp\left(-\dfrac{E}{RT_1}\right)}. \tag{5.115}$$

The shape of the heat generation curve gives rise to the possibility of three intersections with the heat removal line and since the numerical solution might reveal only one of these, a more careful analysis of the process must be made.

Considering a situation in which the feed temperature T_0 can vary over a small range while all of the other parameters remain constant, the possible types of solution are illustrated in Fig. 5.10. For a feed temperature of $T_{0\cdot1}$ the steady state reactor temperature is $T_{1\cdot1}$. A slight change in T_0 from $T_{0\cdot1}$ to $T_{0\cdot2}$ will cause T_1 to change from $T_{1\cdot1}$ to $T_{1\cdot2}$. In fact changes in the feed temperature from $T_{0\cdot1}$ to $T_{0\cdot4}$ bring about changes in the reactor temperature from $T_{1\cdot1}$ to $T_{1\cdot4}$. For a reactor operating in this temperature range, i.e. between points A and B on the curve, gentle fluctuations in T_0 result in gentle fluctuation in the reactor temperature T_1 and the process is stable. However, an increase in the feed temperature beyond $T_{0\cdot4}$ brings about a large change in the reactor temperature. For example, a feed temperature of $T_{0\cdot5}$ results in a reactor operating temperature of $T_{1\cdot9}$. Aris and Amundson (1958) called this temperature 'the reactor ignition temperature'. In the same way it can be deduced that for fluctuations of feed temperature between $T_{0\cdot2}$ and $T_{0\cdot5}$ there would be corresponding fluctuations in the reactor temperature between $T_{1\cdot6}$ and $T_{1\cdot9}$ but that if the feed temperature dropped below $T_{0\cdot2}$ the reactor temperature would drop to below $T_{1\cdot2}$ with a low fractional conversion. Aris and Amundson conversely called this the 'quench temperature'. For the process conditions illustrated, reactor temperatures in the ranges $T_{1\cdot1}$ to $T_{1\cdot4}$ or $T_{1\cdot6}$ to $T_{1\cdot9}$ are

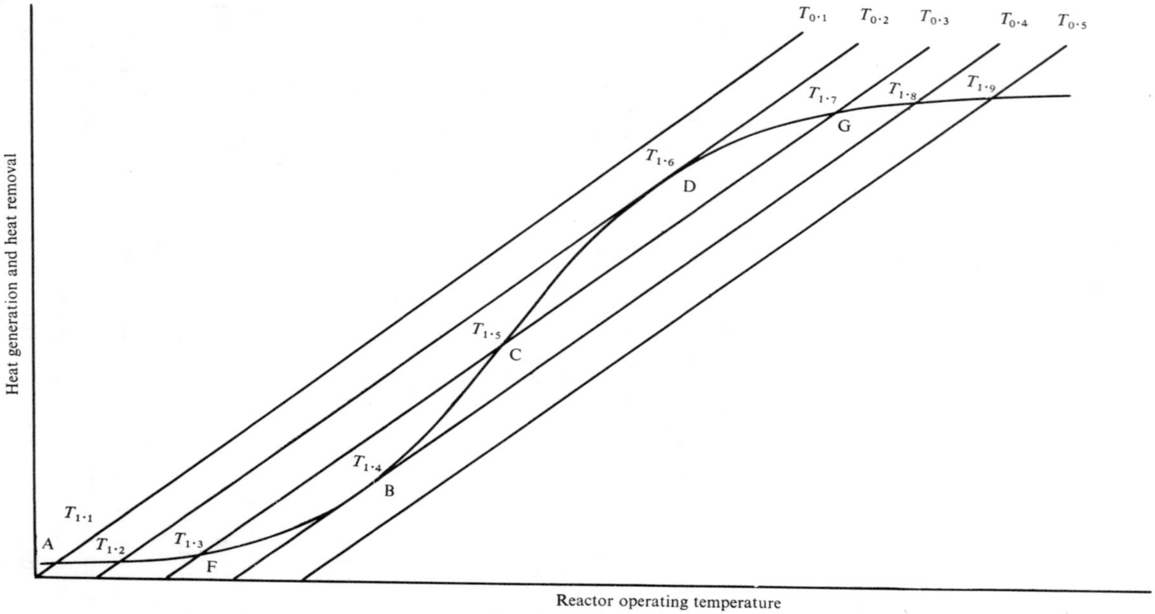

FIG. 5.10. Stable and unstable operating conditions—quench and ignition temperatures

inherently stable and processes operating in this way were described by Van Heerden (1953) as 'autothermal'.

However, reactor operating temperatures between $T_{1\cdot4}$ and $T_{1\cdot6}$ are not inherently stable. The instability in the region B–C–D of the heat generation curve will become apparent by considering a feed temperature $T_{0\cdot3}$. There are then three possible solutions to equation (5.110) at points F, C, and G. Considering the solution at G, an increase in temperature above $T_{1\cdot7}$ would mean a greater rate of heat removal than heat generation, while a decrease in temperature below $T_{1\cdot7}$ would mean that the heat generation rate exceeds the removal rate; either disturbance would result in a restoration of the stable operating temperature $T_{1\cdot7}$.

Similar remarks apply to the stable operating point F. However, if the temperature is increased above $T_{1\cdot5}$ at solution C, the heat generation rate exceeds the removal rate and the temperature continues to rise to $T_{1\cdot7}$ when solution G of equation (5.110) is satisfied. A decrease in temperature below $T_{1\cdot5}$ results in the heat removal rate exceeding that generated and a cooling to $T_{1\cdot3}$. Hence it is possible to distinguish between stable and unstable operating conditions by a comparison of the slopes with respect to temperature of the heat generation and heat removal lines. If the former is the greater, then the process conditions are not inherently stable.

The slope of the heat generation curve may be developed from equation (5.111) assuming the heat of reaction to be independent of temperature. Two temperature dependent terms, k and $(1 - f)$ then remain to be considered. An increase in temperature will increase k and also the fractional conversion 'f'; hence $(1 - f)$ will decrease. The terms k and $(1 - f)$ thus have opposite effects on the slope. Except at high fractional conversions the influence of the $(1 - f)$ term is the smaller and an approximation to the slope may be evaluated assuming that the fractional conversion does not change with temperature. In this way the slope is overestimated and gives conservative conclusions in the assessment of stability. This approximation is not easy to avoid in the case of complex kinetics but a rigorous analytical treatment is possible for simple integral reaction orders. Thus for a general forward reaction, equations (5.111) and (5.114) give:

$$Q_G = \frac{-aV_R c_0^n \exp(-E/RT_1)(1 - f)^n \Delta H^0}{F} \tag{5.116}$$

which for constant 'f' has a slope:

$$\frac{dQ_G}{dT_1} = -\frac{aEV_R c_0^n \exp(-E/RT_1)(1-f)^n \Delta H^0}{RT_1^2 F}. \tag{5.117}$$

Hence an approximate criterion of stability is that:

$$C_P + \frac{UA}{F} > -\frac{aEV_R c_0^n \exp(-E/RT_1)(1-f)^n \Delta H^0}{RT_1^2 F} \tag{5.118}$$

(ΔH^0 to be substituted together with sign) which applied to a first order reaction gives:

$$C_P + \frac{UA}{F} > -\frac{aEV_R c_0^n \exp(-E/RT_1)(1-f)^n \Delta H^0}{RT_1^2 F}. \tag{5.119}$$

This result for a first order reaction is compared below with the rigorous answer obtained by differentiation of equation (5.115) which yields:

$$\frac{dQ_G}{dT_1} = -ac_0 V_R \Delta H^0 \times \left[\frac{E\exp(-E/RT_1)}{RT_1^2[F + aV_R\rho\exp(-E/RT_1)]} - \frac{aEV_R\rho\exp(-E/RT_1)\exp(-E/RT_1)}{RT_1^2[F + aV_R\rho\exp(-E/RT_1)]^2}\right] \tag{5.120}$$

and on simplification:

$$\frac{dQ_G}{dT_1} = -\frac{aEV_R c_0 \exp(-E/RT_1)(1-f)^2 \Delta H^0}{RT_1^2 F}. \tag{5.121}$$

Hence:

$$\frac{\text{Exact Slope}}{\text{Approximate Slope}} = 1 - f \tag{5.122}$$

and since if any reaction takes place $(1 - f)$ must be less than unity, the approximate slope is an overestimate at all values of f.

A numerical example follows to illustrate the above analysis.

Example 6. A solution of acetic anhydride containing 0·22 g mole litre^{-1} is to be hydrolysed continuously in a stirred tank reactor to give an effluent containing an anhydride concentration of 0·04 g mole litre^{-1}. The solution has a specific gravity of 1·05 and is to be fed at the rate of 50 litre min^{-1} into an effective reactor volume of 750 litre. If the external surface area of the reactor available for heat transfer is 5·0 m^2, the ambient temperature 25°C and the surface coefficient of heat transfer 0·5 cal $\text{cm}^{-2}\ \text{h}^{-1}\ \text{degC}^{-1}$, determine the reactor operating temperature and the required feed solution temperature in order that it will be unnecessary to instal a cooling coil for further heat removal.

The kinetic data are first order with respect to anhydride concentration and may be expressed as:

$$k = 0{\cdot}158 \exp\left(18{\cdot}55 - \frac{10980}{RT_1}\right) \text{min}^{-1} \tag{I}$$

or

$$r = 0{\cdot}158\, c_0 (1 - f) \exp\left(18{\cdot}55 - \frac{10980}{RT_1}\right) \text{g moles litre}^{-1}\ \text{min}^{-1}. \tag{II}$$

From the basic C.S.T.R. design equation, the value of velocity constant required to achieve the specified conversion is given by:

$$\frac{0{\cdot}04}{0{\cdot}22} = \frac{1}{1 + \dfrac{k \times 750}{50}} \tag{III}$$

and

$$k = 0{\cdot}30\ \text{min}^{-1} \tag{IV}$$

and hence from equation (I) above, the reactor operating temperature can be fixed at 308·7°K (35·7°C). This rate equation and equations (5.110) and (5.115) may be combined to give:

$$\frac{-0{\cdot}158\, c_0 V_R\, \Delta H^0 \exp\left(18{\cdot}55 - \dfrac{10980}{RT_1}\right)}{F + 0{\cdot}158\, V_R \rho \exp\left(18{\cdot}55 - \dfrac{10980}{RT_1}\right)} = $$

$$= T_1 \left(C_P + \frac{UA}{F}\right) - \left(C_P T_0 + \frac{UAT_S}{F}\right) \tag{V}$$

where $c_0 = 0{\cdot}22$ g mole litre^{-1}
$V_R = 750$ litre [and $V_R\rho = 787\,500$ g]
$\Delta H^0 = -50000\ \text{cal(g mole)}^{-1}$ anhydride

$$\text{Flow} = 50 \text{ litre min}^{-1} \text{ [and } F = 52\,500 \text{ g min}^{-1}]$$
$$C_P = 0{\cdot}7 \text{ cal g}^{-1} \text{ degC}^{-1}$$
$$UA = \frac{0{\cdot}5 \times 5 \times 10^4}{60} = 416{\cdot}7 \text{ cal min}^{-1} \text{ degC}^{-1}$$
$$T_S = 298°\text{K}$$
T_0 to be determined.

Hence the left-hand side of equation (V) becomes:

$$\frac{0{\cdot}158 \times 0{\cdot}22 \times 750 \times 50000 \exp\left(18{\cdot}55 - \dfrac{10980}{1{\cdot}986T_1}\right)}{52\,500 + 0{\cdot}158 \times 787\,500 \exp\left(18{\cdot}55 - \dfrac{10980}{1{\cdot}986T_1}\right)} \text{ cal g}^{-1}.$$

This has been plotted in Fig. 5.11 as a function of T_1 giving the sigmoidal curve. A line of gradient:

$$C_P + \frac{UA}{F} = 0{\cdot}7 + \frac{416{\cdot}7}{52\,500} = 0{\cdot}708 \text{ cal g}^{-1} \text{ degK}^{-1}$$

and passing through $T_1 = 308{\cdot}7$ has been superimposed to represent the heat removal rate. Since the heat removal line is now fixed and its equation contains the feed temperature T_0, the latter may be determined as 296°K.

The relative slopes of the heat removal line and the heat generation curve at 308·7°K confirm that the operating point is an inherently stable one. The heat removal line slope was developed above as 0·708 and application of equation (5.121) gives the rate of change of heat generation with temperature as:

$$0{\cdot}090\,5 \text{ cal(g mole)}^{-1} \text{ degK}^{-1}.$$

5.9 Transient Behaviour of C.S.T.R.'s

The text so far has considered the normal steady state behaviour of well stirred reaction vessels. The various stages of a battery in steady state operation will have a sequence of concentrations and as already discussed there may also be a temperature profile along the battery to obtain desirable products in a complex reaction system. Nevertheless, the temperatures and concentrations in any given tank have been considered to remain constant with time.

It is the purpose of the present section to consider the effect of disturbances in feed composition, throughput or temperature.

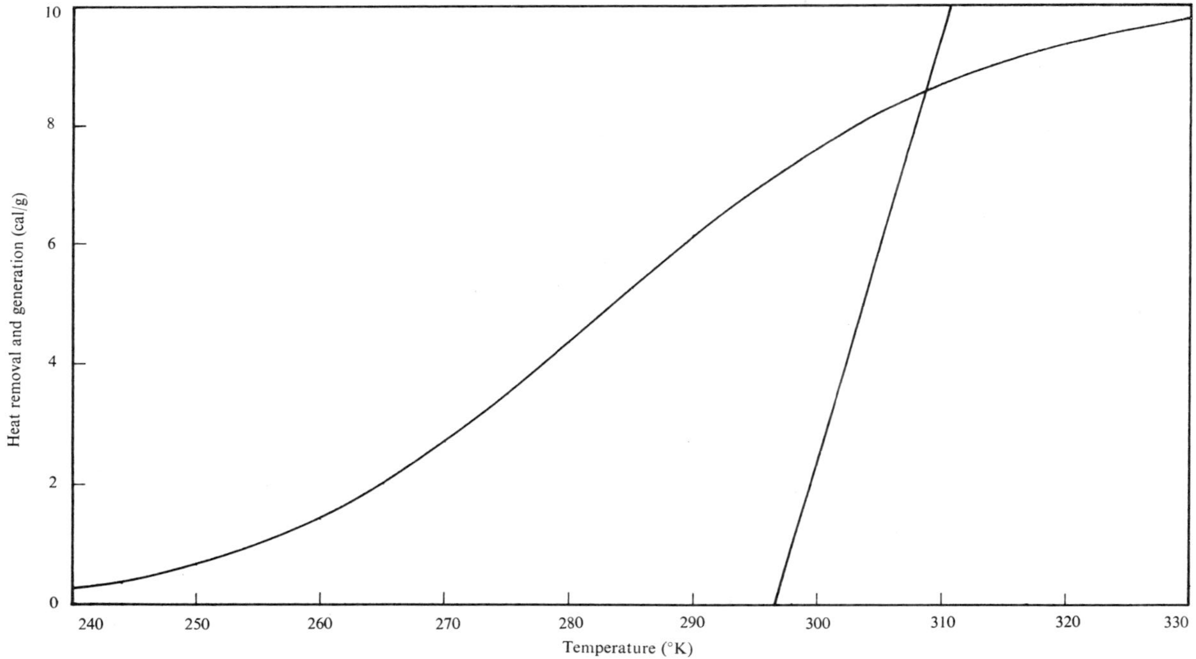

FIG. 5.11. Heat generation and removal plot

Such changes may arise (*i*) because the process plant supplying the reactor is not functioning to specification, (*ii*) because of a failure in the reactor system itself such as steam or coolant failure or (*iii*) during start-up or shut-down periods. The effects of all of these unsteady state conditions must be considered in assessing the reactor performance and some illustrations are given in the following sections.

5.9.1 Effect of change in feed concentration

These initial considerations will assume that the reactor control system is able to maintain the required temperature despite the changed reaction performance and as a result the rate constant will be fixed.

5.9.1.1. *Single tank with first order reaction*

If the feed concentration is a time dependent function $c(t)$, the unsteady state mass balance on the tank may be developed from equation (5.6) by the introduction of the accumulation term and written as:

$$Qc(t) = Qc_1 + kV_Rc_1 + V_R\frac{dc_1}{dt} \tag{5.123}$$

where c_1 is the time variable outlet concentration. This type of equation is conveniently solved using the Laplace transformation which in this instance gives:

$$\left[\frac{V_R}{Q + kV_R} s + 1\right]\bar{C}_1 = \frac{V_R}{Q + kV_R}C_1(0) + \frac{Q}{Q + kV_R}\bar{C} \tag{5.124}$$

where $\bar{c}_1$ is the transform of c_1
$\bar{c}$ is the transform of $c(t)$
and $c_1(0)$ is the initial value of c_1 and has the value:

$$\frac{c_0}{1 + \dfrac{kV_R}{Q}}$$

in which c_0 is the initial feed concentration.

To continue the development, new symbols are introduced for the constants and the equation is rewritten:

$$(Ls + 1)\,\bar{c}_1 = Lc_1(0) + K\bar{c} \tag{5.125}$$

If, for instance, $c(t)$ is made up of the initial concentration c_0 together with a step change of magnitude 'a', then:

$$c = \frac{c_0 + a}{s} \tag{5.126}$$

and

$$c_1 = \mathscr{L}^{-1}\left[\frac{Lc_1(0)}{Ls + 1} + \frac{K(c_0 + a)}{s(Ls + 1)}\right] \tag{5.127}$$

i.e.

$$c_1 = c_1(0)\,\mathrm{e}^{-t/L} + K(c_0 + a)\,(1 + \mathrm{e}^{-t/L}) \tag{5.128}$$

$$c_1 = K(c_0 + a) + \mathrm{e}^{-t/L}\,[c_1(0) - K(c_0 + a)]. \tag{5.129}$$

This equation shows how the outlet concentration will change from $c_1(0)$ at $t = 0$ to $K(c_0 + a)$ at $t = \infty$. This latter value is the potential deviation and will be attained if no corrective control action is taken to counter the effects of the change in feed concentration. The value at infinite time is the new steady state value and may be confirmed by use of equation (5.6) with c_0 replaced by $(c_0 + a)$ and noting the definition of K.

The analysis of such disturbances is normally made in terms of perturbation variables and the above case will now be reconsidered in this way. If the steady state equation is subtracted from the unsteady state one and the constant $c_1(0)$ introduced into the differential term:

$$Q[c(t) - c_0] = Q[c_1 - c_1(0)] + kV_{\mathrm{R}}[c_1 - c_1(0)] + V_{\mathrm{R}}\frac{d}{dt}[c_1 - c_1(0)]. \tag{5.130}$$

The following new variables are now defined

$$c(t) - c_0$$

and

$$c_1 - c_1(0)$$

which are respectively the increase in feed concentration above the initial feed concentration and the increase in outlet concentration above the former steady outlet concentration. These are the perturbation variables and will be abbreviated to simply c and c_1, it being understood that these symbols have this different meaning from their use in equation (5.123). Equation (5.130) then gives a result which appears to be identical with equation (5.123):

$$Qc = Qc_1 + kV_R c_1 + V_R \frac{dc_1}{dt}. \tag{5.131}$$

Transformation of this gives:

$$\left[\frac{V_R}{Q + kV_R} s + 1\right] c_1 = \frac{Q}{Q + kV_R} \bar{c} \tag{5.132}$$

or

$$(Ls + 1)\,\bar{c}_1 = K\bar{c}. \tag{5.133}$$

This result is a simplification of equation (5.124) because the initial value of the perturbation variable c_1 is zero by its definition. The transform of $\bar{c}$ is now a/s

and $$c_1 = \mathscr{L}^{-1} \frac{Ka}{s(Ls + 1)} = Ka(1 - e^{-t/L}). \tag{5.134}$$

Addition of $c_1(0)$ to the result of this equation enables the absolute concentration to be found and may be demonstrated to give the same result as equation (5.129).

This technique of using perturbation variables will be used in the following text as appropriate.

5.9.1.2. Single tank—second order reaction

The unsteady state mass balance is:

$$Qc(t) = Qc_1 + kV_R c_1^2 + V_R \frac{dc_1}{dt}. \tag{5.135}$$

This is an example of a first order non-linear differential equation. Provided the changes in c_1 are not too great it may be converted to a linear differential equation, by subtraction of the corresponding steady state balance as follows:

$$Q[c(t) - c_0] = Q[c_1 - c_1(0)] + kV_R[c_1^2 - c_1(0)^2] + V_R \frac{d}{dt}[c_1 - c_1(0)]. \tag{5.136}$$

In particular the difference of squares may be expressed as:

$$[c_1 + c_1(0)]\,[c_1 - c_1(0)]$$

which for small changes in c_1 may be approximated by:

$$2c_1(0)\,[c_1 - c_1(0)].$$

Hence defining perturbation variables and transforming:

$$\left[\frac{V_R}{Q + 2kV_R c_1(0)}\,s + 1\right]\bar{c}_1 = \frac{Q}{Q + 2kV_R c_1(0)}\,\bar{c}. \tag{5.137}$$

This differential equation may now be treated as for the first order case with slightly different definitions for the constants L and K.

5.1.9.3. *Single tank—general forward kinetics*

The concentration difference term:

$$kV_R[c_1^n - c_1(0)^n]$$

will arise in the analysis of this case, and is linearised by application of Taylor's theorem to c_1^n giving:

$$c_1^n \simeq c_1(0)^n + [c_1 - c_1(0)]\left[\frac{d}{dc_1}(c_1^n)\right]_{c_1(0)} \tag{5.138}$$

i.e.

$$c_1^n \simeq c_1(0)^n + [c_1 - c_1(0)]\,n[c_1(0)]^{n-1}. \tag{5.139}$$

Hence the concentration difference term becomes:

$$nkV_R[c_1(0)]^{n-1}\,[c_1 - c_1(0)]$$

and the results of section (5.9.1.1) may be used with the appropriate modification to L and K.

5.1.9.4. *Multiple tanks—first order reaction*

The unsteady state material balance on tank 'm' is:

$$Qc_{m-1} = Qc_m + kV_{R_m}c_m + V_{Rm}\frac{dc_m}{dt}. \tag{5.140}$$

Putting

$$\theta_m = \frac{V_{Rm}}{Q} \tag{5.141}$$

and the initial value of

$$c_m = c_m(0) \tag{5.142}$$

transformation gives:

$$\left[\frac{\theta_m}{1 + k\theta_m} s + 1\right] \bar{c}_m = \frac{c_m(0)\theta_m}{1 + k\theta_m} + \frac{\bar{c}_{m-1}}{1 + k\theta_m} \tag{5.143}$$

or replacing $1 + k\theta_m$ by λ_m:

$$\bar{c}_m = \frac{c_m(0)}{s + \dfrac{\lambda_m}{\theta_m}} + \frac{\bar{c}_{m-1}}{\theta_m\left(s + \dfrac{\lambda_m}{\theta_m}\right)}. \tag{5.144}$$

If this equation is solved sequentially through the cascade we have for the first vessel:

$$\bar{c}_1 = \frac{c_1(0)}{s + \dfrac{\lambda_1}{\theta_1}} + \frac{\bar{c}_0}{\theta_1\left(s + \dfrac{\lambda_1}{\theta_1}\right)}. \tag{5.145}$$

If the change is a sudden step such that the absolute value of the new input concentration is c_0, then:

$$\bar{c}_1 = \frac{c_1(0)}{s + \dfrac{\lambda_1}{\theta_1}} + \frac{c_0}{\theta_1 s\left(s + \dfrac{\lambda_1}{\theta_1}\right)}. \tag{5.146}$$

For the second vessel:

$$\bar{c}_2 = \frac{c_2(0)}{s + \dfrac{\lambda_2}{\theta_2}} + \frac{\bar{c}_1}{\theta_2\left(s + \dfrac{\lambda_2}{\theta_2}\right)} \tag{5.147}$$

$$= \frac{c_2(0)}{s + \dfrac{\lambda_2}{\theta_2}} + \frac{c_1(0)}{\theta_2\left(s + \dfrac{\lambda_1}{\theta_1}\right)\left(s + \dfrac{\lambda_2}{\theta_2}\right)} + \frac{c_0}{\theta_1\theta_2 s\left(s + \dfrac{\lambda_1}{\theta_1}\right)\left(s + \dfrac{\lambda_2}{\theta_2}\right)}. \tag{5.148}$$

Likewise for the nth vessel:

$$\bar{c}_n = \frac{c_n(0)}{s + \dfrac{\lambda_n}{\theta_n}} + \frac{c_{n-1}(0)}{\theta_n\left(s + \dfrac{\lambda_n}{\theta_n}\right)\left(s + \dfrac{\lambda_{n-1}}{\theta_{n-1}}\right)} + \ldots +$$

$$+\frac{c_2(0)}{\prod_3^n \theta_i \prod_2^n \left(s+\frac{\lambda_i}{\theta_i}\right)}+\frac{c_1(0)}{\prod_2^n \theta_i \prod_1^n \left(s+\frac{\lambda_i}{\theta_i}\right)}$$

$$+\frac{c_0}{s\prod_1^n \theta_i \left(s+\frac{\lambda_i}{\theta_i}\right)}. \qquad (5.149)$$

In the special case where the tanks are of equal size, both λ and θ are constant and if λ/θ is replaced by α:

$$\bar{c}_n = \left[\frac{c_n(0)}{s+\alpha}+\frac{c_{n-1}(0)}{\theta(s+\alpha)^2}+\ldots+\frac{c_1(0)}{\theta^{n-1}(s+\alpha)^n}\right]$$

$$+\left[\frac{c_0}{s\theta^n(s+\alpha)^n}\right] \qquad (5.150)$$

on resolving the term in the second bracket into its partial fractions:

$$\bar{c}_n = \left[\frac{c_n(0)}{s+\alpha}+\frac{c_{n-1}(0)}{\theta(s+\alpha)^2}+\ldots+\frac{c_1(0)}{\theta^{n-1}(s+\alpha)^n}\right]$$

$$+\frac{c_0}{\theta^n}\left[\frac{1}{\alpha^n s}-\left\{\frac{1}{\alpha}\frac{1}{(s+\alpha)^n}+\frac{1}{\alpha^2}\frac{1}{(s+\alpha)^{n-1}}+\ldots+\right.\right.$$

$$\left.\left.+\frac{1}{\alpha^{n-1}}\frac{1}{(s+\alpha)^2}+\frac{1}{\alpha^n}\frac{1}{(s+\alpha)}\right\}\right]. \qquad (5.151)$$

Hence:

$$c_n = e^{-\alpha t}\left[c_n(0)+\frac{t}{\theta}c_{n-1}(0)+\ldots+\left(\frac{t}{\theta}\right)^{n-1}\frac{c_1(0)}{(n-1)!}\right]$$

$$+\frac{c_0}{\lambda^n}\left[1-e^{-\alpha t}\left\{\frac{(\alpha t)^{n-1}}{(n-1)!}+\frac{(\alpha t)^{n-2}}{(n-2)!}+\ldots+\alpha t+1\right\}\right]. \qquad (5.152)$$

Example 6. Acetic anhydride is to be continuously hydrolysed in a series of three stirred tank reactors. The hold-up of each reactor is 1800 cm^3 and initially all contain anhydride solution at concentration 0·21 g moles $litre^{-1}$. A throughput of 600 cm^3 min^{-1} containing 0·135 g moles $litre^{-1}$ of anhydride is then introduced to the system. The first order velocity constant at the operating temperature of 40°C is 0·38 min^{-1}.

Equation (5.152) will be used to examine the approach to the new steady state with the following numerical terms:

$$c_3(0) = c_2(0) = c_1(0) = 0{\cdot}21 \text{ g moles litre}^{-1} \quad \text{(I)}$$
$$c_0 = 0{\cdot}135 \text{ g moles litre}^{-1} \quad \text{(II)}$$
$$\theta = 3 \text{ min} \quad \text{(III)}$$
$$\lambda = 1 + 3 \times 0{\cdot}38 = 2{\cdot}14 \quad \text{(IV)}$$
$$\alpha = 0{\cdot}713 \quad \text{(V)}$$

Hence at time 't' after the feed has started:

$$c_3 = 0{\cdot}21\,e^{-0{\cdot}713t}\left[1 + \frac{t}{3} + \frac{t^2}{18}\right] + \frac{0{\cdot}135}{2{\cdot}14^3}\left[1 - e^{-0{\cdot}713t}\left\{\left(\frac{0{\cdot}713t}{2}\right)^2 + 0{\cdot}713t + 1\right\}\right] \quad \text{(VI)}$$

i.e.

$$c_3 = \frac{0{\cdot}135}{2{\cdot}14^3} + e^{-0{\cdot}713t}\left[0{\cdot}21\left(1 + \frac{t}{3} + \frac{t^2}{18}\right) - \frac{0{\cdot}135}{2{\cdot}14^3} \times \left(1 + 0{\cdot}713t + 0{\cdot}255t^2\right)\right]. \quad \text{(VII)}$$

At the new steady state c_3 will become $0{\cdot}135/2{\cdot}14^3$ (or 0·0138) and the fractional response at time 't' is therefore:

$$\frac{0{\cdot}21 - c_3}{0{\cdot}21 - 0{\cdot}0138}$$

Values of c_3 from equation (VII) then enable a quantitative measure of the approach to the new steady state to be defined. Thus at 5·03 minutes the fractional response is 0·90 while at 9·3 minutes it is 0·99, and the whole system may therefore be considered to have reached its steady state after about ten minutes.

5.9.2 Change in throughput

If Q_0 is the initial throughput, the steady state mass balance for a first order reaction is:

$$Q_0 c_0 = Q_0 c_1(0) + kVc_1(0). \quad (5.153)$$

After a change to Q_1 the unsteady state balance is:

$$Q_1 c_0 = Q_1 c_1 + kVc_1 + V\frac{dc_1}{dt}. \quad (5.154)$$

This equation introduces a further problem in linearisation not encountered in the previous examples, i.e. the product $Q_1 c_1$ of two time dependent quantities now arises.

This product may be approximated by taking the linear terms of a Taylor expansion as:

$$Q_1 c_1 \simeq Q_0 c_1(0) + Q_0[c_1 - c_1(0)] + c_1(0)[Q_1 - Q_0]. \quad (5.155)$$

Hence on subtraction of equations (5.154) and (5.155):

$$c_0(Q_1 - Q_0) = Q_0[c_1 - c_1(0)] + c_1(0)[Q_1 - Q_0] + kV[c_1 - c_1(0)] + V\frac{dc_1}{dt} \quad (5.156)$$

$$\left[\frac{V}{Q_0 + kV}s + 1\right]\bar{c}_1 = \frac{c_0 + c_1(0)}{Q_0 + kV}\bar{Q} \quad (5.157)$$

where $\bar{c}_1$ and $\bar{Q}$ are the transforms of the perturbation variables:

$c_1 - c_1(0)$ and

$Q_1 - Q_0$.

For changes of throughput with more complicated reaction mechanisms, the reaction term would also have to be linearised as in the previous sections.

5.9.3 Changes in temperature

It has been assumed in the foregoing dynamic situations that there would not be any resultant change in the temperature of the reaction mixture. This restriction is now relaxed and consequently it is necessary to include an overall energy balance and also to take into account the variation of reaction rate with temperature as well as with concentration.

Two situations will be examined. In the one, reactor throughput will be held constant but both feed temperature and reactant feed concentration may vary with time and the effect on the reactor operating temperature and composition will be deduced. In the other, reactor throughput and the feed conditions will all be held constant and equations governing the return to steady state after an arbitrary disturbance of reactor operating conditions will be deduced. The form of response in both situations may be either underdamped, overdamped, stable or unstable, and the identical criteria which apply in both situations will be presented as a final section.

5.9.3.1 *Disturbances in reactor feed conditions*

The unsteady state mass balance in terms of a reaction rate r is:

$$Qc(t) = Qc_1 + Vr_1 + V\frac{dc_1}{dt} \qquad (5.158)$$

and on subtracting the corresponding steady state equation from this:

$$Q[c(t) - c_0] = Q[c_1 - c_1(0)] + V[r_1 - r_0] + V\frac{dc_1}{dt}. \qquad (5.159)$$

It will be assumed for the present illustration that 'r' is given by:

$$a\exp(-E/RT)c.$$

Then for an initial steady state reactor temperature $T_1(0)$,

$$r_0 = a\exp\{-E/RT_1(0)\}c_1(0) \qquad (5.160)$$

and with T_1 as the transient reactor temperature, r_1 may be obtained by linearisation as:

$$r_1 = r_0 + a\exp\{-E/RT_1(0)\}[c_1 - c_1(0)] + \frac{aE}{RT_1(0)^2}c_1(0)\exp\{-E/RT_1(0)\}[T_1 - T_1(0)]. \qquad (5.161)$$

Hence identifying constants K_1 and K_2 which are functions of the initial conditions, equation (5.161) yields:

$$r_1 - r_0 = K_1[C_1 - c_1(0)] + K_2[T_1 - T_1(0)]. \qquad (5.162)$$

Defining the perturbation variables:

$$c_1 \equiv c_1 - c_1(0)$$

$$c \equiv c(t) - c_0$$

and

$$T_1 \equiv T_1 - T_1(0)$$

equations (5.159) and (5.162) may be combined for a zero initial value of c_1 to give:

$$Q\bar{c} = Q\bar{c}_1 + V[K_1 c_1 + K_2 T_1] + Vs\bar{c}_1 \tag{5.163}$$

or

$$\left[\frac{V}{Q + VK_1} s + 1\right] c_1 = \frac{Q}{Q + VK_1} \bar{c} - \frac{VK_2}{Q + VK_1} \bar{T}_1 \tag{5.164}$$

or

$$[L_1 s + 1]\,\bar{c}_1 = K_3\bar{c} - K_4\bar{T}_1. \tag{5.165}$$

For the energy balance, let

T_0 be the steady state feed temperature,
$T(t)$ be the transient feed temperature,
T_S be the temperature of the heat transfer medium (assumed fixed in this exercise),
U_1 be the product of the overall heat transfer coefficient and the heat transfer area,
ρ be the fluid density,
and C_p be the fluid specific heat.

With ΔH to be substituted together with its algebraic sign, the steady state energy balance is:

$$\rho Q c_p T_0 - \Delta H V r_0 + U_1[T_S - T_1(0)] = \rho Q c_p T_1(0) \tag{5.166}$$

and the unsteady state one:

$$\rho Q C_p T(t) - \Delta H V r_1 + U_1 [T_S - T_1] = \rho Q C_p T_1 + V\rho C_p \frac{dT_1}{dt}. \tag{5.167}$$

Subtracting and transforming with a zero initial value of the perturbation $T_1 \equiv T_1 - T_1(0)$, these equations give:

$$\left[\frac{V\rho C_p}{Q\rho C_p + U_1 + \Delta H V K_2} s + 1\right] \bar{T}_1 = \frac{\rho Q C_p}{Q\rho C_p + U_1 + \Delta H V K_2} \bar{T} - \frac{\Delta H V K_1}{Q\rho C_p + U_1 + \Delta H V K_2} \bar{c}_1 \tag{5.168}$$

wherein $\bar{T} = T(t) - T_0$

or

$$[L_2 s + 1]\bar{T}_1 = K_5\bar{T} - K_6\,\bar{c}_1. \tag{5.169}$$

Equations (5.165) and (5.169) are simultaneous equations for the reactor conditions $\bar{T}_1$ and $\bar{c}_1$ as functions of the feed stream disturbances $\bar{T}$ and $\bar{c}$. Independent expressions for $\bar{c}_1$ and $\bar{T}_1$ may be isolated as:

$$\left[s^2 + \left(\frac{1}{L_1} + \frac{1}{L_2}\right) s + \frac{1 - K_4K_6}{L_1L_2}\right]\bar{c}_1 = \frac{K_3(L_2s + 1)}{L_1L_2}\bar{c}_1 - \frac{K_5K_6}{L_1L_2}\bar{T} \qquad (5.170)$$

and

$$\left[s^2 + \left(\frac{1}{L_1} + \frac{1}{L_2}\right) s + \frac{1 - K_4H_6}{L_1L_2}\right]\bar{T}_1 = \frac{K_5(L_1s + 1)}{L_1L_2}\bar{T} - \frac{K_3K_6}{L_1L_2}\bar{c}. \qquad (5.171)$$

5.9.3.2 *Return to steady state after a disturbance*

The fixed feed conditions to the reactor are T_0 and c_0 and the steady state operating conditions of the reactor are T_1^0 and c_1^0. At zero time the actual reactor operating conditions will be taken as $T_1(0)$ and $c_1(0)$ and equations are deveoped for the transient changes from $T_1(0)$ to T_1^0 and $c_1(0)$ to c_1^0.

The unsteady state mass balance is:

$$Qc_0 = Qc_1 + Vr_1 + V\frac{dc_1}{dt} \qquad (5.172)$$

and linearising r_1 relative to the final steady state conditions:

$$Qc_0 = Qc_1 + V[r_0 + K_1(c_1 - c_1^0) + K_2(T_1 - T_1^0)] + V\frac{dc_1}{dt} \qquad (5.173)$$

or

$$\left[\frac{V}{Q + VK_1}D + 1\right]c_1 = \frac{Qc_0 - V(r_0 - K_1c_1^0 - K_2T_1^0}{Q + VK_1} - \frac{VK_2}{Q + VK_1}T_1 \qquad (5.174)$$

or
$$[L_1D + 1]\ c_1 = K_7 - K_4T_1. \qquad (5.175)$$

Absolute values of c_1 and T_1 will be used in the present development, although the steady state equation could be subtracted and perturbation variables then introduced. On the chosen basis, equation (5.175) is transformed for the non-zero initial condition of $c_1 = c_1(0)$ into:

$$(L_1s + 1)\bar{c}_1 = L_1c_1(0) + K_7/s - K_4\overline{T}_1. \qquad (5.176)$$

Similarly the transient energy balance is:

$$\rho Qc_pT_0 = \rho Qc_pT_1 + V\Delta H\,[r_0 + K_1(c_1 - c_1^0) + K_2(T_1 - T_1^0)] - U_1(T_S - T_1) + V\rho\, c_p\frac{dT_1}{dt} \qquad (5.177)$$

which on re-arrangement gives:

$$\left[\frac{V\rho C_p}{\rho QC_p + V\Delta HK_2 + U_1}D + 1\right]T_1 = \frac{\rho QC_pT_0 + U_1T_S - V\Delta H\,[r_0 - K_1\,c_1^0 - K_2T_1^0]}{\rho QC_p + V\Delta HK_2 + U_1} - \frac{V\Delta HK_1}{\rho QC_p + V\Delta HK_2 + U_1}c_1 \qquad (5.178)$$

or
$$[L_2D + 1]T_1 = K_8 - K_6C_1. \qquad (5.179)$$

On transformation this gives:

$$[L_2s + 1]\,\overline{T}_1 = L_2T_1(0) + K_8/s - K_6c_1. \qquad (6.180)$$

The simultaneous equations for $\bar{c}_1$ and $\overline{T}_1$ yield:

$$\left[s^2 + \left(\frac{1}{L_1} + \frac{1}{L_2}\right)s + \frac{1 - K_4K_6}{L_1L_2}\right]\bar{c}_1 = sc_1(0) + \frac{L_1c_1(0) - K_4L_2T_1(0) + L_2K_7}{L_1L_2} + \frac{K_7 - K_4K_8}{L_1L_2s} \qquad (5.181)$$

and:

$$\left[s^2 + \left(\frac{1}{L_1} + \frac{1}{L_2}\right)s + \frac{1-K_4K_6}{L_1L_2}\right]\overline{T}_1 = sT_1(0) + \frac{L_2T_1(0) - K_6L_1c_1(0) + L_1K_8}{L_1L_2} + \frac{K_8 - K_6K_7}{L_1L_2}s. \qquad (5.182)$$

If the roots of the quadratic are denoted by m_1 and m_2, the solution of equation (5.181) may be written as:

$$c_1 = \frac{c_1(0)}{m_1 - m_2}\left[m_1\,\mathrm{e}^{m_1 t} - m_2\,\mathrm{e}^{m_2 t}\right]$$
$$+ \frac{L_1(\qquad - K_4 L_2 T_1(0) + L_2 K_1}{L_1 L_2 (m_1 - m_2)}\left[\mathrm{e}^{m_1 t} - \mathrm{e}^{m_2 t}\right]$$
$$+ \frac{K_7 - K_4 K_8}{m_1 m_2 L_1 L_2}\left[1 + \frac{1}{m_1 - m_2}\left(m_2\,\mathrm{e}^{m_1 t} - m_1\,\mathrm{e}^{m_2 t}\right)\right]. \tag{5.183}$$

However, the actual form of solution depends upon whether m_1 and m_2 are positive, negative, real or complex. The criteria for this are examined in the following section.

5.9.3.3 *Criteria for the various forms of response*

In the above sections equations (5.170), (5.171), (5.181) and (5.182) have yielded identical quadratic forms for the one side of the equation. It is the nature of the roots of this quadratic which will determine whether the response will be stable or unstable and whether overdamped or underdamped.

The first test is to decide whether the roots are real or complex. For real roots:

$$\left(\frac{1}{L_1} + \frac{1}{L_2}\right)^2 - \frac{4(1 - K_4 K_6)}{L_1 L_2} > 0 \tag{5.184}$$

and if this criterion is satisfied, the response will not be oscillatory. For the exponentials of the time domain solution to give a stable response (i.e. a decay with time) both roots of the quadratic must be negative, which will be the case if:

$$\frac{1}{L_1} + \frac{1}{L_2} > 0 \tag{5.185}$$

and

$$\frac{1 - K_4 K_6}{L_1 L_2} > 0. \tag{5.186}$$

Thus if both of these conditions are satisfied in addition to that of equation (5.184) the response will be non-oscillatory and stable, as in curves A. B or C of Fig. 5.12.

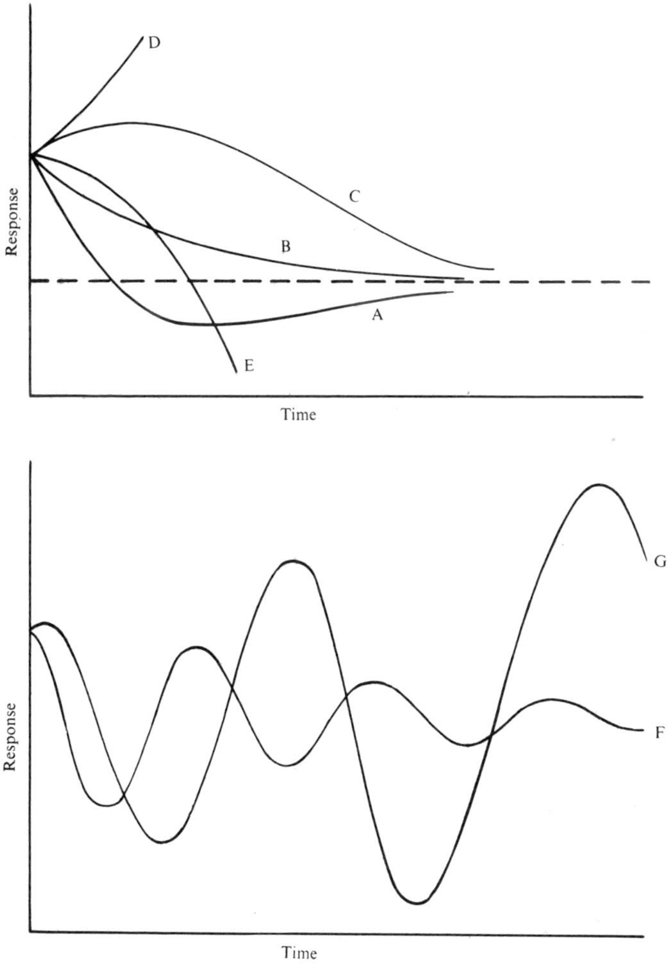

FIG. 5.12. Stable and unstable forms of response

Should both roots of the quadratic be positive, then:

$$\frac{1}{L_1} + \frac{1}{L_2} < 0 \tag{5.187}$$

and

$$\frac{1 - K_4 K_6}{L_1 L_2} > 0 \tag{5.188}$$

and should one be positive and the other negative, then:

$$\frac{1 - K_4K_6}{L_1L_2} < 0 \qquad (5.189)$$

while $\frac{1}{L_1} + \frac{1}{L_2}$ may take either sign.

In either event, if there is at least one positive root, the response will diverge with time and is unstable, as in curves D or E.

Returning to equation (5.184); if the value of the left-hand side is negative, then the roots will be complex and the response oscillatory. If the real part of the complex roots is negative, the oscillation will decay and give a stable response (curve F), while a positive value leads to divergence (curve G).

The criteria of equations (5.184), (5.185) and (5.186) will now be expanded into the various physical parameters from which they were defined:

(*i*) $\frac{1}{L_1} + \frac{1}{L_2} > 0$

becomes:

$$\frac{Q + VK_1}{V} + \frac{\rho QC_p + U_1 + \Delta HVK_2}{\rho C_p V} > 0 \qquad (5.190)$$

i.e.

$$2 + \frac{VK_1}{Q} + \frac{U_1}{\rho C_p Q} > -\frac{\Delta HVK_2}{\rho C_p Q}. \qquad (5.191)$$

By the physical nature of the system all terms in this inequality, with the exception of ΔH, have positive values. Thus endothermic reactions (see also equation 5.194) must lead to a stable response, while one of the conditions for an exothermic reaction to be stable is:

$$-\Delta H < \frac{\rho QC_p}{VK_2}\left[2 + \frac{VK_1}{Q} + \frac{U_1}{\rho C_p Q}\right]. \qquad (5.192)$$

(*ii*) $\frac{1 - K_4K_6}{L_1L_2} > 0$ gives:

$$\left[\frac{Q}{V} + K_1\right]\left[\frac{Q}{V} + \frac{U_1}{\rho C_p Q} + \frac{\Delta HK_2}{\rho C_p}\right]\left[1 - \frac{V^2\Delta HK_1K_2}{(Q + VK_1)(\rho C_p Q + U_1 + \Delta HVK_2)}\right] > 0 \qquad (5.193)$$

from which:

$$1 + \frac{U_1}{\rho C_p Q}\left[1 + \frac{VK_1}{Q}\right] + \frac{VK_1}{Q} > -\frac{V\Delta H K_2}{\rho C_p Q}. \quad (5.194)$$

Thus the second condition to be satisfied alongside (5.192) for a stable exothermic reaction is:

$$-\Delta H < \frac{\rho C_p Q}{VK_2}\left[1 + \frac{U_1}{\rho C_p Q}\left(1 + \frac{VK_1}{Q}\right) + \frac{VK_1}{Q}\right]. (5.195)$$

(*iii*)
$$\left(\frac{1}{L_1} + \frac{1}{L_2}\right)^2 - \frac{4(1 - K_4 K_6)}{L_1 L_2} > 0$$

may now be simplified utilising (5.191) and (5.194) to give:

$$\left(\frac{VK_1}{Q}\right)^2 + \left(\frac{U_1}{\rho C_p Q}\right)^2 + \left(\frac{V\Delta H K_2}{\rho C_p Q}\right)^2 > \frac{2K_1 V U_1}{\rho C_p Q^2} - \frac{2K_1 K_2 \Delta H V^2}{\rho C_p Q^2} - \frac{2K_2 \Delta H V U_1}{(\rho C_p Q)^2}. \quad (5.196)$$

Thus an endothermic reaction is almost certain to have a non-oscillatory response. However, for an exothermic reaction, the whole of the right-hand side of (5.196) will be positive and may well exceed the left-hand side and hence give rise to an oscillatory response.

Use of the above criteria for determination of stability is confined to the effect of small disturbances about a steady state such that the linearised equations remain applicable. For larger disturbances, Liapunov's second method or the technique of phase-plane analysis would be utilised (Aris and Amundson, 1958, 1958a; Warden, Aris and Amundson, 1964).

Problems

1. Acetic anhydride is to be hydrolysed in three C.S.T.R.'s operating in series. If the volume of each reactor is 1800 ml and the feed rate is 582 ml min^{-1}, estimate the degree of hydrolysis accomplished in each reactor, and the total fraction hydrolysed at the exit of each reactor.

The reaction can be assumed to be first order with the rate expressed by the equation:

$$r = 0{\cdot}158c$$

where c is the concentration of acetic anhydride in g moles cm^{-3}.

ANS. 0·328, 0·549 and 0·697.

2. Substance A is to be converted into substance B in solution. The kinetics are first order ($k = 0{\cdot}18$ h^{-1}) with respect to A and the reverse reaction can be neglected.

Find the residence times for 95% conversion of A for the cases of (*i*) 3 well stirred tanks of equal volume in series, (*ii*) a large well stirred tank followed by a second one of 10% this capacity.

ANS. (*i*) 28·5 hr, (*ii*) 57·1 h .

3. Show that the outlet concentrations from a continuous well stirred reactor effecting the first order reactions:

$$A \xrightarrow{k_1} B \xrightarrow{k_2} C$$

are

$$c_A = \frac{c_0}{1 + k_1\theta}$$

$$c_B = \frac{k_1\theta c_0}{(1 + k_1\theta)(1 + k_2\theta)}$$

and

$$c_C = \frac{k_1 k_2 \theta^2 c_0}{(1 + k_1\theta)(1 + k_2\theta)}$$

where θ is the mean residence time and the concentration of A in the feed is c_0 with B and C zero.

Show that for c_B to be at a maximum, θ should be $(k_1 k_2)^{-\frac{1}{2}}$ and that the maximum value is then:

$$\frac{c_0}{2\left(\frac{k_2}{k_1}\right)^{\frac{1}{2}} + \left(\frac{k_2}{k_1}\right) + 1}.$$

4. The first order reaction scheme:

$$A \xrightarrow{k_1} B \underset{k_3}{\overset{k_2}{\rightleftharpoons}} C$$

is to be carried out in a continuous well stirred reactor. The feed Q ft^3 h^{-1} contains A at a concentration c_0 lb mole ft^{-3} and the reactor hold up is V ft^3. Show that the outlet concentration of B is:

$$\frac{k_1\theta(1 + k_3\theta)\,C_0}{(1 + k_1\theta)(1 + k_2\theta + k_3\theta)}$$

where $\theta = V/Q$.

Write down an energy balance for the system assuming reaction A → B to be exothermic and reaction B → C endothermic.

5. Show that for a first order irreversible reaction taking place at given throughput Q and fractional conversion f_2 in a series of two tanks at different temperatures, the minimum total reactor volume is:

$$\frac{Q}{k_2}\left[\frac{2K}{(1-f_2)^{\frac{1}{2}}} - (1 + k^2)\right]$$

and that the ratio of tank volumes is given by:

$$\frac{V_1}{V_2} = K\left[\frac{(1-f_2)^{\frac{1}{2}} - K(1-f_2)}{K(1-f_2)^{\frac{1}{2}} - (1-f_2)}\right]$$

where k_1 and k_2 are the velocity constants in the first and second tanks of volume V_1 and V_2 respectively and $K = (k_2/k_1)^{\frac{1}{2}}$.

6. Show that for 'n' completely mixed stages to be preferable in continuous operation to a batch system, using all 'n' tanks in each case for a first order reaction having a large equilibrium constant, the change-round time of the latter will have to be greater than:

$$\frac{n}{k}\left[\left(\frac{c_0}{c_1}\right)^{1/n} - 1 - \frac{1}{n}\operatorname{Ln}\frac{c_0}{c_1}\right]$$

where c_0 and c_1 are the initial and final reactant concentrations.

For a given reaction in a two stage system compare these minimum times for fractional conversions 0·90 and 0·99.

ANS. $T_{0\cdot90} = 0{\cdot}151\,T_{0\cdot99}$

7. Styrene (A) and Butadiene (B) are to be polymerised in a series of C.S.T.R.'s each of volume 26·5 m^3. If the initial concentration of styrene is 0·795 kg mole m^{-3} and of butadiene is 3·55 kg mole m^{-3} with a feed rate of 19·7 ton h^{-1} estimate the total number of reactors required for polymerisation of 85% of the limiting reactant.

Density of reaction mixture $\rho = 0{\cdot}87$ g cm^{-3}

rate equation: $r = kc_Ac_B$

where $k = 10^{-5}$ litre (g mole)$^{-1}$ sec^{-1}

Molecular Weight of styrene 104

Molecular Weight of butadiene 54.

ANS. = 20 tanks

8. In a proposed chemical process 1000 lb h^{-1} of a pure liquid A is fed continuously to the first of two equal sized C.S.T.R.'s operating in series. If both reactors are to operate at the same temperature so that the chemical reaction

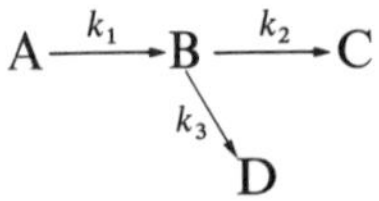

will take place, estimate the size of reactor that will give the maximum yield of the product B if $k_1 = k_2 = k_3 = 0{\cdot}1$ min^{-1}.

ANS. $\theta = 6{\cdot}5$ min $V_R = 17{\cdot}5$ ft^3

9. A series of three well stirred reactors of equal volume, and operating at the same temperature, is effecting an irreversible first order reaction under steady state conditions. Because of the failure in the second tank, it has to be taken out of service and the output from the first tank is then fed directly into the third tank with the system throughput remaining constant. Develop an expression for the change of system exit concentration with time. Find the potential value of the exit concentration and evaluate the ultimate percentage change in fractional conversion.

What difficulty would you anticipate in the solution of this problem had the reaction been second order?

ANS. $$c_3 = \frac{c_0}{(1+k\theta)^2}\left\{1 - \frac{k\theta}{1+k\theta}\exp\left[-\left(\frac{1}{\theta}+k\right)t\right]\right\};$$

$$c_3 = \frac{c_0}{(1+k\theta)^2}; \quad \frac{100}{3(1+k\theta)+(k\theta)^2}$$

10. A single C.S.T.R. of volume V_R is filled with pure solvent. A flow of Q ft^3 h^{-1} of a solution containing $c_{A,0}$ mole ft^{-3} of A and $c_{B,0}$ mole ft^{-3} of B is then initiated. Inside the reactor the reversible reaction

$$\mathrm{A} \underset{k_2}{\overset{k_1}{\rightleftharpoons}} \mathrm{B}$$

takes place. Derive expressions giving the composition of the stream leaving the reactor as a function of time. Hence show that the ratio of the final concentration of A to the maximum possible final concentration is given by:

$$\frac{c_{\mathrm{A}}}{c_{\mathrm{A}e}} = \frac{1 + \dfrac{Q}{Vk_2}\left[\dfrac{c_{\mathrm{A0}}}{c_{\mathrm{A0}} + c_{\mathrm{B0}}}\right]}{1 + \dfrac{Q}{Vk_2}\left[\dfrac{1}{1 + K}\right]}.$$

ANS.

$$c_{\mathrm{A}} = \frac{c_{\mathrm{A0}}\left(\dfrac{Q}{V} + k_2\right) + c_{\mathrm{B0}}k_2}{\dfrac{Q}{V} + k_1 + k_2} - \frac{k_2(c_{\mathrm{A0}} + c_{\mathrm{B0}}) \exp\left[-\dfrac{Qt}{V}\right]}{k_1 + k_2}$$

$$+ \frac{\dfrac{Q}{V}(c_{\mathrm{B0}}k_2 - c_{\mathrm{A0}}k_1) \exp - \left[\dfrac{Q}{V} + k_1 + k_2\right]t}{(k_1 + k_2)\left(\dfrac{Q}{V} + k_1 + k_2\right)}$$

$$c_{\mathrm{B}} = \frac{c_{\mathrm{B0}}\left(\dfrac{Q}{V} + k_1\right) + c_{\mathrm{A0}}k_1}{\dfrac{Q}{V} + k_1 + k_2} - \frac{k_1(c_{\mathrm{A0}} + c_{\mathrm{B0}}) \exp\left[-\dfrac{Qt}{V}\right]}{k_1 + k_2}$$

$$+ \frac{\dfrac{Q}{V}(c_{\mathrm{A0}}k_1 - c_{\mathrm{B0}}k_2) \exp - \left[\dfrac{Q}{V} + k_1 + k_2\right]t}{(k_1 + k_2)\left(\dfrac{Q}{V} + k_1 + k_2\right)}$$

11. A well stirred reactor is effecting a first order exothermic reaction A → B under the following conditions:

Mean residence time	1·0 min
Feed conc.	1·0 g mole litre^{-1}
Feed temperature	350°K
Coolant temperature	350°K
Velocity constant	$\exp\left[25 - \dfrac{10000}{T}\right] \mathrm{min}^{-1}$

$$\frac{UA}{\rho V C_p} = 1{\cdot}0\ \mathrm{min}^{-1}$$

$$\frac{\Delta H}{\rho C_p} = -200\ \mathrm{degK\ litre(g\ mole)}^{-1}$$

Show that the two stable operating conditions are:

$T = 354°K$, $c = 0{\cdot}964$ g mole litre^{-1}
$T = 441°K$, $c = 0{\cdot}0885$ g mole litre^{-1}

and that there is an unstable operating point at:

$T = 400°K$, $c = 0{\cdot}500$ g mole litre^{-1}.

12. Show that the concentration of A leaving the nth stage of a series of well stirred reactors effecting the first order reaction:

$$A \underset{k_2}{\overset{k_1}{\rightleftarrows}} B$$

is:
$$\frac{a_0}{k_1 + k_2}\left\{k_1\left[\frac{1}{1+\theta(k_1+k_2)}\right]^n + k_2\right\}.$$

where the feed mixture contains A at concentration a_o.

Write down the concentration of B leaving the nth tank and develop the corresponding expressions when the feed also contains B at concentration b_o.

REFERENCES

ARIS, R., and AMUNDSON, N. R., 1958. An analysis of chemical reaction stability and control I. The possibility of local control with perfect or imperfect control mechanisms. *Chem. Engng Sci.*, 7, 121.

—— and —— 1958. II. The evolution of proportional control. *Chem. Engng Sci.*, 7, 132.

DENBIGH, K. G. 1965. *Chemical reactor theory*. Cambridge Univ. Press, London.

ELLIS, S. R. M., JEFFREYS, G. V., and WHARTON, J. T. 1964. Raschig synthesis of hydrazine. Investigation of chloramine formation reaction. *Ind. Engng, Chem., Proc. Des. Dev. Qtrly.*, **3**, 18.

JEFFREYS, G. V., and JENSON, V. G. 1965. *Mathematical methods in chemical engineering*. Academic Press, New York.

—— and WHARTON, J. T. 1964. Raschig synthesis of hydrazine. The stability of chloramine in the reaction solution. *J. app. Chem.*, **14**, 203.

—— and —— 1965. Raschig synthesis of hydrazine. Formation of hydrazine from chloramine. *Ind. Engng Chem. Proc. Des. Dev. Qtrly*, **4**, 71.

STRAIN, M. 1961. *Mathematical methods for technologists*. English Univ. Press, London.

VAN HEERDEN, C. 1953. Autothermic processes. Properties and reaction design. *Ind. Engng Chem.*, **45**, 1242.

WARDEN, R. B., ARIS, R., and AMUNDSON, N. R. 1964. An analysis of chemical reaction stability and control. VIII. The direct method of Lyapunov. Introduction and applications to simple reactions in stirred vessels. *Chem. Engng Sci.*, **19**, 149.

——, —— and ——. 1964. IX. Further investigations into the direct method of Lyapunov. *Chem. Engng Sci.*, **19**, 173.

6

PLUG FLOW AND LAMINAR FLOW TUBULAR REACTORS

6.1 Introduction to Tubular Reactors

Consider a tubular reactor operating under steady state conditions of feed rate and feed composition. For convenience let the reactor operate isothermally so that the specific reaction rate is constant at all radial and axial positions. At any point in the general plane AB of Fig. 6.1 the concentration and rate of chemical reaction will be constant and independent of time. The flow profile of the fluid passing through a tubular reactor may take the form shown in the left-hand side of the figure. Therefore, it will be seen that fluid elements at different distances from the tube axis will have different times of passage through the reactor, with those near to the wall remaining for the longest time and those on the axis for the shortest time. Hence the extent of reaction will vary from a maximum in the fluid elements near to the wall to a minimum along the tube axis. This will give rise to radial concentration and temperature gradients with the need to consider the consequent mass and heat transfer effects. These conditions are pronounced for example in polymerisation reactions where highly viscous fluids cause laminar flow conditions to be established. In contrast, particularly with gaseous reactions, turbulent flow conditions may give a velocity profile as in the centre of Fig. 6.1. Since there is only slight variation of velocity in this case, the idealised plug flow approximation shown on the right of Fig. 6.1 is a justifiable representation. Under plug flow conditions the tubular reactor performance is equivalent to the corresponding batch process operated for a time equal to the time of passage of the plug.

The plug flow model of a tubular reactor implies uniform velocity and concentration across the whole radius at any point along the length of the reactor. It also implies no longitudinal diffusional mixing of either reactants or products between the plugs at different positions along the reactor. If such longitudinal diffusion cannot be

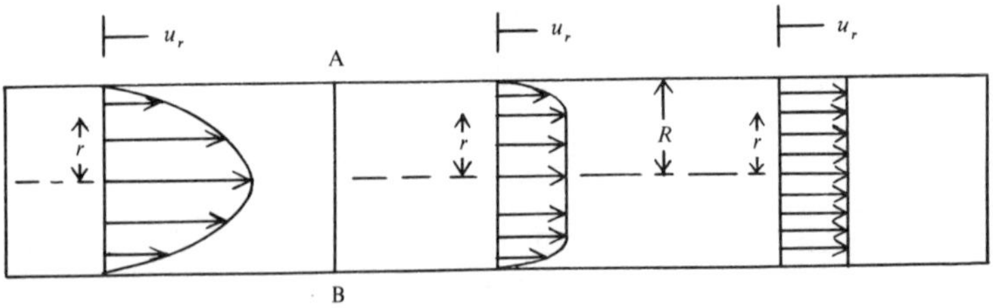

FIG. 6.1. Tubular reactor velocity profiles

neglected a further tubular reactor model has to be considered and forms the subject of section (7.4).

6.2 Laboratory Tubular Reactors

While a knowledge of the reaction kinetics as determined in batch reactors may be combined with a knowledge of velocity profiles and diffusion coefficients to develop a reactor model, the many uncertainties make it desirable to obtain kinetic data from relevant flow equipment. Furthermore, the numerous variables involved render difficult the derivation of a rate equation from observations on a tubular reactor operating with a significant change of fractional conversion. In a small laboratory reactor the temperature and pressure can readily be maintained constant. If then the operating conditions are arranged so that composition changes are small, the reaction rate is virtually constant and its mechanism can be more readily evaluated. Operated in this way the apparatus is called a 'laboratory differential reactor'. However, it frequently happens that small changes in the feed and product composition cannot be detected with sufficient accuracy and in addition the heat of reaction may be so high that isothermal conditions cannot be maintained between reactor inlet and outlet. In these circumstances it is more reasonable to use a larger reactor and accept the significant change of composition. The outlet conversion then represents the integrated value of the point differential reaction rates throughout the reactor, and the apparatus is called a 'laboratory integral reactor'. The results from the integral reactor can be interpreted either by graphical differentiation of the conversion versus space velocity curve or by integration of an assumed rate equation between the observed limits and comparison of the resultant conversions with the experimental values.

6.3 Basic Equations for the Continuous Plug Flow Reactor

6.3.1 Development of Equations

Consider a homogeneous chemical reactor of effective volume V_R and constant cross-sectional area A operating continuously on a total feed of Q_0 volumes per unit time containing reactant at a concentration c_0 moles per unit volume. Let the reactant concentration be c in an element of length 'δl' situated at a distance 'l' along the reactor (Fig. 6.2) and let the total flow into this element be Q.

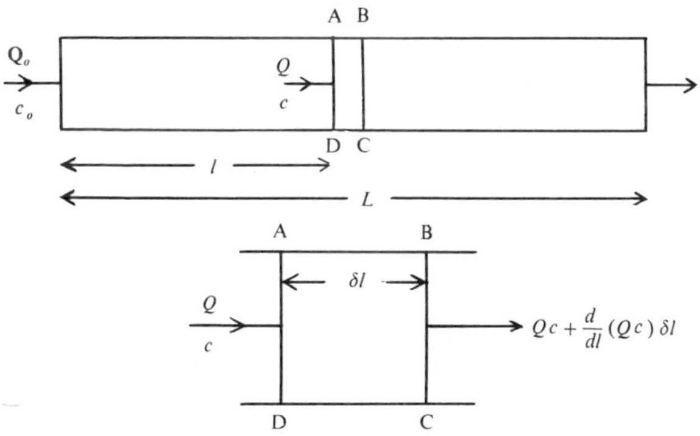

FIG. 6.2. Plug flow reactor mass balance

For a reaction rate of r moles per unit volume per unit time, a mass balance over the element gives:

$$Qc = Qc + \frac{d}{dl}(Qc)\,\delta l + Ar\delta l \tag{6.1}$$

or

$$\frac{d}{dl}(Qc) + Ar = 0. \tag{6.2}$$

In general both Q and c are functions of 'l' and the equation is satisfactorily developed by introduction of the fractional conversion 'f':

$$f = \frac{Q_0c_0 - Qc}{Q_0c_0} = 1 - \frac{Qc}{Q_0c_0} \tag{6.3}$$

so that

$$Q_0c_0\,df = Ardl \tag{6.4}$$

or

$$Fdf = Ardl \tag{6.5}$$

where F is the moles per unit time of inlet reactant.

Hence for a reactor of total length L:

$$Q_0 c_0 \int_0^f \frac{df}{r} = A \int_0^L dl = AL = V_R \tag{6.6}$$

i.e.

$$\frac{V_R}{Q_0} = c_0 \int_0^f \frac{df}{r} \tag{6.7}$$

or in terms of reactant mass feedrate

$$\frac{V_R}{F} = \int_0^f \frac{df}{r}. \tag{6.8}$$

In dealing with heterogeneous gas-solid systems in which the reaction rate is likely to be expressed as: r moles converted per unit mass of solid per unit time it is convenient to express the differential mass balance in terms of an incremental mass dW of solid. Thus equations (6.4) and (6.5) become:

$$Q_0 c_0 df = r dW \tag{6.9}$$

and

$$Fdf = r dW \tag{6.10}$$

and the total mass hold-up of solid W, either:

$$\frac{W}{Q_0} = c_0 \int \frac{df}{r} \tag{6.11}$$

or

$$\frac{W}{F} = \int \frac{df}{r}. \tag{6.12}$$

For an isothermal plug flow reactor, the evaluation of the integrals in equations (6.7), (6.8), (6.11) or (6.12) now depends upon expressing the rate as a function of fractional conversion. Thus considering a first order reaction and using the data of Example (2) in Chapter 5, equation (6.7) yields:

$$\frac{100}{Q_0} = c_0 \int_0^{0\cdot 9} \frac{df}{1 \times c_0(1-f)}$$

$$= \text{Ln } 10$$

or

$$Q_0 = 43{\cdot}5 \text{ ft}^3 \text{ min}^{-1}.$$

If the operation is not isothermal then the temperature term in the rate expression must also be expressed in terms of fractional conversion by an energy balance.

6.3.2 Space time and space velocity

The term $[V_R/Q_0]$ of equation (6.7) was also seen in Chapter 5 to arise in the analysis of continuous well stirred reactors. The dimensions of this term are time and it is accordingly called the 'nominal holding time' or 'space time' for the reaction process. Thus by definition *space time* is the time required to process one reactor volume of feed normally measured at the feed conditions, although the N.T.P. volume is sometimes specified. This means that, for a space time of 15 min, the reactor processes to the required conversion one reactor volume of feed every 15 min or four reactor volumes per hour. The reciprocal of space time $[Q_0/V_R]$ is known as space velocity.

By definition, *space velocity* is the permissible feed rate per unit volume of reactor for the required conversion. These definitions were proposed by Hougen and Watson (1947). They are more rational criteria than time for the design and analysis of any continuous chemical reactor, because at any point within a continuous reactor the composition of the reaction mixture is constant. For example at any plane AB in Fig. 6.1 the concentration profile of A is constant and independent of time, although the concentration does change along the length of the reactor. Furthermore $[V_R/Q_0]$ expresses a relationship with conversion 'f' which is independent of size or shape or feed rate as long as axial diffusion is negligible and the plug flow assumption maintained. Hence it is applicable to evaluating the correlating reaction rate data for different flow rates or to following the conversion along the length of the reactor. However, space time should not be confused with mean residence time in a reactor. Thus if in a gas phase reaction the number of moles of product leaving the reactor differs from the number of moles of reactant entering, the space time and mean residence time will be different. The use of the space time concept and a comparison with the mean residence time is illustrated in the following example.

Example 1. Acetaldehyde is to be decomposed in a plug flow reactor at 791°K and 1 atm with a 50% conversion of the pure feed. The specific reaction rate at this temperature is known to be 0·33 litre (g mole)$^{-1}$ sec^{-1} and the reaction is second order irreversible with respect to acetaldehyde concentration, i.e.:

$$r = 0{\cdot}33\, c^2 \text{ g moles litre}^{-1}\text{ sec}^{-1} \tag{I}$$

where c is measured in g moles litre^{-1} and the stoichiometric decomposition is given by:

$$CH_3CHO \rightarrow CH_4 + CO \tag{II}$$

Using equation (6·7) the space time may be found by evaluation of the integral:

$$\frac{V_R}{Q_0} = c_0 \int_0^f \frac{df}{r}$$

in which Q_0 is the volumetric feed rate of acetaldehyde.

For a molal feed rate F of acetaldehyde, the molal flow rates of the various components after a conversion 'f' are:

CH_3CHO	$F(1-f)$
CH_4	Ff
CO	Ff
Total	$F(1+f)$

Then using the ideal gas law the local volumetric flow rate is:

$$\frac{RTF(1+f)}{P}$$

so that the acetaldehyde concentration c is given by:

$$c = F(1 - f) \times \frac{P}{RTF(1 + f)} \text{ g moles litre}^{-1} \qquad \text{(III)}$$

with appropriate choice of units for R, T and P.

Using this expression for 'c' in the rate equation, the integral to be evaluated is:

$$\frac{V_R}{Q_0} = \left(\frac{RT}{P}\right)^2 \frac{c_0}{0{\cdot}33} \int_0^f \left(\frac{1 + f}{1 - f}\right)^2 df \text{ sec} \qquad \text{(IV)}$$

or since

$$c_0 = \frac{P}{RT} \qquad \text{(V)}$$

$$\frac{V_R}{Q_0} = \frac{RT}{0{\cdot}33P} \int_0^f \left(\frac{1 + f}{1 - f}\right)^2 df \text{ sec} \qquad \text{(VI)}$$

$$= \frac{RT}{0{\cdot}33P}\left[\frac{4}{1 - f} - 4 \operatorname{Ln} \frac{1}{1 - f} - (1 - f)\right]_0^f \qquad \text{(VII)}$$

$$= \frac{RT}{0{\cdot}33P}\left[\frac{4}{1 - f} - 4 \operatorname{Ln} \frac{1}{1 - f} + (f - 4)\right]. \qquad \text{(VIII)}$$

In this particular instance:

$$T = 791°\text{K} \qquad P = 1 \text{ atm} \qquad f = 0{\cdot}5$$

and $R = 0{\cdot}082$ litre atm (g mole)$^{-1}$ degK^{-1}.

Hence $$\frac{V_R}{Q_0} = 197\,[8 - 2{\cdot}77 - 3{\cdot}5] = 340 \text{ sec.} \qquad \text{(IX)}$$

So far in this example no mention has been made of reactor size or throughput, and it is seen that this value of space time is a function of the prevailing rate expression and the required fractional conversion. The value 340 seconds can be utilised to specify the required size of plug flow reactor to convert 50% of any nominated throughput of pure acetaldehyde at 791°K. Thus a feed rate of 10 litre sec^{-1} would require 3400 litres of reactor volume.

For this kinetic equation, the variation of space time with conversion is obtained by substituting a series of values for f into equation (VIII) and the results are given in Table 6.1.

TABLE 6.1

f	Space time (seconds)
0·25	85·2
0·30	116
0·35	154
0·40	202
0·45	262
0·50	340
0·60	578
0·70	1030
0·80	2040
0·90	5455
0·95	12800
0·98	35700
0·99	74600
0·999	782000

The mean residence time in the reactor is found by firstly considering the time of passage dt through a reactor element as:

$$\frac{\text{distance}}{\text{velocity}} \text{ or } \frac{\text{distance} \times \text{cross-sectional area}}{\text{volumetric flow rate}}$$

i.e.

$$dt = \frac{dl \times A \times P}{F(1+f)RT}. \qquad \text{(X)}$$

Now Adl is eliminated from equation (X) and r substituted as above to give:

$$dt = \frac{Fdf\,.\,P}{F(1+f)RT \times 0{\cdot}33}\left(\frac{RT}{P}\right)^2 \frac{(1+f)^2}{(1-f)^2} \qquad \text{(XI)}$$

$$\int_0^t dt = \frac{RT}{0{\cdot}33P}\int_0^f \frac{(1+f)\,df}{(1-f)^2}\text{ sec} \qquad \text{(XII)}$$

$$t = \frac{RT}{0{\cdot}33P}\left[\frac{2}{1-f} - \text{Ln}\frac{1}{1-f}\right]_0^f \qquad \text{(XIII)}$$

$$= \frac{RT}{0{\cdot}33P}\left[\frac{2}{1-f} - 2 - \text{Ln}\frac{1}{1-f}\right]. \qquad \text{(XIV)}$$

Substituting the values for R, T, P and f as before:

$$t = 197\,[4 - 2 - 0{\cdot}693] = 257\text{ sec}. \qquad \text{(XV)}$$

6.4 Isothermal Plug Flow Reactors

Since it is difficult practically to maintain a tubular reactor at a constant temperature, an isothermal analysis is an idealisation. However, the results obtained may be of value for an initial assessment before making more exact and lengthy calculations.

In the previous section, an example of an isothermal forward reaction was considered and extension is made below first to a reverse reaction whose kinetics are expressed in terms of partial pressure and then to a gas–solid reaction with adsorption–desorption equilibrium.

Example 2. Butadiene is to be dimerised in a tubular reactor at 1 atm and 911°K:

$$2C_4H_6 \rightleftharpoons C_8H_{12} \tag{I}$$

The reactor feed is a mixture of butadiene and steam in the mole ratio of 2:1, the diluent steam having been introduced to preheat the butadiene. The forward reaction is second order and the reverse reaction first order. At 911°K the specific reaction rate of the forward reaction is 114·6 g moles litre^{-1} h^{-1} atm^{-2} and the equilibrium constant is 1·27 atm^{-1}. Estimate the volume of tubular reactor required to achieve 95% of the equilibrium conversion of a feed of 50 kg mole h^{-1} of butadiene, and specify the space velocity.

From the dimensions of the rate constant, the rate equation may be formulated as:

$$r = 114{\cdot}6\left[p_B^2 - \frac{p_D}{1{\cdot}27}\right] \text{g mole h}^{-1}\text{ litre}^{-1}. \tag{II}$$

These partial pressures in atm may be expressed in terms of the fractional conversion 'f' by considering a feed of 2 moles of butadiene. Then at a conversion 'f', the mixture consists of:

Moles C_4H_6	$2(1-f)$
Moles C_8H_{12}	f
Moles H_2O	1
Moles total	$3-f$

so that at 1 atm operating pressure:

$$p_B = \frac{2(1-f)}{3-f}\text{ atm} \tag{III}$$

and $$p_D = \frac{f}{3 - f}\text{ atm} \tag{IV}$$

The equilibrium conversion f_e may be found by setting equation (II) equal to zero with these values of p_B and p_D substituted.

i.e. $$\frac{4(1 - f_e)^2}{(3 - f_e)^2} = \frac{f_e}{1{\cdot}27(3 - f_e)} \tag{V}$$

giving $$f_e = 0{\cdot}502. \tag{VI}$$

Hence the upper limit of integration is:

$$0{\cdot}95 \times 0{\cdot}502 = 0{\cdot}478. \tag{VII}$$

It is convenient to use the basic design equation (6.11) in which F will be 50 000 g moles h^{-1}.
Hence:

$$\frac{V_R}{50000} = \frac{1}{114{\cdot}6}\int_0^{0{\cdot}478} \frac{(3 - f)^2\,df}{4(1 - f)^2 - 0{\cdot}788f(3 - f)} \tag{VIII}$$

or $$V_R = \frac{50000}{114{\cdot}6 \times 4}\int_0^{0{\cdot}478} \frac{9 - 6f + f^2}{(1 - f)^2 - 0{\cdot}197f(3 - f)}\,df \text{ litres.} \tag{IX}$$

Dividing out and resolving into partial fractions:

$$V_R = 109\int_0^{0{\cdot}478}\left[0{\cdot}835 + \frac{1{\cdot}30}{f - 1{\cdot}664} - \frac{4{\cdot}5}{f - 0{\cdot}5}\right]df. \tag{X}$$

Integration between the limits gives:

$$V_R = 109\left[0{\cdot}835 \times 0{\cdot}478 + 1{\cdot}30\,\text{Ln}\,\frac{1{\cdot}664 - 0{\cdot}478}{1{\cdot}664} - 4{\cdot}5\,\text{Ln}\,\frac{0{\cdot}5 - 0{\cdot}478}{0{\cdot}5}\right] \tag{XI}$$

$$= 109\,[0{\cdot}40 - 0{\cdot}44 + 8{\cdot}52] \tag{XII}$$

$$= 925 \text{ litres } (32{\cdot}7 \text{ ft}^3). \tag{XIII}$$

If this were to be accommodated in 4 in I.D. piping, the reactor length 'L' would be given by:

$$\left(\frac{\pi}{4}\right)\left(\frac{4}{12}\right)^2 L = 32{\cdot}7 \tag{XIV}$$

i.e. $$L = 374 \text{ ft.} \tag{XV}$$

The space velocity is found by expressing the total feed of 75000 g mole hr^{-1} as litres sec^{-1}.

i.e. $$V = \frac{nRT}{P} = \frac{75000 \times 0{\cdot}082 \times 911}{1 \times 3600} = 1560 \text{ litre sec}^{-1} \tag{XVI}$$

giving a space velocity of 1·69 litre sec^{-1} litre^{-1} and a space time of 0·592 sec.

Example 3. The isothermal rate equation for the gas–solid reaction:

$$C + CO_2 \rightarrow 2CO \tag{I}$$

in terms of gas phase partial pressures is:

$$r = \frac{k_1 p_{CO_2}}{1 + k_2 p_{CO} + k_3 p_{CO_2}} \tag{II}$$

moles of CO_2 converted per unit time per unit mass of coke in the reactor.

The gas stream fed to the reactor is:

CO_2	n_0	moles per unit time
CO	m_0	moles per unit time
N_2	N	moles per unit time

An expression is to be developed for the required hold-up of coke assuming plug flow of gas through the reactor at pressure P. Application of equation (6.12) gives:

$$\frac{W}{n_0} = \frac{1}{k_1} \int_0^f \left[\frac{1}{p_{CO_2}} + k_2 \frac{p_{CO}}{p_{CO_2}} + k_3 \right] df \tag{III}$$

and it remains to express the partial pressures as functions of 'f'. If 'n' moles of CO_2 have reacted at a position corresponding to the conversion 'f', then the moles of the three components are:

CO_2	$n_0 - n$
CO	$m_0 + 2n$
N_2	N
Total	$N + n_0 + m_0 + n$

Since $f = n/n_0$, the partial pressures of CO_2 and CO are:

$$p_{CO_2} = \frac{n_0(1-f)P}{N + m_0 + n_0(1+f)} \quad \text{and} \quad p_{CO} = \frac{(m_0 + 2n_0 f)P}{N + m_0 + n_0(1+f)} \tag{IV}$$

Hence

$$\frac{W}{n_0} = \frac{1}{k_1}\int_0^f \left[\frac{N + m_0 + n_0(1+f)}{n_0(1-f)P} + k_2\frac{(m_0 + 2n_0 f)}{n_0(1-f)} + k_3\right] df. \tag{V}$$

The leading terms of the integral may be rearranged to give:

$$\frac{W}{n_0} = \frac{1}{k_1}\int_0^f \left[\left(\frac{N + m_0 + n_0}{n_0 P} + \frac{k_2 m_0}{n_0}\right)\frac{1}{1-f} + \left(\frac{1}{P} + 2k_2\right)\frac{f}{1-f} + k_3\right] df \tag{VI}$$

giving:

$$W = \frac{1}{k_1}\left[-\left(\frac{N + m_0 + n_0}{P} + k_2 m_0\right)\text{Ln}\,(1-f) - n_0\left(\frac{1}{P} + 2k_2\right)\left(f + \text{Ln}\,\{1-f\}\right) + k_3 n_0 f\right] \tag{VII}$$

or

$$W = \frac{f n_0}{k_1}\left[k_3 - \left(\frac{1}{P} + 2k_2\right)\right] - \frac{\text{Ln}\,(1-f)}{k_1} \times \left[\frac{N + m_0 + 2n_0}{P} + k_2(m_0 + 2n_0)\right]. \tag{VIII}$$

6.5 The Non-isothermal Plug Flow Reactor

Not only is it difficult to operate a tubular reactor isothermally but it may also be undesirable to operate in this manner as discussed in section (6.7). The non-isothermal reactor may be operated either adiabatically or there may be some heat transfer. In this latter case the heat transfer conditions may be arranged to impose a predetermined temperature profile on the reactor. Whatever the reasons for non-isothermal operation, an energy balance over an increment of length is combined with the incremental mass balance of equations (6.4) or (6.5) to analyse the reactor. The following energy balance for a flow system was developed in Chapter 2:

$$q = \Delta H^0_{T_0} + \int_{T_0}^{T_2} (C_p)_P \, dT - \int_{T_0}^{T_1} (C_p)_R \, dT \,. \qquad (2.12)$$

Application of this to a general flow through an element of a plug flow reactor Fig. 6.2 requires:

(*i*) q to be replaced by $UA_s dl(T_s - T)_m$
where U is the overall heat transfer coefficient between the heating medium and the reactor contents
A_s is the heat transfer surface area per unit length of reactor,
T_s is the temperature of the heat transfer medium,
and
$(T_s - T)_m$ is the mean temperature driving force.

(*ii*) $\Delta H^\circ_{T_0}$ to be replaced by $rAdl\Delta H$
where r is the rate and is a function of both fractional conversion and temperature,
A is the reactor cross-sectional area,
and ΔH now stands for the heat of reaction at the datum temperature.

(*iii*) $\int_{T_0}^{T_2} (C_p)_P dT$ to be replaced by $\int_{T_0}^{T_2} \sum F_P (C_p)_P \, dT$

where F_P represents all products and inert components leaving the element
and $(C_p)_P$ their specific heats.

(*iv*) $\int_{T_0}^{T_1} (C_p)_R \, dT$ to be replaced by $\int_{T_0}^{T_1} \sum F_R (C_p)_R dT$ where F_R represents all components (reactants and inerts) entering the element and $(C_p)_R$ their specific heats.

Hence equation (2.12) applied to this situation gives:

$$UA_s dl(T_s - T)_m = rAdl\Delta H + \int_{T_0}^{T_2} \sum F_P (C_p)_P dT - \int_{T_0}^{T_1} \sum F_R (C_p)_R dT \quad (6.13)$$

and the integrals can be re-arranged to give:

$$UA_s dl(T_s - T)_m = rAdl\Delta H + \int_{T_0}^{T_1} \left[\sum F_P(C_p)_P - \sum F_R(C_p)_R\right] dT + \int_{T_1}^{T_2} \sum F_P(C_p)_P dT. \qquad (6.14)$$

If the reaction in question and the specific heats are such that:

$$\sum F_P(C_p)_P = \sum F_R(C_p)_R \qquad (6.15)$$

then equation (6.14) reduces to:

$$UA_s dl(T_s - T)_m = rAdl\Delta H + \int_{T_1}^{T_2} \sum F_P(C_p)_P dT \qquad (6.16)$$

Furthermore, if the change in temperature:

$$T_2 - T_1 \equiv \Delta T \qquad (6.17)$$

over a finite increment of reactor Δl is small, the values of C_p in the integral between T_1 and T_2 will be essentially constant and equations (6.14) and (6.16) respectively become:

$$UA_s \Delta l(T_s - T)_m = rA\Delta l\Delta H + \int_{T_0}^{T_1} \left[\sum F_P(C_p)_P - \sum F_R(C_p)_R\right] dT + \sum F_P(C_p)_P \Delta T \qquad (6.18)$$

and

$$UA_s \Delta l(T_s - T)_m = rA\Delta l\Delta H + \sum F_P(C_p)_P \Delta T. \qquad (6.19)$$

6.5.1 Adiabatic operation

In the absence of heat transfer, the left-hand side of equations (6.13), (6.14), (6.16), (6.18) and (6.19) may be set at zero. In particular for a finite length of reactor element equations (6.18) and (6.19) become:

$$\int_{T_0}^{T_1} \left[\sum F_P(C_p)_P - \sum F_R(C_p)_R\right] dT + \sum F_P(C_p)_P \Delta T = -rA\Delta l\Delta H \qquad (6.20)$$

and

$$\sum F_P(C_p)_P \Delta T = -rA\Delta l\Delta H \qquad (6.21)$$

in either of which the substitution:

$$rA\Delta l = F\Delta f \qquad (6.22)$$

based on equation (6.5), may be made. On making this substitution the unique connection between fractional conversion and temperature in adiabatic operation is established, and as many corresponding values of f and T may be evaluated as is desirable for the definition of the reactor. The resulting values of 'r' enable the necessary lengths of increment to be evaluated. The procedure is illustrated in the following example.

Example 4. For the conversion of acetaldehyde into methane and carbon monoxide:

$$CH_3CHO \rightarrow CH_4 + CO$$

$$\Delta H = -4{\cdot}55 \text{ kcal (g mole)}^{-1}$$

and

$$r = 0{\cdot}00033\, c^2 \exp\left(31{\cdot}71 - \frac{25\,080}{T}\right) \text{g mole cm}^{-3} \text{ sec}^{-1}.$$

Considering a feed of F mole sec^{-1} of pure acetaldehyde, the rate of reaction at fractional conversion 'f' is:

$$r = \frac{0{\cdot}00033}{R^2T^2}\left(\frac{1-f}{1+f}\right)^2 \exp\left(31{\cdot}71 - \frac{25080}{T}\right) \text{g mole cm}^{-3} \text{ sec}^{-1}$$

while for a reactor increment having 'f' as the inlet fractional conversion and $f + \Delta f$ as the outler value, the moles of the three components are:

	Inlet	Outlet
CH_3CHO	$F(1-f)$	$F(1-f-\Delta f)$
CH_4	Ff	$F(f+\Delta f)$
CO	Ff	$F(f+\Delta f)$

and if their respective specific heats as functions of temperature are denoted C_{P1}, C_{P2}, and C_{P3}, the integral of equation (6.20) becomes:

$$\int_{T_0}^{T_1} (-F\Delta f C_{P1} + F\Delta f C_{P2} + F\Delta f C_{P3})\, dT$$

or

$$F\,\Delta f \int_{T_0}^{T_1} (C_{P2} + C_{P3} - C_{P1})\, dT.$$

The term $\sum F_P(C_P)_P \Delta T$ may be written as:

$$F[(1-f-\Delta f)C'_{P1} + (f+\Delta f)(C'_{P2} + C'_{P3})]\,\Delta T$$

where C'_{P1}, C'_{P2} and C'_{P3} are now the point values of specific heat evaluated at the prevailing temperature in the increment, and without error could be evaluated at the increment inlet temperature.

With the elimination of $rA\Delta l$, equation (6.20) yields:

$$\Delta T = \frac{\Delta f\left[-\Delta H - \int_{T_0}^{T_1} (C_{P2} + C_{P3} - C_{P1})\,dT\right]}{(1 - f - \Delta f)C'_{P1} + (f + \Delta f)(C'_{P2} + C'_{P3})}.$$

Using the specific heat values from Table 2.2, the following connection between fractional conversion and temperature in successive increments may be established; the increment size is $\Delta f = 0{\cdot}05$ except for the final one of 0·04.

TABLE 6.2

Inlet conversion	Outlet conversion	Inlet T °K	Outlet T °K	Increment length cm
0	0·05	780·0	790·5	57·9
0·05	0·10	790·5	801·1	47·6
0·10	0·15	801·1	811·7	39·6
0·15	0·20	811·7	822·4	33·3
0·20	0·25	822·4	833·1	28·3
0·25	0·30	833·1	843·8	24·5
0·30	0·35	843·8	854·6	21·4
0·35	0·40	854·6	865·4	19·1
0·40	0·45	865·4	876·2	17·3
0·45	0·50	876·2	887·1	16·0
0·50	0·55	887·1	898·1	15·1
0·55	0·60	898·1	909·1	14·7
0·60	0·65	909·1	920·2	14·7
0·65	0·70	920·2	931·3	15·4
0·70	0·75	931·3	942·5	16·9
0·75	0·80	942·5	953·7	19·9
0·80	0·85	953·7	965·0	26·1
0·85	0·90	965·0	976·4	40·5
0·90	0·95	976·4	987·8	86·3
0·95	0·99	987·8	997·0	553·3

The final column in the table is developed by evaluating the rate at both inlet and outlet of an element taking into account the values of both fractional conversion and temperature; the arithmetic average of these two rates is used in equation (6.22) with $F = 0{\cdot}1$ g mole sec^{-1} and $A = 1500$ cm^2 to obtain Δl in cm. It is of interest to note the relatively large increments of length needed initially because of the low temperatures, and at high conversion because of the low reactant concentration; this accounts for the sigmoidal shape of the plot of these results shown in Fig. 6.3.

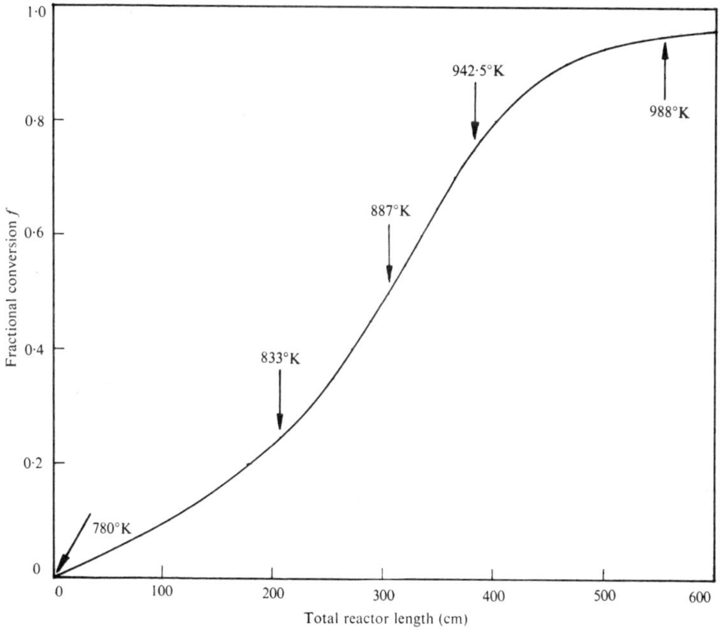

FIG. 6.3. Adiabatic exothermic plug flow reactor

6.5.2 General non-isothermal operation

The connection between fractional conversion and temperature is no longer unique and the mass and energy balances have to be solved by trial and error. It is usual to nominate the size of reactor increment and to then iterate the calculation of the resulting increments of temperature and fractional conversion. The mass and energy balances written in terms of finite changes have been developed above and are:

$$F\Delta f = Ar\Delta l \qquad (6.22)$$

and

$$\int_{T_0}^{T_1} \left[\left(\sum F_{\mathrm{P}}(C_p)_{\mathrm{P}} - \sum F_{\mathrm{R}}(C_p)_{\mathrm{R}}\right] dT + \sum F_{\mathrm{P}}(C_p)_{\mathrm{P}}\Delta T + rA\Delta l\Delta H = UA_{\mathrm{s}}\Delta l(T_{\mathrm{s}} - T)_{\mathrm{m}} \qquad (6.23)$$

where

$$r = \phi(f, T). \tag{6.24}$$

The steps in the iteration of the nth increment for a chosen value of Δl are as follows and a computer flow diagram for the reactor evaluation is shown in Fig. 6.4.

(*i*) From the values f_n and T_n at the inlet to the nth, increment, use (6.24) to calculate r_n at entry to the increment.

(*ii*) Assuming r_n to apply throughout the increment use (6.22)

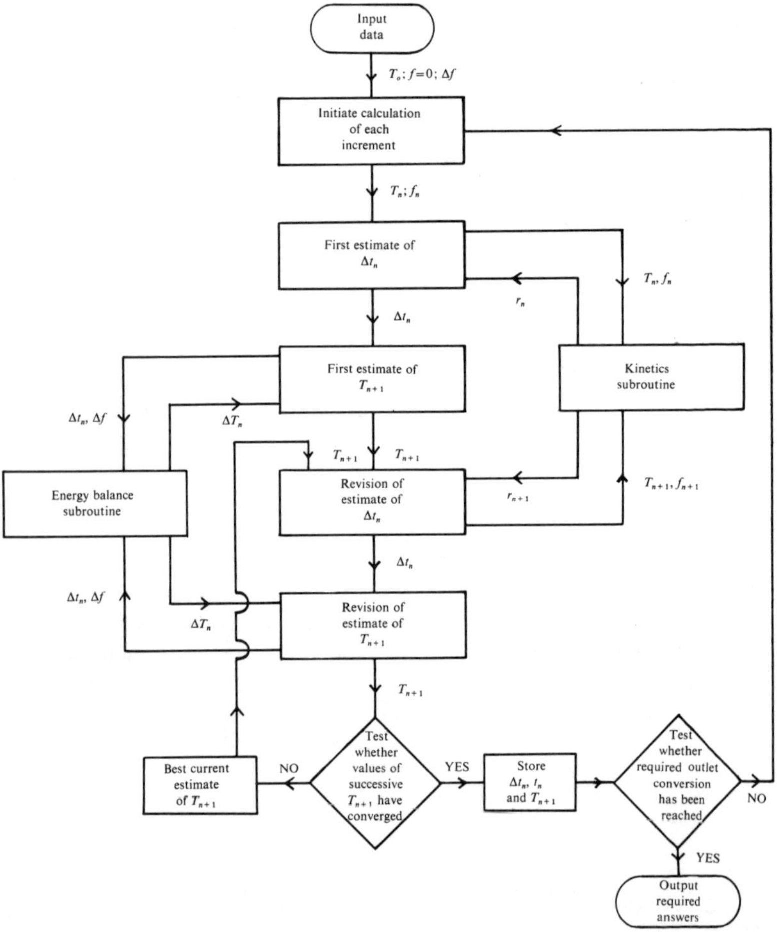

FIG. 6.4. Computer flow diagram for non-isothermal plug flow reactor

with $r = r_n$ to obtain a first estimate of Δf and then f_{n+1} from:

$$f_{n+1} = f_n + \Delta f. \tag{6.25}$$

Also making whatever assumptions are relevant to the mean heat transfer driving force, i.e. the temperature difference $(T - T_s)_m$, use (6.23) with $r = r_n$ to obtain a first estimate of ΔT and hence T_{n+1} from:

$$T_{n+1} = T_n + \Delta T. \tag{6.26}$$

(*iii*) From these estimates of f_{n+1} and T_{n+1}, use the rate equation (6.24) to obtain an estimate of r_{n+1}. Then take an average rate:

$$r_{n+\frac{1}{2}} = \tfrac{1}{2}(r_{n+1} + r_n) \tag{6.27}$$

to revise the estimates of f_{n+1} and T_{n+1}.

(*iv*) Continue to revise $r_{n+\frac{1}{2}}$, f_{n+1} and T_{n+1} until a satisfactory level of convergence is achieved.

(*v*) Use the finally accepted answers of f_{n+1} and T_{n+1} to restart the above procedure for the calculation of the next increment exit values f_{n+2} and T_{n+2}.

Example 5. Ethylene and propylene are formed by thermal decomposition of n-butane in a tubular reactor. For the conditions specified below and the listed data, it is required to estimate the length of 4 in I.P.S. (0·296 ft I.D.) piping needed to effect 25% conversion.

(*a*) Feed is 101 lb h^{-1} ft^{-2} of butane gas at 510°C.

(*b*) Operating pressure P is 1 atm.

(*c*) The following three reactions occur in the ratio 10:4:1.

$$C_4H_{10} \rightarrow CH_4 + C_3H_6 \qquad \Delta H = 16670 \tag{I}$$

$$C_4H_{10} \rightarrow C_2H_6 + C_2H_4 \qquad \Delta H = 22150 \tag{II}$$

$$C_4H_{10} \rightarrow H_2 + C_4H_8 \qquad \Delta H = 28670 + 5T \tag{III}$$

(*d*) The kinetic equation proposed by Echols and Pease (1939) for the thermal decomposition of n-butane is:

$$-\frac{dp}{dt} = k_1 p^{1\cdot 5} + \frac{k_2 p^2}{p_0 - p} \tag{IV}$$

where p is the butane pressure
and p_0 is the initial butane pressure.

The rate constants in units of minutes and millimetres of mercury are:

T°C	480	490	500	510	520	535
$k_1 \times 10^4$	0·642	1·043	1·80	2·85	4·50	8·30
$k_2 \times 10^4$	0·182	0·338	0·625	1·126	1·95	4·77

(*e*) Thermal data:
C_p for butane 6.84 + 0·05T

Heat transfer coefficient: $$U = \frac{1{\cdot}65}{D}\left[\frac{wC_p}{l}\right]^{0\cdot 33} [\bar{k}]^{0\cdot 67}$$

where $\bar{k}$ is the thermal conductivity of the gas. For the prevailing properties:

$$U = 0{\cdot}021\, l^{-0\cdot 33} \tag{V}$$

Temperature of heat transfer medium 833°K.

Solution. Since each of the decomposition reactions generates two moles of gas per mole of butane reacted, a material balance based on a feed of n_0 moles butane at conversion f gives:

Butane: $n_0(1 - f)$ moles
Reaction products $2n_0 f$ moles
Total $n_0(1 + f)$ moles

Hence $$P = \left(\frac{1 - f}{1 + f}\right) P \tag{VI}$$

and $$\frac{dp}{dt} = \frac{-2P}{(1 + f)^2}\frac{df}{dt}. \tag{VII}$$

Eliminating dp/dt between this expression and the kinetic equation and noting that for the feed of pure butane $p_0 = p$:

$$\frac{df}{dt} = (1 + f)^{\frac{1}{2}} (1 - f)^{\frac{3}{2}} \frac{k_1 p^{\frac{1}{2}}}{2} + \frac{k_2(1 + f)(1 - f)^2}{4f}. \tag{VIII}$$

The total gas volume passing through the element of length dl is:

$$n_0(1 + f)\frac{RT}{P}$$

and so for a cross-sectional area A, the time of passage dt is given by:

$$dt = \frac{PAdl}{n_0(1 + f)RT} \tag{IX}$$

and hence:

$$\frac{df}{dl} = \frac{(1-f)^{\frac{3}{2}}}{(1+f)^{\frac{1}{2}}} \cdot \frac{k_1 A P^{\frac{3}{2}}}{2n_0 RT} + \frac{(1-f)^2}{f} \cdot \frac{k_2 A P}{4n_0 RT}. \qquad \text{(X)}$$

Substituting numerical values for the terms which are independent of conversion and temperature as follows:

$$A = 0{\cdot}25\pi \times (0{\cdot}296)^2 = 0{\cdot}0687 \text{ ft}^2$$

$$P = 760 \text{ mmHg}$$

$$n_0 = \frac{101 \times 0{\cdot}0687}{60 \times 58} = 2 \times 10^{-3} \text{ lb mole min}^{-1}$$

$$R = 998{\cdot}9 \text{ ft}^3 \text{ mmHg lb mole}^{-1} \text{ degK}^{-1}$$

gives:

$$\frac{df}{dl} = \frac{357\,k_1}{T}\frac{(1-f)^{1\cdot 5}}{(1+f)^{0\cdot 5}} + \frac{6{\cdot}53\,k_2}{T}\frac{(1-f)^2}{f}. \qquad \text{(XI)}$$

The energy changes in the reaction mixture in an element of reactor are:

$$2 \times 10^{-3}(6.84 + 0{\cdot}05T)\,dT + 2 \times 10^{-3}\,df \times \left[\frac{10}{15} \times 16670 + \frac{4}{15} \times 22150 + \frac{1}{15}(28670 + 5T)\right].$$

The rate of heat transfer from the jacket is:

$$\pi \times 0{\cdot}296 \times U(833 - T)\,dl.$$

Equating these expressions and re-arranging gives:

$$\frac{dT}{dl} = 465U\frac{833 - T}{6{\cdot}84 + 0{\cdot}05T} - \frac{56730 + T}{3(6{\cdot}84 + 0{\cdot}05T)}\frac{df}{dl} \qquad \text{(XII)}$$

At reactor entry $T = 783°\text{K}$ giving $k_1 = 2{\cdot}85 \times 10^{-4}$ and $k_2 = 1{\cdot}126 \times 10^{-4}$.

Let $f = 0{\cdot}01$

$$\frac{df}{dl} = 2{\cdot}24 \times 10^{-4} \qquad \text{from XI}$$

and

$$\frac{dT}{dl} = 10{\cdot}6. \qquad \text{from XII}$$

Then first trial of f at 0·5 ft of tube is

$$f_{0\cdot 5} = f_0 + \left(\frac{df}{dl}\right)_0 0\cdot 5 = 0\cdot 01 + 2\cdot 24 \times 10^{-4} \times 0\cdot 5 = 0\cdot 0101$$

$$T_{0\cdot 5} = T_0 + \left(\frac{dT}{dl}\right)_0 0\cdot 5 = 783 + 10\cdot 6 \times 0\cdot 5 = 788\cdot 3°\text{K}.$$

The gradients at $l = 0\cdot 5$ are then

$$\left(\frac{df}{dl}\right)_{0\cdot 5} = 2\cdot 82 \times 10^{-4}; \left(\frac{dT}{dl}\right)_{0\cdot 5} = 9\cdot 42.$$

Then $\left(\frac{df}{dl}\right)_{av} = 2\cdot 53 \times 10^{-4}$ and $\left(\frac{dT}{dl}\right)_{av} = 10\cdot 01$

giving $f_{0\cdot 5} = 0\cdot 0101$ and $T_{0\cdot 5} = 788°\text{K}$

The above procedure is repeated until $f = 0\cdot 25$, giving a result of $l = 350$ ft for the reactor length.

The mole fraction composition of the exit gas is:

$H_2 = 0\cdot 0133$	$C_3H_6 = 0\cdot 1338$
$CH_4 = 0\cdot 1338$	$C_4H_8 = 0\cdot 0133$
$C_2H_4 = 0\cdot 0535$	$C_4H_{10} = 0\cdot 5980$
$C_2H_6 = 0\cdot 0535$	

6.6 Laminar Flow Reactors

Consider a tubular reactor which is carrying out a liquid reaction process with the liquid flow characteristics such that the flow is in the laminar regime. The fluid then passes through the tube in the form of streamlines with no radial diffusion and the relative velocities of these streamlines generate a parabolic velocity profile as shown in Fig. 6.5. Accepting the concept of the existence of streamlines implies that fluid elements at different distances from the tube axis have different times of passage through the reactor, but that all fluid in a streamline at a particular radius has the same residence time. Hence the laminar velocity profile in Fig. 6.5a can be approximated by a series of annuli in each of which the velocity is constant as illustrated in Fig. 6.5b.

Each annulus can be considered to be a plug flow tubular reactor having its own space velocity. The velocities of the fluid elements

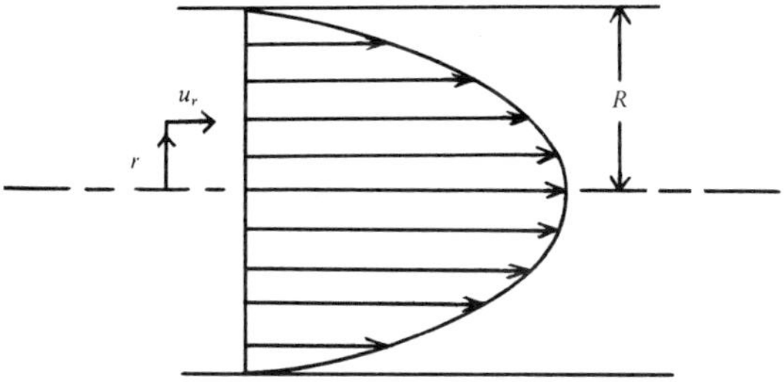

(a)

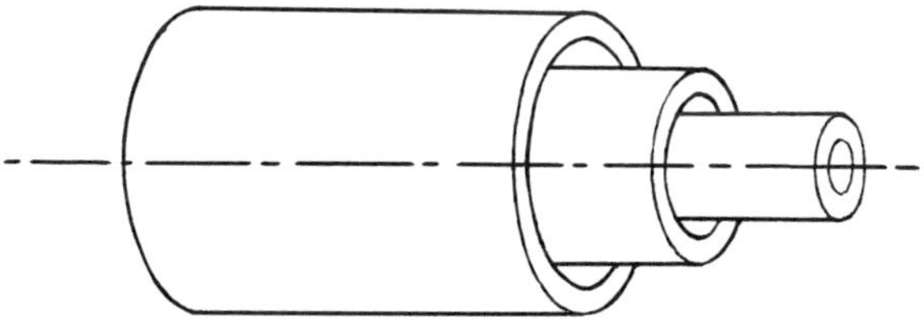

(b)

FIG. 6.5. Laminar flow reactor

at different radii are given by the parabolic velocity profile for fully developed laminar flow:

$$u_r = u_0 \left(1 - \frac{r^2}{R^2}\right) \tag{6.28}$$

where u_r is the velocity at any radius 'r',
u_0 is the velocity at the tube axis,
and R is the tube radius.

From elementary fluid mechanic considerations this velocity profile entails a total volumetric flow Q given by

$$Q = \frac{\pi u_0 R^2}{2} \tag{6.29}$$

and a mean velocity u_m related to the maximum velocity at the tube axis by:

$$u_m = \frac{u_0}{2}. \tag{6.30}$$

If t is the residence time of fluid at radius 'r', we have for a reactor of length L:

$$t = \frac{L}{u_r} = \frac{L}{u_0\left(1 - \dfrac{r^2}{R^2}\right)} \tag{6.31}$$

from which:

$$dt = \frac{2rLdr}{u_0R^2\left(1 - \dfrac{r^2}{R^2}\right)^2} = \frac{2u_0t^2}{L} \cdot \frac{rdr}{R^2} \tag{6.32}$$

and

$$rdr = \frac{LR^2}{2u_0} \cdot \frac{dt}{t^2}. \tag{6.33}$$

The flow through an annular element at radius 'r' is:

$$2\pi u_r r\,dr \quad \text{or} \quad \frac{2\pi Lr\,dr}{t} \tag{6.34}$$

and the fractional conversion achieved will be the same as for a batch after time 't' which may be deduced from equation (3.33) for a second order reaction:

$$f = \frac{ktc_0}{1 + ktc_0} \tag{6.35}$$

and so the amount of material converted in the annulus by a second order reaction is:

$$\frac{2\pi Lr\,dr}{t} \cdot \frac{ktc_0}{1 + ktc_0}.$$

which on substitution of $r\,dr$ from equation (6.33) gives:

$$\frac{\pi L^2R^2kc_0}{u_0} \cdot \frac{dt}{t^2(1 + ktc_0)}.$$

If the mean fractional conversion over the exit plane is f_m:

$$\frac{\pi R^2u_0}{2} f_m = \frac{\pi L^2R^2kc_0}{u_0} \int_{t_0}^{\infty} \frac{dt}{t^2(1 + ktc_0)} \tag{6.36}$$

with the integral to be evaluated from the minimum time of passage t_0 on the tube axis to infinite time at the tube wall.
Hence:

$$f_m = \frac{2kc_0L^2}{u_0^2}\int_{t_0}^{\infty}\left[\frac{1}{t^2} - \frac{kc_0}{t} + \frac{k^2c_0^2}{1 + ktc_0}\right]dt. \tag{6.37}$$

Integrating and replacing L^2/u_0^2 by t_0^2:

$$f_m = 2kc_0t_0^2\left[kc_0 \,\text{Ln}\,\frac{1 + ktc_0}{t} - \frac{1}{t}\right]_{t_0}^{\infty}. \tag{6.38}$$

As $t \to \infty$, the ratio within the logarithm tends to kc_0 and so:

$$f_m = 2kc_0t_0\left[1 + kc_0t_0 \,\text{Ln}\,\frac{kc_0t_0}{1 + kc_0t_0}\right] \tag{6.39}$$

or in terms of t_m which from (6·30) is $2t_0$:

$$f_m = kc_0t_m\left[1 + \frac{kc_0t_m}{2}\,\text{Ln}\,\frac{kc_0t_m}{2 + kc_0t_m}\right]. \tag{6.40}$$

In general for second order reactions, the difference between plug reactor and streamline reactor performance is a function of kc_0t_m and some comparative values are given in Table 6.3.

TABLE 6.3

kc_0t_m	f-Plug	f-Streamline
1·0	0·50	0·45
2·0	0·67	0·61
10·0	0·91	0·87
25·0	0·96	0·94

A specific calculation of the lengths of plug flow and laminar flow reactors is given below in example 6.

For a first order reaction, the fractional conversion for a batch after time t is:

$$f = 1 - e^{-kt} \tag{6.41}$$

and combining this with the flow through an annulus and integrating as for the second order case leads to:

$$f_m = 2t_0^2\int_{t_0}^{\infty}\left(\frac{1}{t^3} - \frac{e^{-kt}}{t^3}\right)dt \tag{6.42}$$

The second term of this integral does not have an analytical solution and must either be evaluated numerically or reduced on integration by parts to an integral of the form:

$$\int_1^\infty \frac{e^{-t/K}}{t}\, dt$$

and then evaluated from tables of the special function $Ei(K)$.

Example 6. Butyl acetate is to be produced in a tubular reactor operating isothermally at 100°C using sulphuric acid as catalyst. The feed contains 4·97 moles butanol per mole of acetic acid and the catalyst concentration is 0·032% H_2SO_4 by weight. It is to be fed at the rate of 1000 lb h^{-1} to the reactor, and a 50% conversion is required. Estimate the lengths of reactor required for this conversion assuming plug flow conditions and also laminar flow-conditions. The following data may be used:

(*i*) The rate equation (Leyes and Othmer, 1945) is $r = kc_A^2$ where r = rate in g moles acid/cm³ min.
$k = 17{\cdot}4$ cm³ (g mole)$^{-1}$ acetic acid min^{-1}.

(*ii*) Specific gravity of feed is 0·75 and is assumed to apply throughout the conversion.

(*iii*) Reactor tube 4·0 inches I.D.

(*iv*) Viscosity of mixture 0·54 c.p.

Solution. Reaction rate $r = kc_A^2 = kc_{A0}^2(1-f)^2$.
The basic plug flow design equation is:

$$\frac{V_R}{Q} = c_{A0}\int_0^{0\cdot5}\frac{df}{r} = \frac{c_{A0}}{kc_{A0}^2}\int_0^{0\cdot5}\frac{df}{(1-f)^2} \qquad \text{I}$$

or

$$\frac{V_R}{Q} = \frac{1}{kc_{A0}}\left[\frac{1}{1-f} - 1\right] = \frac{1}{kc_{A0}}(2-1) = \frac{1}{kc_{A0}} \qquad \text{II}$$

where c_{A0} is:

$$\frac{0{\cdot}75}{(4{\cdot}97 \times 74) + (1 \times 60)} = 0{\cdot}0018 \text{ g mole acetic acid cm}^{-3}.$$

Then $$\frac{V_R}{Q} = \frac{1}{17{\cdot}4 \times 0{\cdot}0018} = 32 \text{ minutes}$$

$\therefore$ $$V_R = \frac{1000 \times 32}{60 \times 0{\cdot}75 \times 62{\cdot}4} = 0{\cdot}356 \times 32 = 11{\cdot}5 \text{ ft}^3$$

and $$\frac{\pi\ 4^2}{4\times144} l = 11{\cdot}5 \text{ or } l = 120 \text{ ft}$$

i.e. a plug flow reactor 120 feet long would be required.

$$Re = \left(\frac{1000}{\frac{\pi\ 4^2}{4\times144}}\right)\frac{4}{12} \times \frac{1}{0{\cdot}54 \times 2{\cdot}42} = 3100$$

While the flow is therefore in the transition region between laminar and turbulent the mean conversion f_m will be assumed to be given by the laminar result:

$$f_m = kc_0t_m \left[1 + \frac{kc_0t_m}{2} \ln\left(\frac{kc_0t_m}{2 + kc_0t_m}\right)\right] \qquad \text{III}$$

where $$kc_0t_m = 17{\cdot}4 \times 0{\cdot}0018 \times 32 = 1{\cdot}0$$

so that: $$f_m = 1 + \frac{2{\cdot}303}{2} \log\left(\tfrac{1}{3}\right) = 0{\cdot}45.$$

That is, the above reactor would be incapable of attaining the required conversion. To obtain the residence time for 0·5 conversion equation (III) must be solved by trial and error.
Thus let $t_m = 40$ minutes, i.e. $kc_0t_m = 1{\cdot}25$.

$$f_m = 1{\cdot}25 \left[1 + 1{\cdot}439 \log\left(\frac{1{\cdot}25}{3{\cdot}25}\right)\right] = 0{\cdot}501.$$

That is $t_m = 40$ minutes is the solution.

$$\frac{V_R}{Q} = 40 \text{ or } V_R = 14{\cdot}5 \text{ ft}^3 \equiv 167 \text{ ft of } 4{\cdot}0 \text{ in I D tube.}$$

6.7 Optimum Operating Temperatures

6.7.1 Introduction

For irreversible reactions, the rate of reaction increases with temperature at all levels of fractional conversion. Hence the greatest temperature which can be tolerated by the materials of construction and without the promotion of unwanted side reactions should be employed in order to maximise the reactor performance. The reactor may then be readily assessed by equations (6.7) or (6.8) using a fixed value of k for the selected temperature.

For endothermic reversible reactions the forward reaction rate constant increases with temperature and because the equilibrium constant also increases with temperature it will be seen from equation (6.43) below that the overall rate of reaction increases monotonously with temperature for any composition of reaction mixture. Hence as with irreversible reactions the maximum feasible temperature for the other practical considerations should be selected.

$$r = k_1\left[C_f - \frac{C_r}{K}\right] \tag{6.43}$$

where C_f is a function of the reactant concentrations
and C_r of the product concentrations.

The equilibrium constant for reversible exothermic reactions decreases with temperature and so the terms k_1 and $[C_f - (C_r/K)]$ of equation (6.43) change in opposite directions as temperature is changed at any mixture composition. For each possible composition there will therefore be a temperature at which the rate is a maximum.

For a liquid phase reaction of the form:

$$\mathrm{A} \rightleftharpoons \mathrm{B} \tag{6.44}$$

with first order kinetics operating with a feed of pure A at concentration a, equation (6.43) may be expressed as:

$$r = k_1 a\left[(1 - f) - \frac{f}{K}\right]. \tag{6.45}$$

The variation of rate with temperature is illustrated in Fig. 6.6 for both endothermic and exothermic reactions. In each case:

$$k_1 = 3 \times 10^7 \exp\left(-\frac{11\,600}{RT}\right)\text{min}^{-1} \tag{6.46a}$$

$$a = 1 \tag{6.46b}$$

$$f = 0{\cdot}5. \tag{6.46c}$$

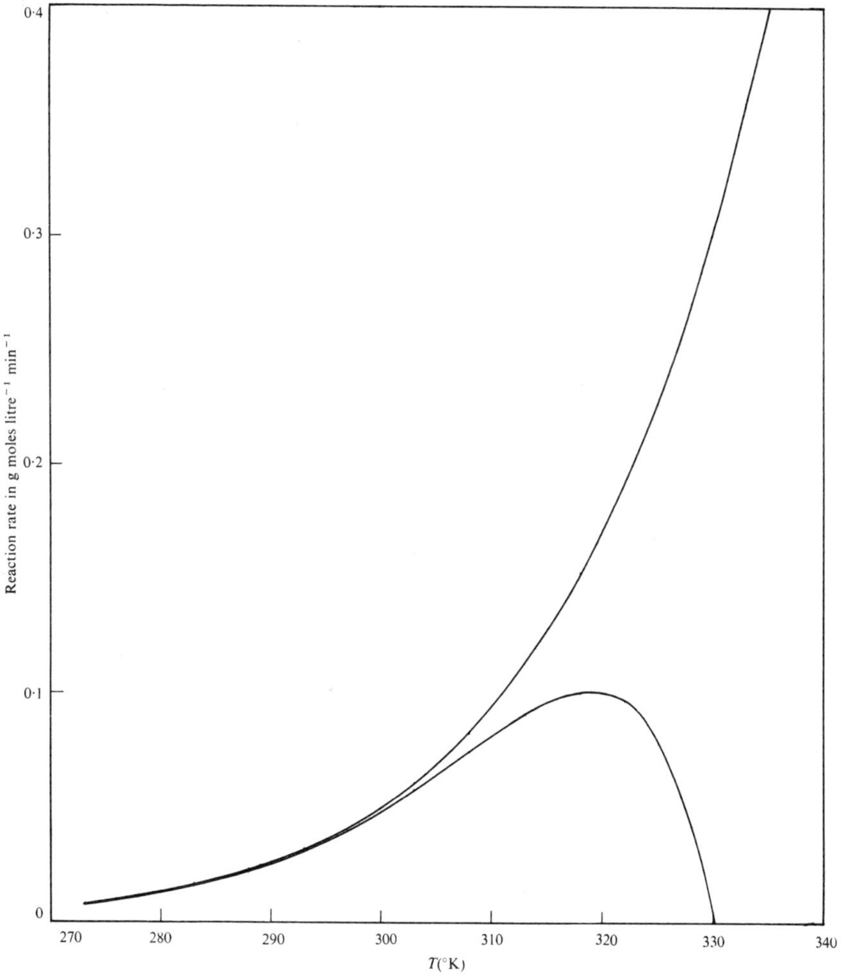

FIG. 6.6. Variation of overall reaction rate with temperature exothermic and endothermic reactions

The equilibrium constant was expressed by rearranging equation (2.68) as:

$$K = K_D \exp\left[-\frac{\Delta H}{R}\left(\frac{1}{T} - \frac{1}{T_D}\right)\right] \qquad (6{\cdot}47a)$$

where K_D is a known equilibrium constant at a datum temperature

T_D. For the curves in Fig. 6.6, the particular values of the constants are:

$$K_D = 19 \text{ at } T_D = 298 \tag{6.47b}$$

while $$\Delta H = +5000 \text{ for the endothermic reaction} \tag{6.47c}$$

and $$\Delta H = -18000 \text{ for the exothermic reaction.} \tag{6.47d}$$

6.7.2 Optimum temperature for isothermal operation

Considering a reversible first order reaction with a feedstock of A at concentration a and B at concentration b, equations (6.7) and (6.43) may be combined to give:

$$\frac{V}{Q} = \frac{a}{k_1}\int_0^f \frac{df}{a(1-f) - \dfrac{b+af}{K}} \tag{6.48}$$

and hence:

$$\frac{V}{Q} = \frac{1}{k_1\left(1+\dfrac{1}{K}\right)} \operatorname{Ln} \frac{a - \dfrac{b}{K}}{\left(a - \dfrac{b}{K}\right) - a\left(1+\dfrac{1}{K}\right)f} \tag{6.49}$$

from which:

$$f = \frac{Ka - b}{a(1+K)}\left[1 - \exp\left\{-\frac{k_1 V(1+K)}{QK}\right\}\right] \tag{6.50}$$

or if there is no product in the feedstream:

$$f = \frac{K}{1+K}\left[1 - \exp\left\{-\frac{k_1 V(1+K)}{QK}\right\}\right]. \tag{6.51}$$

If the reactor size and throughput are fixed, then the fractional conversion attainable will vary with the operating temperature.

From the data used in Fig. 6.6. for the exothermic reaction, plots of f versus temperature for values of $V/Q = 5$, 10 and 15 are shown in Fig. 6.7. This figure shows how the optimum temperature varies with mean residence time and also how the reactor yield may be enhanced by correct selection of temperature.

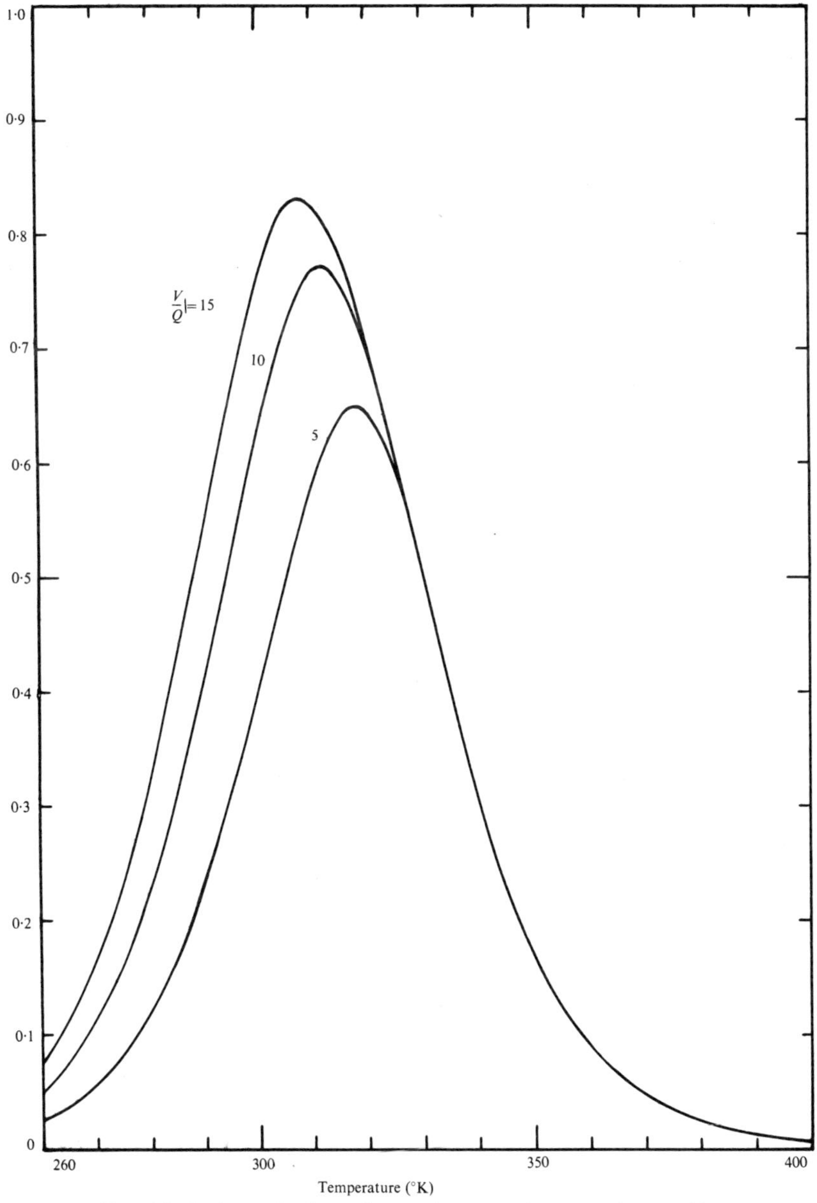

FIG. 6.7. Optimum isothermal operation temperature at given residence time

6.7.3 Optimum temperature profiles

It has been demonstrated above that for exothermic reactions there is a temperature at which the reaction rate is a maximum for each level of fractional conversion. The reaction rate may be maintained at a maximum throughout the course of the reaction provided the temperature is adjusted progressively along the length of the reactor to be optimal for the local conversion at each particular point.

First order kinetics will again be considered for the reaction:

$$A \rightleftharpoons B \tag{6.44}$$

For a feed of A at concentration a and B at b and a liquid phase reaction, the concentrations of A and B after conversion f are:

$$a(1-f) \quad \text{and} \quad b + af$$

and the reaction rate is:

$$r = k_1\left[a(1-f) - \frac{b+af}{K}\right]. \tag{6.52}$$

However, for a gas phase reaction at constant pressure the concentrations will depend upon the temperature T at the point under consideration and the concentrations of A and B will be:

$$a(1-f)\frac{T_0}{T} \quad \text{and} \quad (b+af)\frac{T_0}{T}$$

where T_0 is the inlet reactor temperature and the rate is then:

$$r = \frac{k_1 T_0}{T}\left[a(1-f) - \frac{b+af}{K}\right]. \tag{6.53}$$

On expressing k_1 and K as functions of temperature the turning point of equation (6·52) is found to occur at:

$$T = \frac{T_D}{1 + \dfrac{RT_D}{\Delta H}\operatorname{Ln}\dfrac{a(1-f)}{b+af}\cdot\dfrac{E_1 K_D}{E_1 - \Delta H}}. \tag{6.54}$$

The fractional conversion f may be extracted from this equation to give the explicit expression:

$$f = \frac{\dfrac{E_1 K_D}{E_1 - \Delta H} - \dfrac{b}{a}\exp\left[\dfrac{\Delta H}{R}\left(\dfrac{1}{T} - \dfrac{1}{T_D}\right)\right]}{\dfrac{E_1 K_D}{E_1 - \Delta H} + \exp\left[\dfrac{\Delta H}{R}\left(\dfrac{1}{T} - \dfrac{1}{T_D}\right)\right]}. \tag{6.55}$$

Similar treatment of equation (6·53) enables the relationship between f and T to be established when variations of concentration with temperature are important. Then:

$$\frac{b + af}{a(1 - f)} = K\left(\frac{RT - E_1}{RT - E_2}\right) = \frac{K_D(RT - E_1)}{(RT - E_2)\exp\left[\dfrac{\Delta H}{R}\left(\dfrac{1}{T} - \dfrac{1}{T_D}\right)\right]} \tag{6.56}$$

While it is possible to obtain an explicit expression for f from this equation, a corresponding expression for T cannot be extracted and an iteration procedure is required.

The stages in the specification of a plug flow reactor having optimum temperature profile are shown in Fig. 6.8 and are as follows:

(*i*) Establish the fractional conversion for which the inlet temperature is the optimum using equations such as (6.55) or (6.56). The reactor will operate isothermally at the inlet temperature until this fractional conversion is attained. Although this is not the optimum temperature with regard to reaction rate, it is assumed that the reactants enter at the maximum temperature permissible with respect to materials of construction and the avoidance of side reactions.

(*ii*) Use an equation for isothermal operation such as (6.7) to estimate V/Q for the fractional conversion calculated in stage (*i*).

(*iii*) Choose an increment of fractional conversion and evaluate the corresponding increments of V/Q using the trapezium rule applied to equation (6.22). The two values of rate should be those which are optimal at the inlet and outlet conversions of the particular increment under consideration, and are evaluated by first using equation (6.54) or an iteration on (6.56) to find the local optimum temperatures.

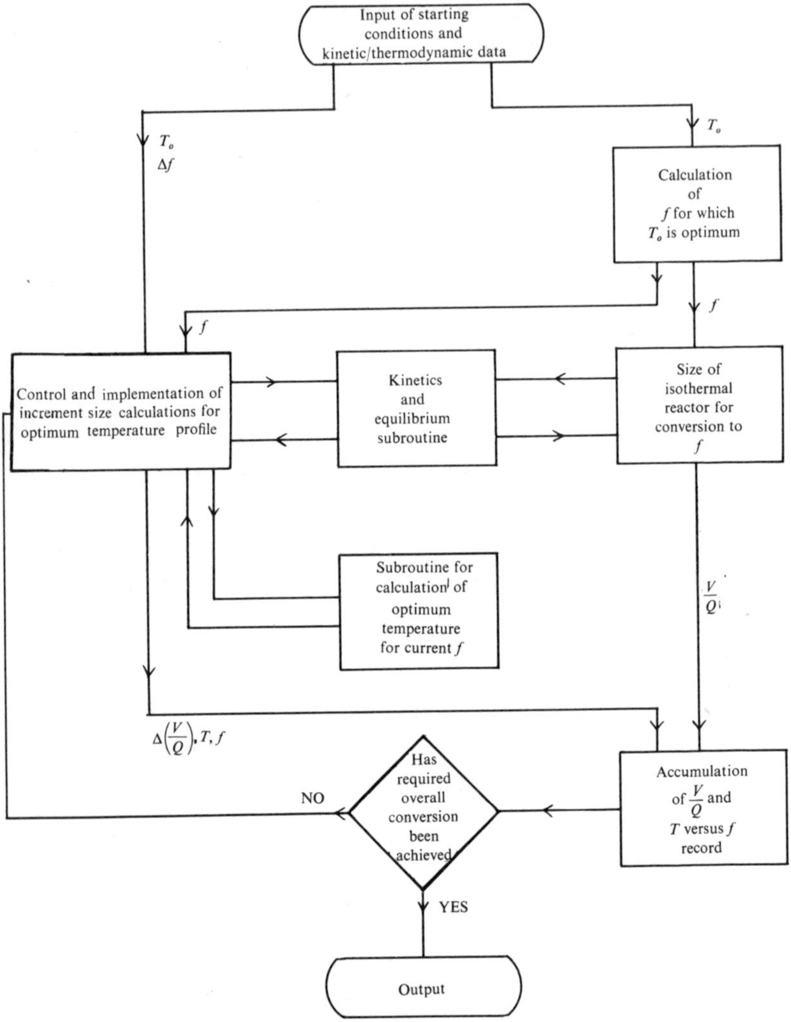

FIG. 6.8. Digital computer block diagram for calculation of optimum temperature profile

Applying the data of section 6.7.1 and the equations developed in the present section to a feed of pure A in a liquid phase reaction, it was found that a feed entering at 60°C will react to $f = 0{\cdot}233$ before the optimum temperature is below 60°C. The step by step calculation in increments of 0·01 was then started at $f = 0{\cdot}23$ with

$T = 333°K$ and evaluation of the optimum temperature $T = 332{\cdot}6°K$ for $f = 0{\cdot}24$. The results of these computations are summarised in Table 6.4 with groupings of 5% in the changes of fractional conversion over part of the range. The temperature profile is plotted in Fig. 6.9.

TABLE 6.4

f	$T°K$	V/Q	$\sum(V/Q)$
0·23	333·0	0·451	0·451
0·24	332·6	0·030	0·481
0·25	331·9	0·031	0·512
0·30	328·9	0·182	0·695
0·35	326·2	0·227	0·922
0·40	323·7	0·283	1·205
0·45	321·3	0·352	1·557
0·50	319·1	0·440	1·997
0·55	316·8	0·554	2·551
0·60	314·6	0·706	3·257
0·65	312·3	0·917	4·174
0·70	309·8	1·224	5·398
0·75	307·2	1·696	7·094
0·80	304·2	2·486	9·580
0·85	300·7	3·863	13·443
0·90	296·2	7·193	20·636
0·95	289·1	18·131	38·767
0·96	287·0	8·000	46·767
0·97	284·3	12·395	59·162
0·98	280·6	22·670	81·832
0·99	274·6	62·257	144·089

A further calculation was carried out in which the optimum temperature was allowed to be operative from 1% conversion and thereby assuming that temperatures in excess of 60°C would be acceptable. The progress of the initial part of the conversion until coincidence with the above results is given in Table 6.5 and the corresponding temperature profile is shown in Fig. 6.10. It is seen that in the initial stages of the conversion the rate is high whichever of the two temperature profiles is followed and the effect on the total reactor size needed for high conversions is very small.

The heat removal rate per unit length of reactor will vary considerably over the length of the reactor. In the isothermal region the heat removal is dependent upon the heat of reaction but in the falling temperature region the sensible heat removal from the reaction mixture also has to be taken into account. The relative

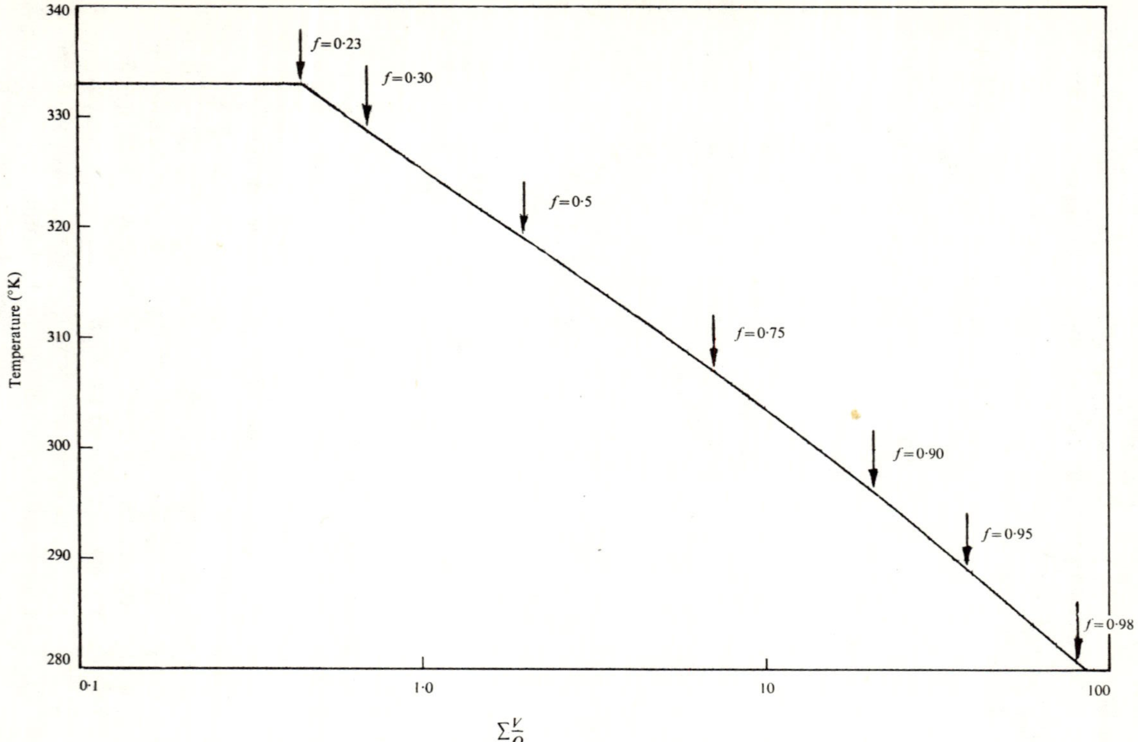

FIG. 6.9. Optimum temperature profile with initial temperature limitation

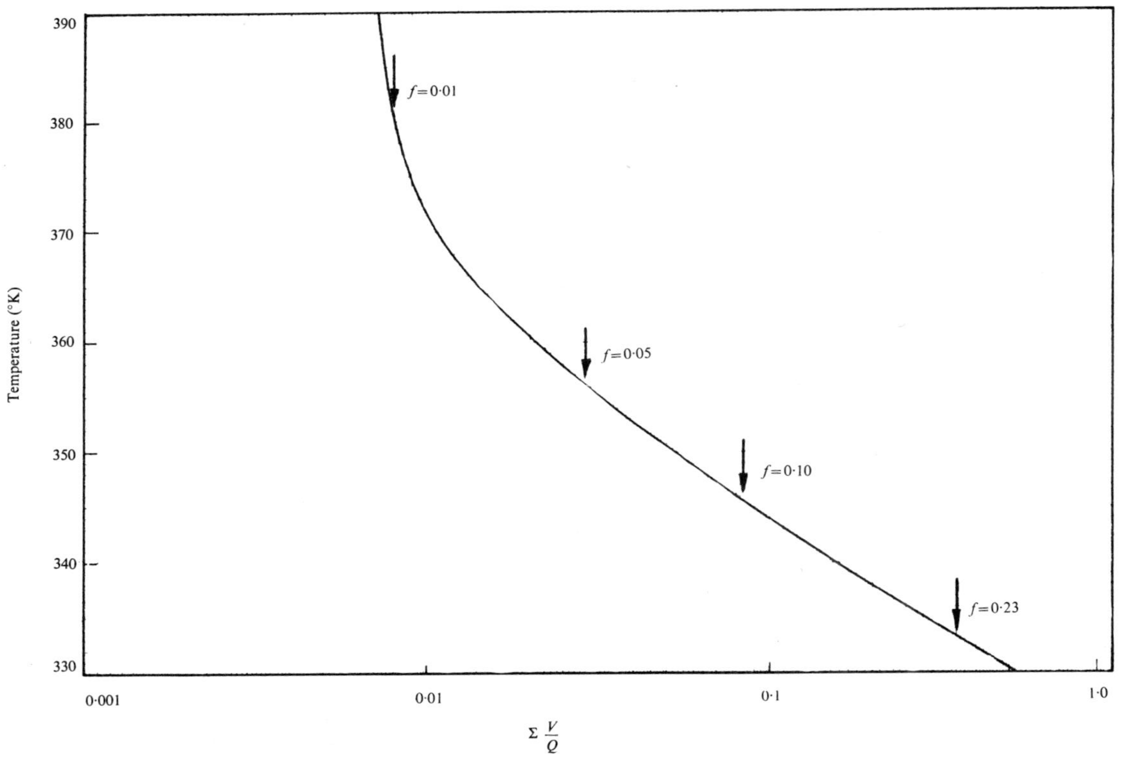

FIG. 6.10. Initial part of temperature profile without temperature limitation

TABLE 6.5

f	T°K	V/Q	$\sum(V/Q)$
0·01	380·6	0·008	0·008
0·02	369·7	0·003	0·011
0·03	363·5	0·005	0·016
0·04	359·3	0·006	0·022
0·05	356·0	0·007	0·029
0·10	345·8	0·053	0·082
0·15	339·8	0·079	0·161
0·20	335·4	0·110	0·271
0·23	333·2	0·081	0·352

rates of the heat removal for the above reaction have been calculated on the assumption that each mole of total reactant and product is in solution at a concentration of 0·002 g mole cm^{-3} and that the thermal properties of the solution are those of water.

For the isothermal region of the reactor, a mass balance over an increment gives:

$$kc_0\left[(1-f)-\frac{f}{K}\right]A\,dl = Qc_0\,df \tag{6.57}$$

where A is the reactor cross-sectional area, and Q the throughput of solution.

The rate of heat generation per unit length of reactor H_g at any position is:

$$H_g = Qc_0\Delta H\frac{df}{dl} \tag{6.58}$$

which on combining with equation (6.57) gives:

$$H_g = Akc_0\Delta H\left[(1-f)-\frac{f}{K}\right]. \tag{6.59}$$

For the region with a falling temperature profile, the mass balance over a finite section having r_1 and r_2 as the terminal reaction rates is:

$$\frac{A\Delta l}{Q} = \frac{\Delta f}{2}\left[\frac{1}{r_1}+\frac{1}{r_2}\right] \tag{6.60}$$

and hence the rate of heat generation per unit length of reactor is:

$$H_g = 2Ac_0\,\Delta H\left[\frac{1}{r_1}+\frac{1}{r_2}\right]^{-1} \tag{6.61}$$

while the sensible heat load per unit length H_s is:

$$H_s = Q\rho C_p \frac{\Delta T}{\Delta l}. \tag{6.62}$$

Numerical values from equations (6.59), (6.61) and (6.62) are presented in Table 6.6 for a reactor of diameter 10 cm, a throughput of 60 litre min^{-1} and the thermodynamic data from section 6.7.1. For these values, the length of reactor in metres is also specified in the final column of the table.

TABLE 6.6

f	T°K	H_g cal cm^{-1} min^{-1}	H_s cal cm^{-1} min^{-1}	Length (metres)
0	333	2042	—	0
0·05	333	1809	—	0·556
0·10	333	1575	—	1·205
0·15	333	1342	—	1·94
0·20	333	1108	—	2·83
0·23	333	968	—	3·5
0·23–0·24	332·8	945	1165	3·7
0·295–0·305	328·9	698	1102	5·3
0·395–0·405	323·7	449·7	601·9	9·2
0·495–0·505	319·1	288·5	360·2	15·2
0·595–0·605	314·6	177·8	224·7	24·8
0·695–0·705	309·8	106·8	147·3	41·1
0·795–0·805	304·2	47·2	83·7	72·5
0·895–0·905	296·2	13·9	41·6	156·4
0·98 –0·99	280·6–274·6	0·45	7·53	1101·2

Problems

1. By consideration of an irreversible first order reaction show that the hold-up in a single well mixed stage would be nearly 2·5 times that in a plug flow system operating at 80% conversion for the same throughput.

2. A first order non-reversible reaction (velocity constant 1 min^{-1}) is to be effected with a throughput of 10000 ft^3 h^{-1}. Two continuous reactors are available, the one (50 ft^3 hold-up) being suitable for plug flow operation and the other (200 ft^3 hold-up) for well mixed conditions. Compare the obtainable conversions (*a*) if the reactors are operated in series and (*b*) if the flow is equally divided between the reactors operating in parallel.

ANS. (*a*) $f = 0{\cdot}664$ (*b*) $f = 0{\cdot}579$

3. The flow through a plug-flow reactor effecting a first order forward reaction is increased by 20% and in order to maintain the fractional conversion at its former value it is decided to increase the reactor operating temperature. If the reaction has an activation energy of 4 kcal (g mole)$^{-1}$ and the initial temperature is 150°C, find the new operating temperature. Would the required elevation of temperature be different if the reactor were well mixed?

ANS. 169°C; No.

4. Butene is to be thermally cracked according to the reaction:

$$C_4H_8 \rightarrow C_4H_6 + H_2$$

for which the kinetics are first order with respect to butene and the reverse reaction may be neglected, i.e. having:

$$r = kp_{C_4H_8} \text{ lb moles cracked h}^{-1} \text{ litre}^{-1}$$

where $p_{C_4H_8}$ is the partial pressure of butene.
The feed consists of n moles of steam per mole of butene. Show that for a fractional conversion 'f' of N lb mole hr^{-1} of butene at an operating pressure P, the required plug flow reactor volume is:

$$-\frac{N}{kP}\left[f + (2 + n)\,\text{Ln}\,(1 - f)\right] \text{ litres}$$

5. A plug flow isothermal reactor is effecting the exothermic liquid phase reaction:

$$A + B \underset{k_2}{\overset{k_1}{\rightleftharpoons}} C + D$$

The feed contains only reactants A and B each at concentration c_0 and the kinetics are given by

$$k_1 c_A c_B - k_2 c_C c_D.$$

Show that the volume of reactor for a volumetric throughput Q and conversion f is given by:

$$\frac{K^{\frac{1}{2}}Q}{2k_1 c_0}\log_e \frac{K^{\frac{1}{2}}(f - 1) - f}{K^{\frac{1}{2}}(f - 1) + f}$$

where K is the equilibrium constant.

Hence, if the temperature dependance of k_1 and K may be represented by:

$$K = A \exp\left(-\Delta H/RT\right)$$

and $$k_1 = B\exp(-\Delta E/RT)$$

show that for a specified conversion and throughput there is an optimum size of reactor when the isothermal operating temperature is governed by the following equation:

$$\exp(\Delta H/2RT)\frac{2\Delta E - \Delta H}{2A^{\frac{1}{2}}\Delta H}\log_e\frac{A^{\frac{1}{2}}(f-1)\exp(-\Delta H/2RT) - f}{A^{\frac{1}{2}}(f-1)\exp(-\Delta H/2RT) + f} = \frac{f(f-1)}{A(f-1)^2\exp(-\Delta H/RT) - f^2}$$

6. A plug flow reactor operating isothermally at pressure P atm is to effect the gas phase reaction:

$$a\mathrm{A} + bB \rightarrow cC$$

for which the kinetics may be represented by:

$$r = kp_{\mathrm{A}}p_{\mathrm{B}} \text{ moles of A converted ft}^{-3}\,\mathrm{h}^{-1}.$$

The reactor feed contains only A and B with B in excess at a molar ratio of B:A = n.

Develop an expression for the size of reactor needed for a given fractional conversion of A for a feed F mole h^{-1} of mixture.

ANS. $$V = \frac{F}{k(n+1)P^2}\int_0^f \frac{[(1+n) + f(q-p-1)]^2}{(1-f)(n-pf)}\,df$$

where $p = \dfrac{b}{a}$ and $q = \dfrac{c}{a}$

7. It is proposed to carry out to within 90% of equilibrium, a gas phase reaction:

$$\mathrm{A} \rightleftarrows \mathrm{B},$$

which is first order and unimolecular, by a single passage through a plug flow reactor at 500°C and 1 atm pressure. The pure reactant A enters at 350°C and the available data is as follows:

$$\Delta H^\circ_{298} = -20\text{ kcal (g mole)}^{-1} \text{ and independent of } T$$

$$\Delta G^\circ_{298} = -13\text{ kcal (g mole)}^{-1}.$$

Forward velocity constant is 0·111 h^{-1} at 300°C with an activation

energy $+30$ kcal mole^{-1}. The relevant mean specific heats of A and B are 8·3 and 12·4 Btu (lb mole)$^{-1}$ degF^{-1} respectively.
What are the implications of this proposal?

ANS. Time 63·6 sec.; 21 800 Btu to be removed per lb mole of A fed.

8. An organic compound undergoes pyrolysis by passing it in plug flow through a continuous heated tube in a furnace. It is anticipated that this reaction which is endothermic will be carried out isothermally at 675°C. The rate constant (sec^{-1}) is given by:

$$\mathrm{Ln}\, k = 34{\cdot}34 - \frac{68\,000}{RT}$$

when T is the reaction temperature °K.

The furnace temperature may be taken as 1000°C and the overall heat transfer coefficient as 6 Btu h^{-1} ft^{-2} degF^{-1}. The heat load on the furnace is 2000000 Btu h^{-1} and the economic pass yield is 30%. If the reaction is of the first order calculate the dimensions of the reaction tube if the volumetric throughput measured at reaction conditions is 150 ft^3 sec^{-1}.

ANS. $D = 2{\cdot}545$ ft, $L = 71{\cdot}1$ ft

9. The decomposition of phosphine takes place according to the stoichiometric equation:

$$4PH_3 \rightarrow P_4 + 6H_2$$

The reaction is irreversible and endothermic. It is first order, the rate being proportional to the phosphine concentration and k in sec^{-1} varies with temperature thus:

$$\log_{10} k = -\frac{18963}{T} + 2\log_{10} T + 12{\cdot}13.$$

It is proposed to produce phosphorus by the decomposition of phosphine at a feed rate of 100 lb h^{-1} in a tubular reactor operating at atmospheric pressure. The highest temperature that can be used in the available material of construction is 680°C, at which temperature the phosphorus is a vapour.

Estimate the conversion expected in a tubular reactor of volume 100 ft^3 operating

(*a*) Isothermally at 680°C.
(*b*) Adiabatically with an inlet temperature of 680°C.

Heat of reaction = 5665 cal (g mole)$^{-1}$ phosphine
Mean $C_p\,(P_4)$ = 14·9 cal degC^{-1} (g mole)$^{-1}$
Mean $C_p\,(PH_3)$ = 12·6 cal degC^{-1} (g mole)$^{-1}$
Mean $C_p\,(H_2)$ = 7·2 cal degC^{-1} (g mole)$^{-1}$

ANS. $f = 0{\cdot}68$ $\quad f = 0{\cdot}13$

10. The chemical reaction:

$$C + CO_2 \rightarrow 2CO$$

is to be effected in a plug flow reactor by a reaction whose rate (lb moles CO_2 converted per hour per ft^3 of coke) is given by the expression:

$$r = \frac{6 \times 10^8\, p_{CO_2} \exp\left(\dfrac{-24\,700}{T}\right)}{1 + 3{\cdot}6\, p_{CO} + 0{\cdot}8\, p_{CO_2}}.$$

The reaction is endothermic and heat is to be supplied through the reactor wall from a heating medium whose temperature will remain constant. The feed gas stream is pure CO_2 at a rate of M moles h^{-1} and the reaction is at essentially atmospheric pressure. Formulate mass and energy balances for the estimation of conversion and temperature change over an increment of reactor length. Hence prepare a flow diagram for a digital computer programme for the calculation of one such increment to an accuracy of 0·2% on the fractional conversion and 0·1% on the absolute temperature at the end of the increment.

11. It has been shown in example (6) that 167 ft of 4 in diameter tube would be suitable for the 50% conversion under laminar flow conditions of a throughput of 0·365 ft^3 min^{-1} containing 0·0018 g mole cm^{-3} of reactant. For the given velocity constant of 17·4 cm^3 g mole^{-1} min^{-1} evaluate the point fractional conversions:

(i) on the tube axis at 83·5 ft and 167 ft.
(ii) at the reactor exit at radii 0·5, 1·0 and 1·5 in.
(iii) at the coordinate l = 83·5 ft, r = 1 in.

REFERENCES

ECHOLS, L. S. and PEASE, R. N. 1939. Kinetics of the decomposition of n-butane, I. Normal decomposition. *J. Am. chem. Soc.*, **61**, 208.

HOUGEN, O. A., and WATSON, K. M. 1947. *Chemical process principles. Part* 3. *Kinetics and catalysis*. Wiley, New York.

LEYES, C. E., and OTHMER, D. F., 1945. Esterification of butanol and acetic acid. *Ind. Engng Chem.*, **37**, 968.

7

FLOW CHARACTERISTICS AND THEIR EFFECTS ON THE PERFORMANCE OF CONTINUOUS REACTORS

7.1 Introduction

In the preceding chapters, the analysis of two specific types of continuous reactor has been considered. These extreme types are:

(*i*) The continuous stirred tank reactor in which the feed is assumed to mix completely with the reactor contents immediately on entry; and

(*ii*) The plug flow tubular reactor in which no longitudinal mixing of fluid occurs during passage through the reactor.

Thus in the former, complete mixing of feed and contents is encouraged while in the latter, no mixing at all is desired. In actual operation neither of these conditions is ever realised, although in many cases the operating conditions are such that one of them may be an acceptable assumption for analytical or design purposes. When the idealised cases are not applicable it is necessary to be able to allow for both minor and major deviations. Thus in order to make a precise analysis of a reactor, a detailed knowledge of the flow characteristics is required. One approach is the establishment of the distribution of residence times of fluid elements in their passage through the reactor. A method which is often used to treat turbulent flow conditions in tubular reactors having some deviation from plug flow, represents the backmixing by a longitudinal diffusion coefficient. Yet another example arises from the analysis of laminar flow reactors (sections 6.6 and 7.5) in which the reaction kinetics are combined with the known velocity profile.

Reactors having basically well stirred reactor geometry have been considered to have deviations from ideal behaviour caused by some reactant short-circuiting the reaction zone and further reactant passing through a plug flow region either in series or parallel with the well mixed region. Complete mixing has been shown to be a reasonable assumption in the majority of circumstances and has been confirmed by the experimental results of Manning and Wilhelm (1963) and by Rice, Toor and Manning (1964).

These workers found that with high speed turbine agitators two streams of different composition were completely mixed during the time that it took the liquids to pass through the impeller, i.e. in less than one second. However, away from the impeller, mixing is not so intense and there is evidence that at low agitator speeds there is some segregation. Thus McDonald and Piret (1951) using a colorimeter method found that the time required to attain complete mixing in a C.S.T.R. agitated by a two bladed paddle operating at more than 20 r.p.m. was one minute for all holding times greater than 15 minutes. They showed that when the nominal holding time exceeded 30 minutes the agitation by the paddle was the predominant factor affecting the rate of mixing, but when the holding time was less than 5 minutes the feed flow rate had a greater effect than that of the agitator. However it should be stated that the two bladed paddle is not the ideal agitator for a stirred tank reactor and therefore in a well designed reactor the effects of poor mixing should be small. Hence the assumption of perfect mixing is valid in most cases, but when this is not the case, analysis of imperfect mixing may be carried out by application of the methods proposed by Cholette and Cloutier (1959); these are treated in sections 7.3.5 to 7.3.9 for the detection of imperfections and in section 7.6 for the estimation of reactor performance when such imperfections exist.

The following section shows the range of reactor performance between the extreme conditions of plug flow and complete mixing and hence the range of uncertainty in prediction of reactor performance in the absence of knowledge of the prevailing mixing conditions.

The remainder of the chapter considers in detail the detection of mixing patterns, followed by the combination with kinetics to give reactor design equations.

7.2 Effects of Backmixing in Tubular Reactors

The effects of backmixing on the performance of real reactors can be considered by comparing the sizes of a C.S.T.R. and the corresponding plug flow reactor required to effect a specified reactor performance. Thus for the irreversible nth order reaction:

$$r = k_n c^n{}_A = k(c_{A0})^n (1 - f)^n$$

taking place in a plug flow tubular reactor, the volume of reactor 'V_0' required for the conversion of 'f' is found from:

$$V_0 = \frac{Q}{(c_{A0})^{n-1}} \int_0^f \frac{df}{(1-f)^n}. \tag{7.1}$$

If the same reaction were carried out in a completely mixed reactor, the volume 'V_∞' required for the same conversion would be:

$$V_\infty = \frac{Qc_{A0}f}{r} = \frac{Qc_{A0}f}{kc_{A0}^n(1-f)^n}. \tag{7.2}$$

Comparison of V_0 and V_∞ can be made for the cases of $n = 1$ and $n \neq 1$ as follows:

(*a*) $n = 1$

$$\frac{V_\infty}{V_0} = -\left(\frac{f}{1-f}\right)\left[\frac{1}{\mathrm{Ln}\,(1-f)}\right] \tag{7.3}$$

(*b*) $n \neq 1$

$$\frac{V_\infty}{V_0} = \frac{f}{(1-f)^n}\left[\frac{(n-1)}{\dfrac{1}{(1-f)^{n-1}} - 1}\right] \tag{7.4}$$

or

$$\frac{V_\infty}{V_0} = \frac{(n-1)f}{(1-f)\,[1-(1-f)^{n-1}]}. \tag{7.5}$$

For the special case of a zero order reaction, equation (7.5) leads to the result that $V_\infty = V_0$ so that either type of reactor could be employed equally well whatever the required conversion. For cases where $n > 0$, both equations (7.3) and (7.5) show that the ratio V_∞/V_0 exceeds unity and depends on the conversion level.

Hence, since in all tubular reactors there will be a certain amount of backmixing, the actual volume of a tubular reactor will be larger than that predicted by the assumption of plug flow conditions but not as large as that predicted for complete mixing. Consequently it is necessary to be able to establish the reactor flow characteristics and to predict their effects on its performance in order to obtain a reactor of correct size for the duty envisaged. Corrections of the kind considered here are due to variation in the residence times of different elements of the feed in the reactor and should not be

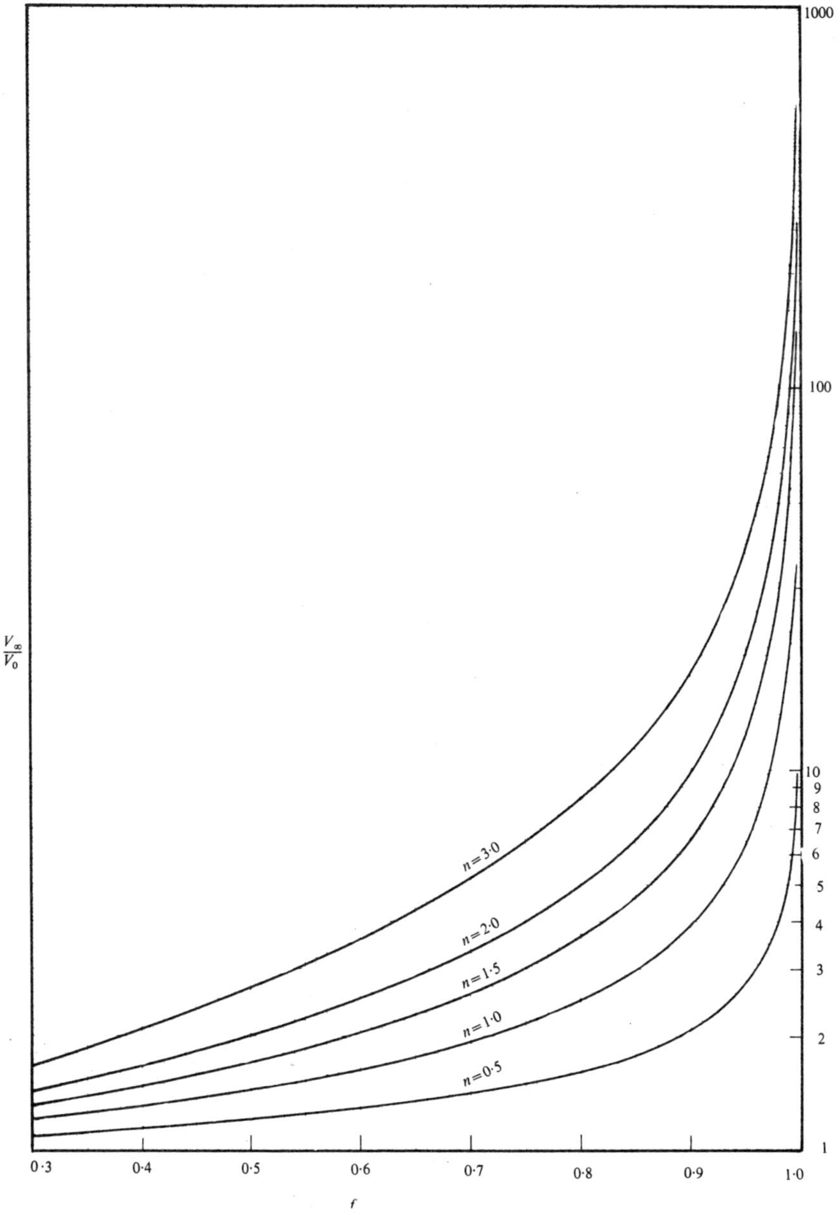

FIG. 7.1. Comparison of C.S.T.R. and plug-flow volumes

confused with the effects of incomplete mixing of the different reactants within a single feed stream. This latter problem was considered in section 7.1 and was shown to be small in the majority of operating conditions encountered in practice. The magnitude of the problem considered here can be assessed by considering equations (7.3) and (7.5) for different conversions for half, first, second and third order reactions. The results are summarised in Fig. 7.1. There it can be seen that for a first order reaction, the volume of an ideal C.S.T.R. would be 1·94 times the volume of a plug flow reactor at 70% conversion. The volume ratio is 3·33 for a second order reaction at the same conversion. At higher conversions, approaching unity, the volume ratio increases very steeply, and may easily reach values in the range 10–100.

7.3 Residence Time Distribution Analysis

In the above section it was shown that the extent of backmixing has a marked effect on the necessary size of reactor for a specified duty.

In the case of a plug flow reactor, all fluid elements spend the same time interval in the reactor and there is therefore no distribution or range of residence times. However, with a well mixed reactor, some fluid will leave almost instantly after entering the reactor and will have a zero residence time, while some further fluid will never be discharged from the reactor and will have an infinite residence time; there will be elements of fluid with the whole range of residence times between these extremes. A practical reactor with some degree of backmixing will have a range of residence times which will not include either zero or infinity. The form of the residence time distribution curve may therefore be used to characterise the reactor and may be combined with first order kinetics to predict reactor performance; for the treatment of other kinetics Zwietering (1959) should be consulted. The residence time distribution of material at the reactor exit can be expressed both as:

(*i*) a point distribution function (E-diagram), and
(*ii*) a cumulative distribution function (F-diagram).

As a point distribution function the fraction of material in the reactor exit stream which has spent a time between t and $(t + dt)$ within the reactor is $E(t)\,dt$ and is termed the external age distribution function. As a cumulative distribution function, the total

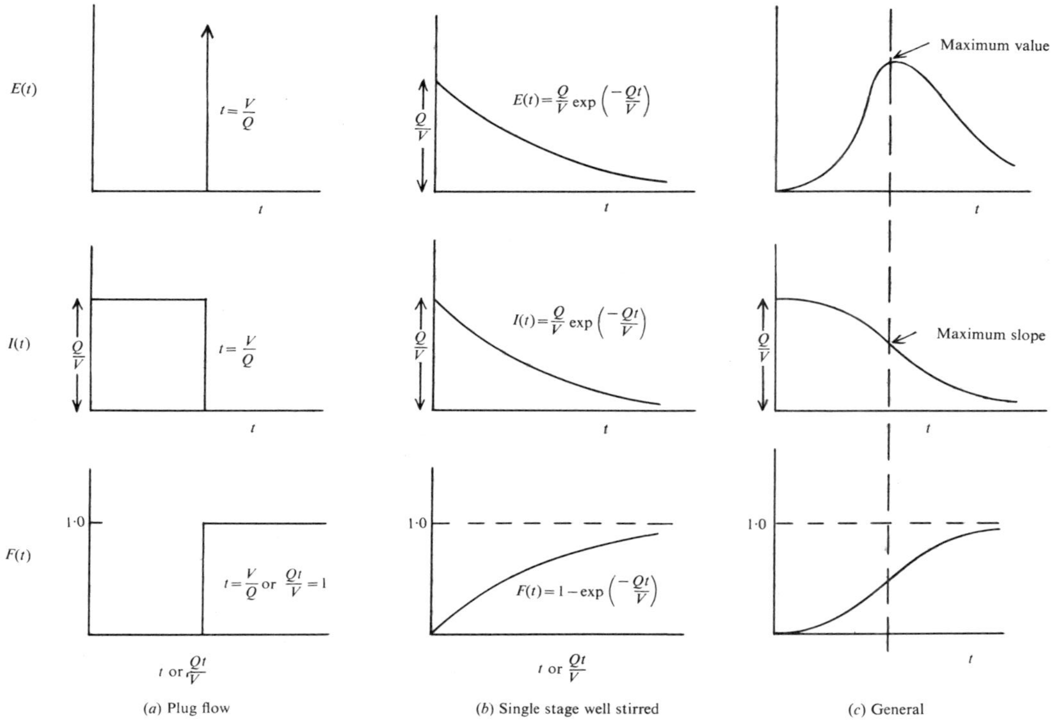

FIG. 7.2. A comparison of $E(t)$, $I(t)$, and $F(t)$ diagrams for plug flow, single well stirred and general systems

material which has spent a time less than t is considered and is therefore deduced by integration of $E(t)\,dt$ over the time interval $0 - t$,

i.e.
$$F(t) = \int_0^t E(t)\,dt. \tag{7.6}$$

In addition to the residence time distribution of material in the reactor exit stream, it is possible to define a distribution of residence times for which the material currently in the reactor has been present there. Thus the fraction of material which has been in the reactor for a time since entry between t and $(t + dt)$ is $I(t)\,dt$ and is termed the internal age distribution function.

It follows from the definitions of $E(t)$, $I(t)$ and $F(t)$ that:

$$\int_0^\infty E(t)\,dt = 1 \tag{7.7}$$

$$\int_0^\infty I(t)\,dt = 1 \tag{7.8}$$

and
$$F(t) = 1 \text{ at } t = \infty. \tag{7.9}$$

The I and E diagrams are not in general identical, and the distinction may be clarified by consideration of Fig. 7.2 in which the corresponding E, I and F diagrams are shown for plug flow, complete mixing of a single stage and an intermediate pattern. Considering the plug flow situation, there is no exit age distribution and the E diagram has an infinitely large ordinate applicable for an infinitely short time and such that the area is unity, equation (7.7). In contrast the material within the vessel will have equal fractions of all ages between zero and the mean residence time, but no material whose age exceeds the mean residence time. The F diagram obtained by integration of the E diagram has a zero value until the mean residence time has elapsed, reaches unity at this value of time and thereafter continues at unity. The identical E and I diagrams obtained for the single well stirred reactor follow from the fact that the exit stream is, by definition, a typical sample of the vessel contents.

The residence time characteristics are established by stimulus response techniques. The stimulus is usually a change in concentration of a solute or tracer material in the feed stream to the system. If the form of the stimulus is an impulse, then the observed response at the system exit is called a C-diagram and will yield the E-diagram on adjustment of the ordinate. A step change stimulus is much more commonly employed than the impulse, and the exit response yields the F-diagram. Impulse testing will be considered briefly in the following section and then step changes will be treated in greater detail.

7.3.1 Impulse testing and the C-diagram

Consider a flow system with a steady volumetric throughput Q. At time $t = 0$, let a mass m of a tracer be injected instantaneously into the feed stream. The tracer could, for example, be a solution of an easily detectable salt or a solution of a compound containing a radioactive isotope, but must not be so large in quantity that the volumetric throughput is disturbed. The concentration $c(t)$ of tracer is observed at the system outlet for a series of time intervals. These data are frequently plotted with $(V/m)c(t)$ as ordinate against dimensionless time Qt/V to give a C-diagram.

Applying the definition of $E(t)$ to the mass m of tracer, a material balance for a time interval dt at the system outlet gives:

$$Qc(t)\,dt = mE(t)\,dt \tag{7.10}$$

and hence

$$E(t) = \frac{Q}{m}c(t). \tag{7.11a}$$

Alternatively $E(t)$ may be plotted as a function of dimensionless time from the C-diagram. Thus if the C-diagram ordinates are represented by $C(t)$:

$$E(t) = \frac{Q}{V}C(t) \tag{7.11b}$$

The form of the F-diagram may be developed from equation (7.6) as:

$$F(t) = \frac{Q}{m}\int_0^t c(t)\,dt. \tag{7.12a}$$

Using the C-diagram:

$$F(t) = \frac{Q}{V}\int_0^t C(t)\,dt$$

$$= \int_0^{\frac{Qt}{V}} C(t)\,d\left(\frac{Qt}{V}\right). \tag{7.12b}$$

The F-diagram may be developed either by integration of the observed exit concentrations with respect to time using equation (7.12a) or by integration of the derived $(V/m)\,c(t)$ values with respect to dimensionless time using equation (7.12b).

The determination of the residence time distribution based on an impulse stimulus can be satisfactory if there is a significant spread of residence times; this is illustrated in Example 1 below. If, however, the system is operating near to plug flow conditions, all of the tracer will leave the system over a short period of time and may not be detected in any of the samples taken. For such situations the step change stimulus is more satisfactory.

Example 1. A vessel of volume 120 ft^3 and operating with a throughput of 8 ft^{3}min^{-1} has a solution containing 800 g of salt added rapidly to the inlet stream. The salt concentration at exit was found to be:

time (min)	0	5	10	15	20	25	30	35
Salt g ft^{-3}	0	3	5	5	4	2	1	0

Before developing the E- and F-diagrams, it is of practical importance to check that the total injected salt may be accounted for in the exit stream. This can be achieved by integration of the left-hand side of equation (7.10). Using Simpson's rule (taking a dummy ordinate at $t = 40$ min), the total effluent salt is:

$$8[\tfrac{5}{3}\{0 + 2(5 + 4 + 1) + 4(3 + 5 + 2 + 0)\}] = 800 \text{ g}.$$

Using $Q/m = 0{\cdot}01$, the following table is developed for the E- and F-diagram ordinates:

t	$c(t)$	$E(t)$	$\int_{t_1}^{t_2} E(t)\,dt$	$F(t)$
0	0	0		0
			0·075	
5	3	0·03		0·075
			0·20	
10	5	0·05		0·275
			0·25	
15	5	0·05		0·525
			0·225	
20	4	0·04		0·750
			0·15	
25	2	0·02		0·900
			0·075	
30	1	0·01		0·975
			0·025	
35	0	0		1·000

The fourth column shows the contribution made during each time interval to the F-diagram, and column five shows the accumulation of of these contributions up to each time observation.

The forms of the C-, E- and F-diagrams for a well stirred vessel will now be developed by consideration of an impulse stimulus. For a single well stirred vessel the transient equation on introduction of tracer is:

$$Qc + V\frac{dc}{dt} = Qc_0. \tag{7.13}$$

In this instance, the product Qc_0 will be the mass 'm' of tracer introduced as an impulse and will therefore have the transform 'm'. Hence for zero tracer in the vessel at time $t = 0$, the transform of equation (7.13) is:

$$Q\bar{c}(s) + Vs\bar{c}(s) = m. \tag{7.14}$$

Hence

$$\bar{c}(s) = \frac{m}{V[s + (Q/V)]} \tag{7.15}$$

and inversion gives:

$$c(t) = \frac{m}{V}\mathrm{e}^{-Qt/V}. \tag{7.16}$$

The c-diagram for a single well mixed reactor is thus an exponentially decreasing function of time. Using equations (7.11) and (7.12) the E- and F-diagrams are:

$$E(t) = \frac{Q}{V}\mathrm{e}^{-Qt/V} \tag{7.17}$$

and

$$F(t) = 1 - \mathrm{e}^{-Qt/V}. \tag{7.18}$$

It is useful at this stage to generalise equation (7.15) for the transmission of an impulse 'm' through a system whose transfer function is $G(s)$. Thus

$$\bar{c}(s) = mG(s) \tag{7.19}$$

and again using equations (7.11) and (7.12):

$$E(s) = Q\,G(s) \tag{7.20}$$

$$E(t) = Q\mathscr{L}^{-1}\,G(s) \tag{7.21}$$

$$F(s) = \frac{Q\,G(s)}{s} \tag{7.22}$$

and
$$F(t) = Q\int_0^t [\mathscr{L}^{-1}\, G(s)]\, dt. \tag{7.23}$$

These latter equations give the relationship between the E- and F-diagrams and the system transfer function, and are not dependent upon their development by consideration of the transmission of an impulse.

7.3.2 Step disturbances and the F-diagram

A step change of a tracer would be imposed by changing from one steady rate of input Q containing no tracer to a second steady input Q containing a concentration c_0 of tracer; Q and c_0 are then maintained at these values throughout the observation period while the concentration c_1 of tracer is observed at the exit. It is demonstrated below that the ratio c_1/c_0 gives the ordinate of $F(t)$ which is normally plotted against dimensionless time Qt/V. Thus considering the mass balance an imposition of an upwards step disturbance in concentration c_0 at fixed flow Q to a system having a general transfer function $G(s)$, the transform of the outlet concentration c_1 is given by:

$$\bar{c}_1(s) = \frac{Qc_0}{s} G(s) \tag{7.24}$$

or
$$\frac{\bar{c}_1(s)}{c_0} = Q\frac{G(s)}{s} \tag{7.25}$$

and hence on comparison with equation (7.22) it is seen that:

$$F(s) = \frac{\bar{c}_1(s)}{c_0} \quad \text{or} \quad F(t) = \frac{c_1}{c_0}. \tag{7.26}$$

As time progresses the ratio c_1/c_0 approaches unity and normally reaches this value after the elapse of a few mean residence times. For mixing patterns near to plug flow the outlet concentration will change from zero to c_0 over a short period of time and the full F-diagram is readily obtained. For systems having a wide distribution of residence times and hence requiring a lengthy experiment to complete the F-diagram, geometrical relationships which exist between portions of any F-diagram may be utilised to estimate the final part of the diagram. Such geometrical features are now developed along with the connection between the F- and I-diagrams.

Consider the material balance of tracer up to a time 't' after the imposition of the step of magnitude c_0 in the tracer input concentration. The total tracer input in this time is:

$$Qc_0t.$$

At an instant of time t_2 the rate at which tracer material is leaving is:

$$Qc_0 \int_0^{t_2} E(t_1)\, dt_1$$

and since all times up to t must be considered, the total tracer which has left the system is:

$$Qc_0 \int_{t_2=0}^{t} \left[\int_{t_1=0}^{t_2} E(t_1)\, dt_1 \right] dt_2.$$

The accumulated tracer in the system is:

$$Vc_0 \int_0^{t} I(t_1)\, dt_1$$

and so the material balance is:

$$Qc_0t = Qc_0 \int_{t_2=0}^{t} \left[\int_{t_1=0}^{t_2} E(t_1)\, dt_1 \right] dt_2 + Vc_0 \int_0^{t} I(t_1)\, dt_1. \quad (7.27)$$

Differentiation with respect to t gives:

$$Qc_0 = Qc_0 \int_0^{t} E(t_1)\, dt_1 + Vc_0 I(t) \quad (7.28)$$

and from the definition of $F(t)$, equation (7.28) gives:

$$1 = F(t) + \frac{V}{Q} I(t). \quad (7.29)$$

Equation (7.29) affords a means of deducing an internal age distribution from the F-diagram. It may be developed as follows to give a useful geometrical result:

Since $$\int_0^{\infty} I(t)\, dt = 1 \quad (7.8)$$

equation (7.29) may be re-arranged to give:

$$\frac{Q}{V} \int_0^{\infty} [1 - F(t)]\, dt = 1. \quad (7.30)$$

The F-diagram is plotted on a dimensionless time scale Qt/V and hence the usual integral is obtained from equation (7.30) by introducing the constant Q/V into the integral,

i.e.
$$\int_0^\infty [1 - F(t)]\, d\left[\frac{Qt}{V}\right] = 1. \tag{7.31}$$

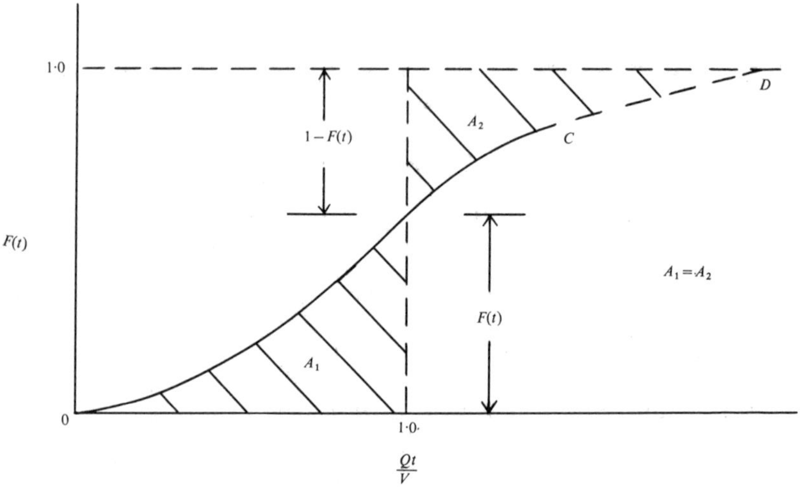

FIG. 7.3. Geometrical check on $F(t)$ diagram

Geometrically this expression means that the area between the F-curve (Fig. 7.3) and the line $F(t) = 1$ is unity. Since the region bounded by the axes and the lines $F(t) = 1$ and $Qt/V = 1$ is also unity, it follows that the two shaded areas A_1 and A_2 must be equal. Thus an F-diagram might be completed approximately by arranging for A_2 to have the same area as A_1 and noting that $F(t)$ must increase monotonously with time. A diagram might be completed in this way by construction of the linear portion CD to have the necessary additional area.

The prediction of reactor performance by a combination of kinetics and the results of residence time distribution experiments may utilise the latter directly in their graphical form or as an analytical expression after fitting to one of several mathematical models. The following sections consider the responses to be expected on imposing a step change in input concentration to a number of such models and hence offer a guide to the recognition of models relevant to particular situations; the combination with kinetic data follows in sections 7.5, 7.6 and 7.7.

The step change may be achieved as already described by a sudden change of the tracer concentration from zero to some known concentration c_0; it is often experimentally more convenient to effect the converse change of abruptly discontinuing the supply of tracer and most of the ensuing mathematical development is on this basis.

Example 2. The following results for outlet to inlet concentration ratio were obtained after the sudden introduction of a continuous stream of tracer to a reactor system of unknown volume. The results give the ordinates of the F-diagram directly but before the F-diagram may be plotted, it is necessary to deduce the mean residence time. The mean residence time may be obtained from an E-diagram by taking the first moment of the distribution about zero time.

i.e.
$$\frac{V}{Q} = \int_0^\infty tE(t)\,dt. \qquad \text{(I)}$$

The E-diagram itself is obtained by taking mean gradients from the F-diagram data, and the various stages of the manipulation are shown in the table.

t hr	c_1/c_0 or F	$E(t)$	$tE(t)$	Qt/V
0	0	0	0	0
0·1	0·023	0·60	0·06	0·197
0·2	0·12	1·235	0·247	0·394
0·3	0·27	1·55	0·465	0·59
0·4	0·43	1·55	0·62	0·787
0·5	0·58	1·35	0·675	0·984
0·6	0·70	1·05	0·63	1·18
0·7	0·79	0·80	0·56	1·38
0·8	0·86	0·53	0·424	1·57(5)
0·9	0·90(5)	0·40	0·36	1·77
1·0	0·94	0·25	0·25	1·97
1·2	0·97	0·125	0·15	2·36
1·4	0·99	0·065	0·091	2·76
1·6	0·996	0·0215	0·0344	3·15
1·8	0·999	0·0075	0·0135	3·54

The column of $tE(t)$ values has been integrated using Simpson's rule for the two equal interval regions from 0–1·0 and 1·0 to 1·8, giving

$$\frac{V}{Q} = \frac{0{\cdot}1}{3}[0{\cdot}25 + 2(1{\cdot}921) + 4(2{\cdot}12)] + \frac{0{\cdot}2}{3}[0{\cdot}2635 + 2(0{\cdot}091) + 4(0{\cdot}1844)]$$

$$= \frac{0{\cdot}1}{3}(12{\cdot}57) + \frac{0{\cdot}2}{3}(1{\cdot}183)$$

$$= 0{\cdot}429 + 0{\cdot}079 = 0{\cdot}508 \text{ hours.}$$

From this value, the volume of the vessel could be determined from the throughput using equation (I) and the final column of dimensionless values of time has been prepared.

7.3.3 The single well stirred reactor E- and F-diagrams

For a general disturbance c_0 in input concentration, the concentration at exit is given by:

$$Qc_1 + V\frac{dc_1}{dt} = Qc_0. \tag{7.32}$$

If c_0 is to be a step upwards from zero to c_0, the transform of this equation is:

$$Q\bar{c}_1 + Vs\bar{c}_1 = Q\frac{c_0}{s} \tag{7.33}$$

and hence

$$\frac{c_1}{c_0} = 1 - e^{-Qt/V}. \tag{7.34}$$

This expression gives the fractional response with time and hence the F-diagram,

i.e.

$$F(t) = 1 - e^{-Qt/V}. \tag{7.35}$$

If c_0 is to be a step downwards from c_0 to zero, the transform of equation (7.32) is:

$$Q\bar{c}_1 + V[s\bar{c}_1 - c_0] = 0 \tag{7.36}$$

$$[Vs + Q]\bar{c}_1 = Vc_0 \tag{7.37}$$

and

$$\frac{c_1}{c_0} = e^{-Qt/V}. \tag{7.38}$$

The fractional response is now given by:

$$\frac{c_0 - c_1}{c_0}, \text{ i.e. by } 1 - \frac{c_1}{c_0}.$$

Hence as before:

$$F(t) = 1 - e^{-Qt/V}. \tag{7.35}$$

The E-diagram equation may be deduced from the result of equation (7.6) by differentiation of equation (7.35) giving:

$$E(t) = \frac{Q}{V} e^{-Qt/V}. \tag{7.39}$$

7.3.4 E- AND F-DIAGRAMS FOR A SERIES OF EQUALLY SIZED WELL MIXED VESSELS

The volume of the whole system will be designated as 'V' and the volume of each of the stages, 'v'. Letting $c_1, c_2, \ldots, c_m, \ldots, c_n$ be the concentrations leaving the successive stages of the cascade, the unsteady state balance on the mth stage is:

$$v\frac{dc_m}{dt} + Qc_m = Qc_{m-1} \tag{7.40}$$

and on repeated application to the series of vessels, the exit of an 'n' stage system has:

$$F(t) \equiv \frac{c_n}{c_0} = 1 - e^{-Qt/v} \times \left[1 + \frac{Qt}{v} + \frac{1}{2!}\left(\frac{Qt}{v}\right)^2 + \ldots + \frac{1}{(n-1)!}\left(\frac{Qt}{v}\right)^{n-1}\right], \tag{7.41}$$

or since:

$$v = \frac{V}{n} \tag{7.42}$$

$$F(t) = 1 - e^{-nQt/V} \times \left[1 + \left(\frac{nQt}{V}\right) + \frac{1}{2!}\left(\frac{nQt}{V}\right)^2 + \ldots + \frac{1}{(n-1)!}\left(\frac{nQt}{V}\right)^{n-1}\right]. \tag{7.43}$$

Similarly:

$$E(t) = \left(\frac{Qt}{v}\right)^{n-1}\left(\frac{Q}{v}\right)\frac{e^{-Qt/v}}{(n-1)!} = \left(\frac{nQt}{V}\right)^{n-1}\left(\frac{nQ}{V}\right)\frac{e^{-nQt/V}}{(n-1)!}. \tag{7.44}$$

The shapes of these curves as n varies are indicated later in Fig. 7.12 along with those for the dispersion model.

7.3.5 Single well stirred vessel with short circuiting and partial stagnation

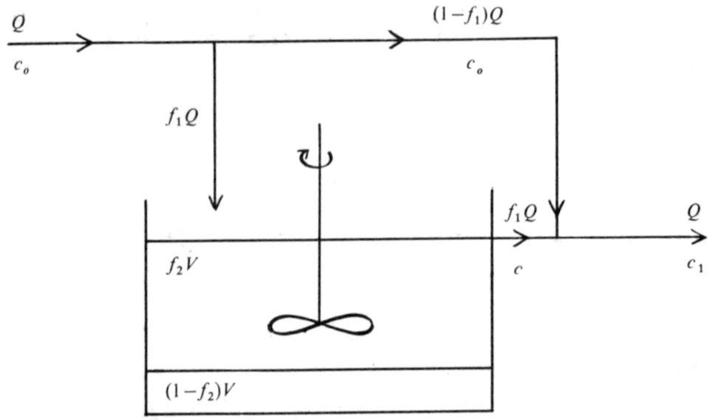

FIG. 7.4. Single well-stirred vessel with short circuiting and partial stagnation

This model (Fig. 7.4) considers the possibility that only a fraction f_1 of the feed will enter the mixing zone while the balance proceeds directly to the vessel outlet. Furthermore, the possibility is considered that only a fraction f_2 of the volume holdup is well stirred, the remainder being stagnant. The equation representing the well mixed zone is then:

$$f_1 Q c_0 = f_1 Q c + f_2 V \frac{dc}{dt}. \tag{7.45}$$

For a step downwards in tracer feed concentration from c_0 to zero:

$$f_1 Q \bar{c} + f_2 V [s\bar{c} - c_0] = 0 \tag{7.46}$$

and

$$\frac{c}{c_0} = \exp\left(-\frac{f_1 Q}{f_2 V} t\right). \tag{7.47}$$

This stream is combined with the short-circuiting stream to give a system outlet concentration c_1 given by:

$$Qc_1 = (1 - f_1)\, Q \times 0 + Q f_1 c_0 \exp\left(-\frac{f_1 Q}{f_2 V} t\right), \tag{7.48}$$

i.e.
$$\frac{c_1}{c_0} = f_1 \exp\left(-\frac{f_1 Q}{f_2 V} t\right) \tag{7.49}$$

$$F(t) = 1 - f_1 \exp\left(-\frac{f_1 Q}{f_2 V} t\right) \tag{7.50}$$

and
$$E(t) = \frac{f_1^2 Q}{f_2 V} \exp\left(-\frac{f_1 Q}{f_2 V} t\right). \tag{7.51}$$

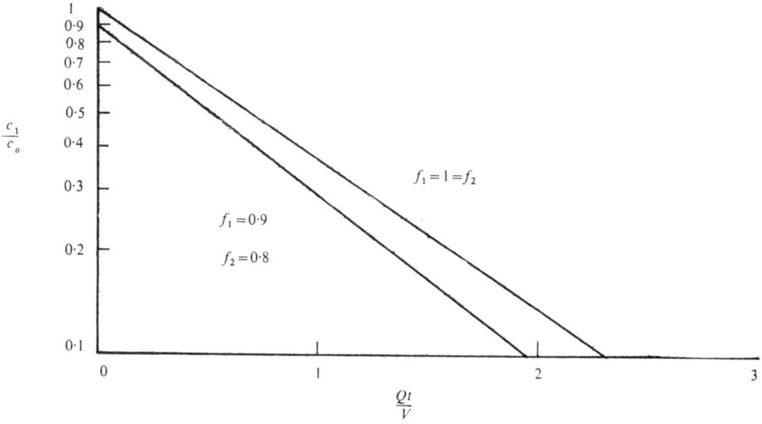

FIG. 7.5. Short circuiting with partial stagnation

The parameters f_1 and f_2 may be obtained by plotting Ln c_1/c_0 against 'Qt/V' (Fig. 7.5); the intercept at $t = 0$ will give Ln f_1 and on allowing for the base ten logarithmic plot, the slope of the line gives $-f_1/f_2$.

The results of equations (7.50) and (7.51) may be reduced to the single well mixed reactor case on putting $f_1 = 1 = f_2$. Likewise either short circuiting or partial stagnation could be discarded individually from these results.

7.3.6 SINGLE VESSEL PARTLY WELL MIXED AND PARTLY PLUG FLOW

A fraction 'f_2' of the hold-up volume is considered well mixed. The throughput Q flows in series through this well mixed region and in plug flow through the remaining volume (Fig. 7.6); the same residence time distribution results whichever of the two regimes is encountered first. Considering the case of a downward step change

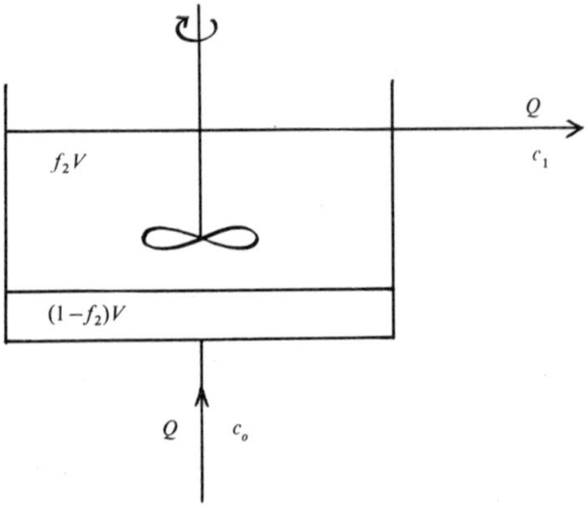

FIG. 7.6. Plug flow and well mixed in series

entering the plug flow region at time $t = 0$, the concentration entering the well mixed region will remain at c_0 for a time:

$$(1 - f_2)\frac{V}{Q}$$

and will then fall to zero; the transform of the input to the well mixed region is then:

$$\frac{c_0}{s} - \frac{c_0}{s}\exp\left\{-s(1 - f_2)\frac{V}{Q}\right\}$$

and the transformed differential equation for this region is:

$$f_2V[s\bar{c}_1 - c_0] + Q\bar{c}_1 = \frac{Qc_0}{s}\left[1 - \exp\left\{-s(1 - f_2)\frac{V}{Q}\right\}\right] \quad (7.52)$$

or:

$$\bar{c}_1 = \frac{c_0}{s + \dfrac{Q}{Vf_2}} + \frac{Qc_0}{Vf_2}\left[\frac{1}{s\left(s + \dfrac{Q}{Vf_2}\right)}\right] \times \left[1 - \exp\left\{-s(1 - f_2)\frac{V}{Q}\right\}\right]. \quad (7.53)$$

Noting that the inverse of the time delay term $\exp[-s(1 - f_2)(V/Q)$

is not applicable until a time $(1 - f_2)(V/Q)$ has elapsed, equation (7.53) may be inverted for the two time periods to give:

For $0 < t < (1 - f_2)(V/Q)$:

$$\frac{c_1}{c_0} = \exp\left(-\frac{Qt}{Vf_2}\right) + \left[1 - \exp\left(-\frac{Qt}{Vf_2}\right)\right] = 1, \tag{7.54}$$

i.e. confirmation that the system outlet remains fixed at c_0 for this period is obtained.
For $t > (1 - f_2)(V/Q)$:

$$\frac{c_1}{c_0} = \exp\left(-\frac{Qt}{Vf_2}\right) + \left[1 - \exp\left(-\frac{Qt}{Vf_2}\right)\right] - \left(1 - \exp\left\{-\frac{Q}{Vf_2}\left[t - (1 - f_2)\frac{V}{Q}\right]\right\}\right), \tag{7.55}$$

i.e.

$$\frac{c_1}{c_0} = \exp\left\{-\frac{1}{f_2}\left[\frac{Qt}{V} - (1 - f_2)\right]\right\}. \tag{7.56}$$

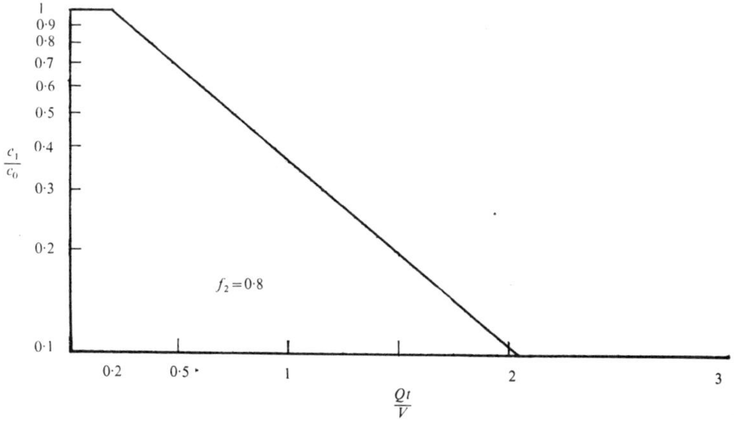

FIG. 7.7. Plug flow and well mixed in series

The result of plotting Ln c_1/c_0 against Qt/V is shown in Fig. 7.7 from which f_2 may be deduced from both the slope and the intercept. Hence for: $0 < t < (1 - f_2)(V/Q)$:

$$F(t) = 0 \tag{7.57}$$

and

$$E(t) = 0 \tag{7.58}$$

and for: $t > (1 - f_2)(V/Q)$:

$$F(t) = 1 - \exp\left(-\frac{1}{f_2}\left[\frac{Qt}{V} - (1 - f_2)\right]\right) \tag{7.59}$$

and
$$E(t) = \frac{Q}{f_2 V}\exp\left(-\frac{1}{f_2}\left[\frac{Qt}{V} - (1 - f_2)\right]\right). \tag{7.60}$$

7.3.7 Single well stirred vessel with parallel regions of plug flow and short circuiting

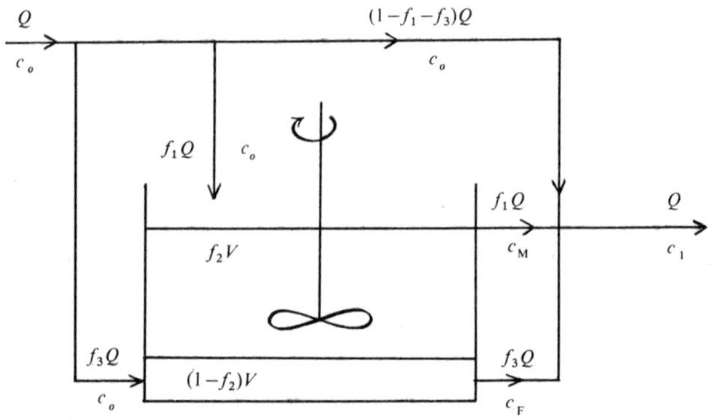

FIG. 7.8. Parallel regions of mixing, plug flow and short circuiting

A fraction f_1 of the feed traverses the well mixed region of volume $f_2 V$ as in an earlier case. A fraction f_3 of the feed (Fig. 7.8) passes through the remaining volume in plug flow, while a fraction $(1 - f_1 - f_3)$ of the feed short circuits.

The results for this model are readily deduced from those of the above models.

Thus for $0 < t < \dfrac{(1 - f_2)V}{f_3 Q}$:

$$c_M = c_0 \exp\left\{-\frac{f_1 Q}{f_2 V} t\right\} \tag{7.61}$$

$$c_F = c_0. \tag{7.62}$$

Hence

$$Qc_1 = (1 - f_1 - f_3)\, Q \times 0 + f_1 Q c_0 \exp\left\{-\frac{f_1 Q}{f_2 V} t\right\} + f_3 Q c_0, \quad (7.63)$$

i.e.

$$\frac{c_1}{c_0} = f_3 + f_1 \exp\left\{-\frac{f_1 Q}{f_2 V} t\right\}. \quad (7.64)$$

For $t > \dfrac{(1 - f_2)V}{f_3 Q}$:

c_M is given by equation (7.61)

and

$$c_F = 0, \quad (7.65)$$

i.e.

$$\frac{c_1}{c_0} = f_1 \exp\left\{-\frac{f_1 Q}{f_2 V} t.\right\} \quad (7.66)$$

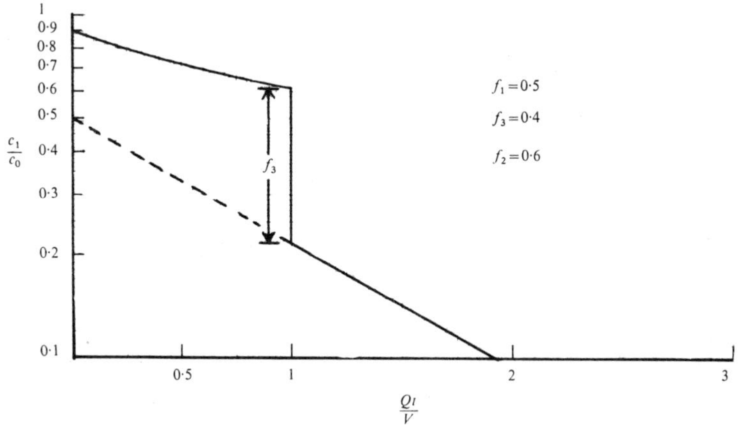

FIG. 7.9. Parallel regions of mixing, plug flow and short circuiting

A plot of Ln c_1/c_0 against Qt/V (Fig. 7.9) shows a discontinuity at:

$$\frac{Qt}{V} = \frac{1 - f_2}{f_3}. \quad (7.67)$$

The slope of the straight line part of the plot gives $-f_1/f_2$ while its extrapolated intercept gives Ln f_1. The value of f_3 is given by the difference between the two values of c_1/c_0 at the discontinuity.

7.3.8 Single well stirred vessel with a series/parallel plug flow region and short circuiting

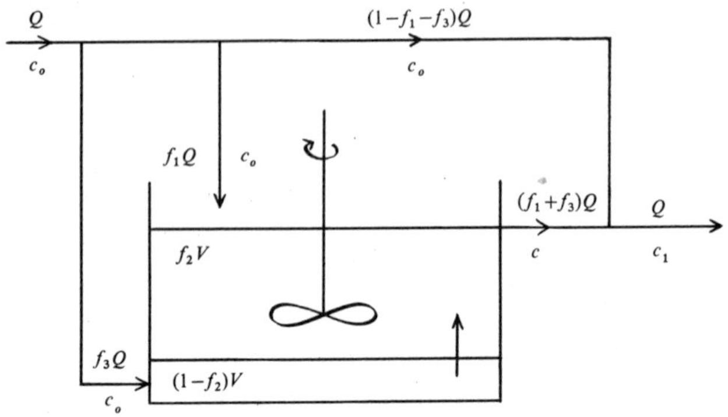

FIG. 7.10. Series/parallel plug flow and short circuiting

Using the symbolism as above, the well mixed region has a fraction f_1 of the feed entering it directly while a fraction f_3 enters it after traversing the plug flow region (Fig. 7.10). The transform of the concentration leaving the plug flow region is:

$$\frac{c_0}{s}\left[1 - \exp\left\{-\frac{s(1-f_2)V}{f_3Q}\right\}\right]$$

and so the transform of the equation representing the well mixed region is:

$$f_2V[s\bar{c} - c_0] + (f_1 + f_3)Q\bar{c} = f_3Q\frac{c_0}{s}\left[1 - \exp\left\{-\frac{s(1-f_2)V}{f_3Q}\right\}\right] \quad (7.68)$$

or

$$\bar{c} = \frac{c_0}{s + \dfrac{(f_1+f_3)Q}{f_2V}} + \frac{f_3Qc_0}{f_2V}\left[\frac{1}{s\left(s + \dfrac{(f_1+f_3)Q}{f_2V}\right)}\right] \times \left[1 - \exp\left\{-\frac{s(1-f_2)V}{f_3Q}\right\}\right]. \quad (7.69)$$

Hence for $0 < t < \dfrac{(1-f_2)V}{f_3Q}$:

$$\frac{c}{c_0} = \exp\left\{-\frac{(f_1+f_3)Qt}{f_2V}\right\} + \frac{f_3}{f_1+f_3}\left(1-\exp\left\{-\frac{(f_1+f_3)Qt}{f_2V}\right\}\right) \quad (7.70)$$

and since:

$$Qc_1 = (f_1 + f_3)Qc \quad (7.71)$$

$$\frac{c_1}{c_0} = f_3 + f_1 \exp\left\{-\frac{(f_1+f_3)Qt}{f_2V}\right\} \quad (7.72)$$

Similarly for $t > \dfrac{(1-f_2)V}{f_3Q}$:

$$\frac{c_1}{c_0} = \left[f_1 + f_3 \exp\left\{\frac{(f_1+f_3)(1-f_2)}{f_2f_3}\right\}\right]\exp\left\{-\frac{(f_1+f_3)Qt}{f_2V}\right\}. \quad (7.73)$$

Again the location of the discontinuity and the slope and intercept of the linear part of the plot give simultaneous equations for f_1, f_2 and f_3.

7.3.9 Single well stirred vessel with circulation through a plug flow region and short circuiting

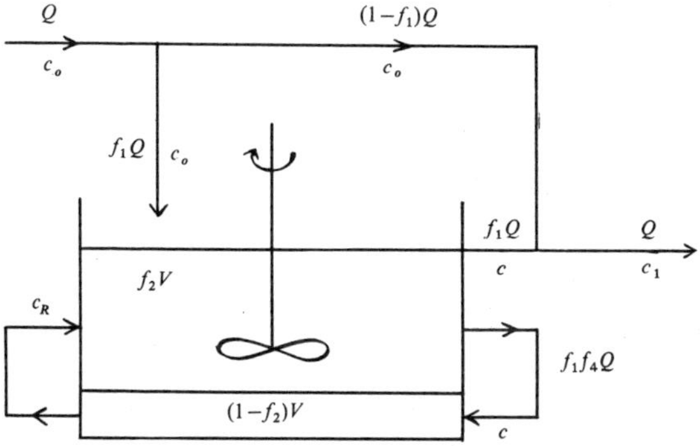

FIG. 7.11. Circulation between the plug flow and well-mixed region and short circuiting

This model (Fig. 7.11) introduces a circulation between the plug flow and well mixed regions; the magnitude of the circulation rate is f_4 times the well mixed region throughput, i.e. $f_4 f_1 Q$.

Let c_R be the tracer concentration recirculated from the plug flow zone; the balance on the well mixed region is then:

$$f_1 f_4 Q c_R = f_1 Q c + f_1 f_4 Q c + f_2 V \frac{dc}{dt} \quad (7.74)$$

which on transformation gives:

$$[s + \alpha]\bar{c} = c_0 + \frac{f_1 f_4 Q}{f_2 V}\bar{c}_R \tag{7.75}$$

where

$$\alpha = \frac{f_1 Q(1 + f_4)}{f_2 V}. \tag{7.76}$$

In the time interval $0 < t < \dfrac{(1 - f_2)V}{f_1 f_4 Q}$

$$\bar{c}_R = \frac{c_0}{s} \tag{7.77}$$

and

$$\bar{c} = \frac{c_0}{s + \alpha} + \frac{f_4 c_0}{1 + f_4}\left[\frac{1}{s\left(\dfrac{s}{\alpha} + 1\right)}\right] \tag{7.78}$$

or

$$\frac{c}{c_0} = e^{-\alpha t} + \frac{f_4}{1 + f_4}(1 - e^{-\alpha t}), \tag{7.79}$$

i.e.

$$\frac{c}{c_0} = \frac{f_4 + e^{-\alpha t}}{1 + f_4} \tag{7.80}$$

and the system outlet concentration c_1 is:

$$\frac{c_1}{c_0} = \frac{f_1[f_4 + e^{-\alpha t}]}{1 + f_4}. \tag{7.81}$$

As time progresses from $[(1 - f_2)V/f_1 f_4 Q]$ to $[2(1 - f_2)V/f_1 f_4 Q]$ the value of c_R is the same as c or $\bar{c}$ given by equations (7.74) and (7.75) for the initial time interval. Hence on introduction of the term:

$$-\frac{c_0}{s}e^{-\beta s}$$

to represent the cessation of the step change c_0 at time β the complete transform of c_R to be used in the solution of equation (7.75) for this second time interval is:

$$\frac{c_0}{s} - \frac{c_0}{s}e^{-\beta s} + \frac{c_0}{s + \alpha}e^{-\beta s} + \frac{f_4 c_0}{1 + f_4}\left[\frac{1}{s\left(\dfrac{s}{\alpha} + 1\right)}\right]e^{-\beta s}$$

where β is the time delay:

$$\frac{(1-f_2)V}{f_1 f_4 Q}.$$

Solution of equation (7.75) for c and then deduction of c_1 gives:

$$\frac{c_1}{c_0} = \frac{f_1 f_4^2}{(1+f_4)^2} + \frac{f_1\,\mathrm{e}^{-\alpha t}}{1+f_4}\left[1 + \mathrm{e}^{\alpha\beta}\left\{\frac{f_4}{1+f_4} - \frac{1-f_2}{f_2} + \frac{f_1 f_4 Q t}{f_2 V}\right\}\right]. \tag{7.82}$$

7.4 Dispersion Models

Of the numerous models used to characterise non-ideal flow patterns in reaction vessels, those which draw on the analogy between mixing and a diffusion process are called 'dispersion models'. Thus consider the plug flow of a fluid through a tubular reactor and let some degree of backmixing be superimposed on the plug flow. The magnitude of the backmixing is independent of position in the vessel so that there is no 'dead space' in the reactor or short circuiting of the fluid element through the reactor. Levenspiel called this type of flow dispersed plug flow. The extent of the superimposed backmixing determines whether the reactor can be classed as a tubular reactor or a C.S.T.R. Furthermore the volume of the real reactor will be between that of the plug flow tubular reactor and the perfectly mixed reactor. Therefore the actual volume of reactor required for a given conversion can be related to the volume required by a plug flow reactor for the same performance by:

$$V_{\text{actual}} = \phi V_0 \tag{7.83}$$

where ϕ is a correction factor dependent on reaction rate, vessel geometry and the intensity of backmixing. Levenspiel and his co-workers characterised the flow with a 'Dispersion Coefficient' which was similar to molecular diffusivity. The superimposed backmixing could bring about longitudinal or radial dispersion. Hence there was a longitudinal dispersion coefficient and a radial dispersion coefficient. Usually mixing perpendicular to the direction of flow was very small and accomplished by molecular diffusion. Since the intensity of backmixing is uniform in a longitudinal

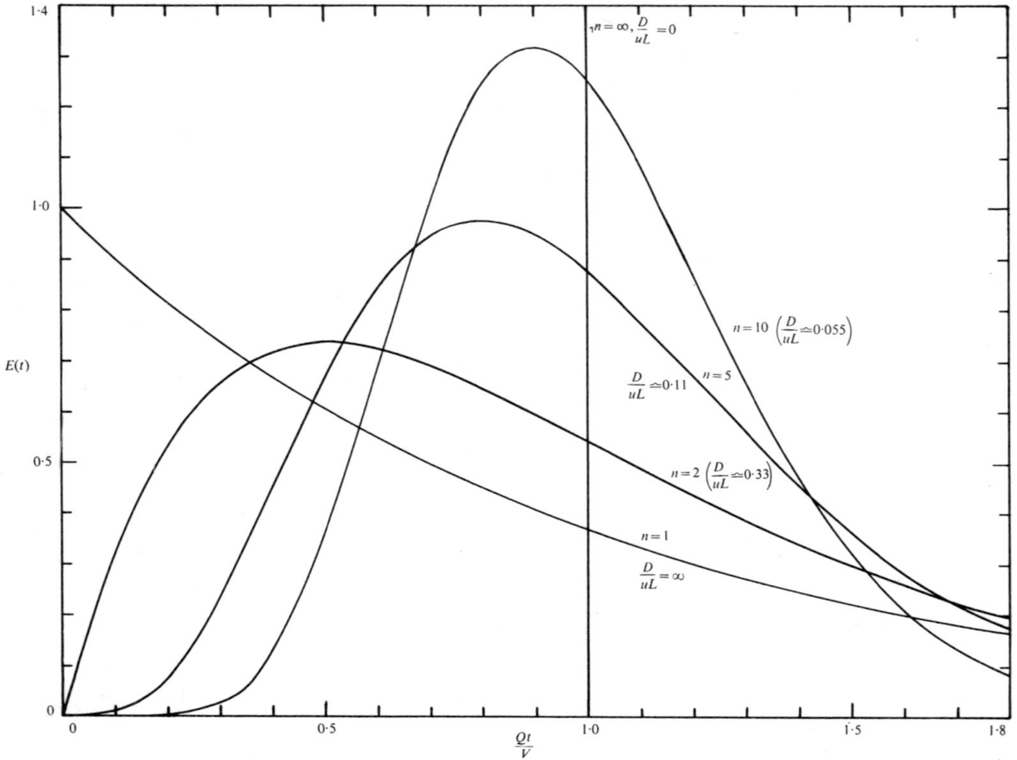

FIG. 7.12a E-diagrams for n equal sized vessels ($q/v = 1$)

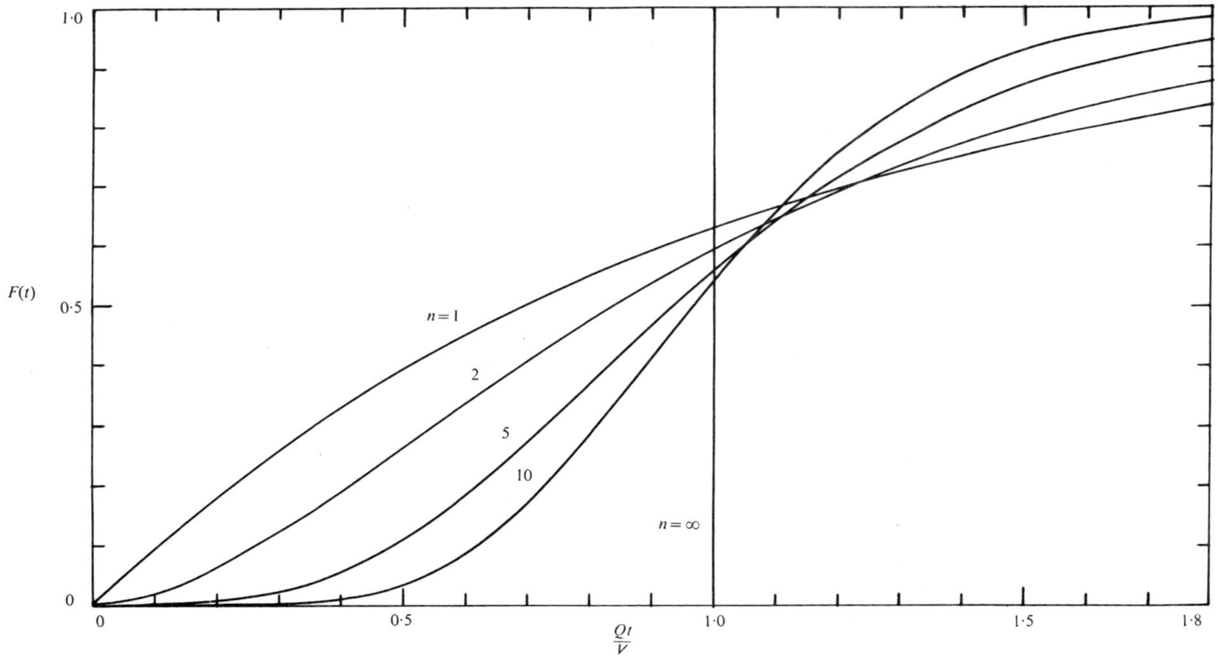

FIG. 7.12b. F-diagrams for n equal sized vessels

direction, the extent can be analysed using the unsteady state diffusion type equation:

$$\frac{\partial c}{\partial t} = D\frac{\partial^2 c}{\partial l^2} \tag{7.84}$$

where D is the dispersion coefficient.

Then for an impulse or step change in the fluid properties, the solution of equation (7.84) will give a family of theoretical E- and F-diagrams having the dispersion coefficient D, the flow velocity u, and a characteristic length L as parameters. The dimensionless group D/uL arises in the analysis and has therefore been found to be a suitable parameter to characterise dispersion; the group is called the 'Dispersion Number' and its reciprocal the 'Péclét' Number'. The results of experimental E- or F-diagram determinations are compared with the family of theoretical curves (Fig. 7.12) to select the one with the best fit, and hence to obtain either a value of dispersion number or the equivalent number of well stirred reactors which represents the non-ideal flow pattern. The matching of theoretical and practical curves may be attempted by comparison of shapes. However, it has been found more satisfactory to match the variance of the theoretical and practical E-diagrams. The variance of the experimental observations may be calculated while the following relationship between variance and dispersion number have been developed by Van der Laan for both open-ended and closed vessels (Fig. 7.13).

(*a*) For a closed vessel with no diffusion across either boundary, the E-diagram variance is:

$$2\left(\frac{D}{uL}\right) - 2\left(\frac{D}{uL}\right)^2(1 - e^{-uL/D}).$$

(*b*) For an open vessel with diffusion across both boundaries it is:

$$2\left(\frac{D}{uL}\right) + 8\left(\frac{D}{uL}\right)^2.$$

(*c*) An open–closed vessel with diffusion across one boundary but not the other gives:

$$2\left(\frac{D}{uL}\right) + 3\left(\frac{D}{uL}\right)^2.$$

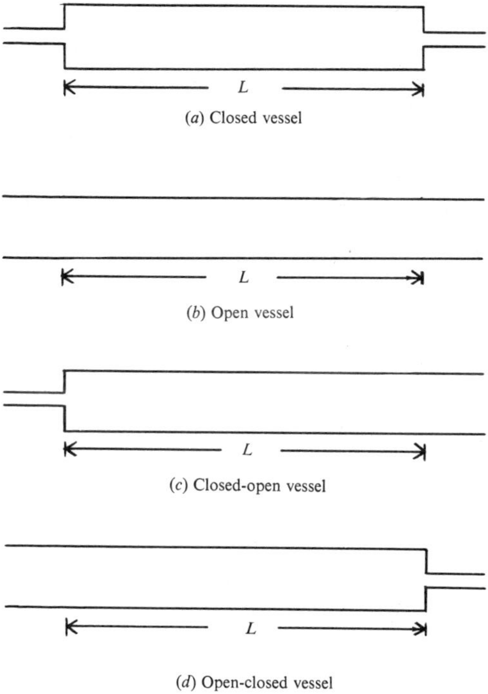

FIG. 7.13. Arrangements of open-ended and closed-ended vessels

When the deviation from plug flow conditions is small and the dispersion number therefore is small, the above results approach each other regardless of the vessel end conditions and the exit distribution is very well approximated by the normal error or Gaussian distribution,

i.e.
$$E(t) = \frac{1}{2\sqrt{\left(\frac{\pi D}{uL}\right)}} \exp\left[-\frac{\left(1-\frac{Qt}{V}\right)^2}{\frac{4D}{uL}}\right] \qquad (7.85)$$

for which the variance is:

$$\frac{2D}{uL}.$$

A general correlation of dispersion number with flow characteristics has been attempted by many workers. Probably the most useful is that presented by Levenspiel (1958) where the radial dispersion number (D/ud_p) in which d_p is the internal diameter of the pipe is plotted against the Reynolds number. This dispersion number can be converted into the longitudinal dispersion number by using the relation

$$\left(\frac{D}{ud_p}\right)\left(\frac{d_p}{L}\right)=\frac{D}{uL} \tag{7.86}$$

where (d_p/L) is the ratio of tube diameter to tube length.

The correlation in terms of Reynolds number, and Schmidt numbers has been taken from Levenspiel, and is shown in Figs. 7.14 and 7.15.

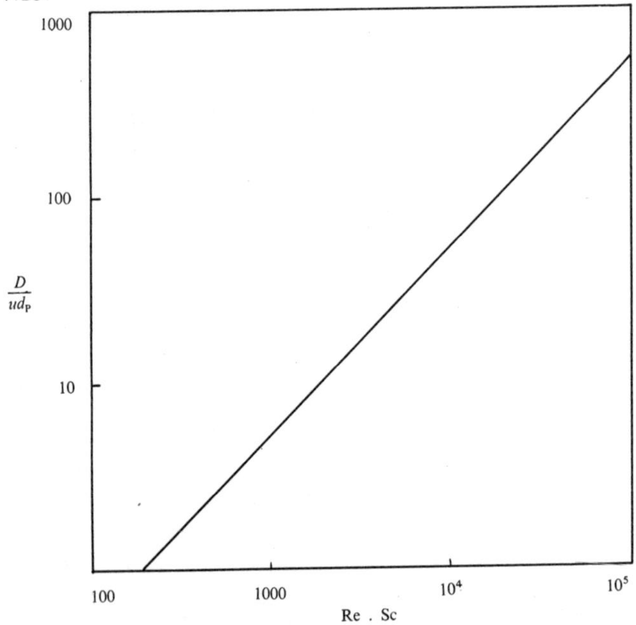

FIG. 7.14. Dispersion of fluid in streamline flow in pipes applicable when

$$(\text{Re})(\text{Sc}) \ll 30\frac{L}{d_p}$$

Example 3. The data of Example 2 in which the mean residence time was estimated will now be used to find the variance of the distribution and hence an estimate of the Péclét number.

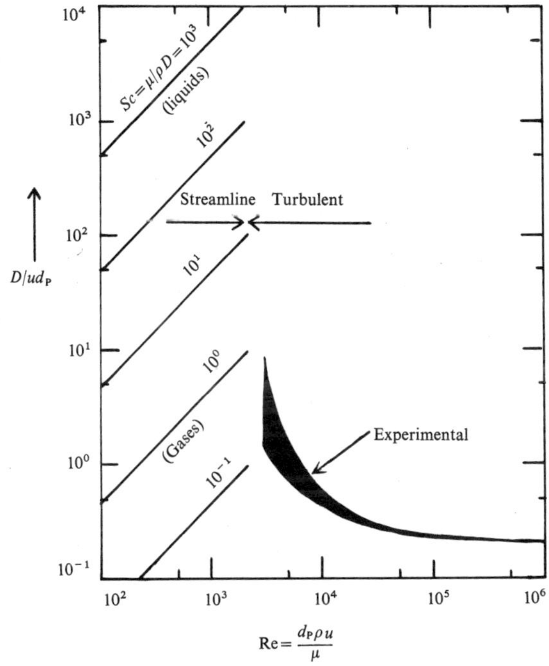

FIG. 7.15. Axial dispersion in pipe flow

For equally spaced observations the variance is given by:

$$\frac{\sum t^2E(t)}{\sum E(t)} - \left(\frac{V}{Q}\right)^2$$

and selecting values at 0·2 h intervals from the table in Example 2 yields the following table:

t	$E(t)$	$t^2E(t)$
0	0	0
0·2	1·235	0·0494
0·4	1·55	0·248
0·6	1·05	0·378
0·8	0·53	0·34
1·0	0·25	0·25
1·2	0·125	0·18
1·4	0·065	0·1275
1·6	0·0215	0·055
1·8	0·0075	0·0243
	$\sum E(t) = 4·8340$	$\sum t^2E(t) = 1·6522$

Hence the variance is:

$$\frac{1{\cdot}6522}{4{\cdot}8340} - (0{\cdot}508)^2 = 0{\cdot}083.$$

Using the following equation on the assumption of a closed vessel:

$$0{\cdot}083 = 2\frac{D}{uL} - 2\left(\frac{D}{uL}\right)^2 [1 - e^{-uL/D}]$$

for small values of D/uL, the exponential term is almost zero and the remaining quadratic then gives:

$$\frac{D}{uL} = 0{\cdot}0436.$$

7.4.1 Error function solution of the diffusion equation

For small deviations from plug flow (Danckwerts, 1953), the concentration at time 't' and any position 'l' is given by:

$$c = \frac{c_0}{2}\left[1 - erf\frac{l}{2\sqrt{(Dt)}}\right] \tag{7.87}$$

and so evaluation of C/C_0 at the tube exit gives:

$$F(t) = \frac{1}{2}\left[1 - erf\frac{L - ut}{2\sqrt{(Dt)}}\right]. \tag{7.88}$$

Using the expressions for mean residence time:

$$\frac{L}{u} \quad \text{and} \quad \frac{V}{Q}$$

equation (7.88) may be re-arranged to:

$$F(t) = \frac{1}{2}\left[1 - erf\frac{1 - \dfrac{Qt}{V}}{2\left(\dfrac{Qt}{V}\dfrac{D}{uL}\right)^{\frac{1}{2}}}\right]. \tag{7.89}$$

The result is seen to be a function of the mean residence time and uL/D the Péclét number (Pe). Further a plot of $F(t)$ against Qt/V is characterised by the value of Pe and so a convenient way of deducing

Pe from it is required. If the above function is differentiated with respect to Qt/V the gradient at any time is:

$$\frac{1}{4\sqrt{\pi}}\left(\frac{uL}{D}\right)^{\frac{1}{2}} e^{\frac{-\left(1-\frac{Qt}{V}\right)^2}{\frac{4Qt}{V}\frac{D}{uL}}} \left[\left(\frac{Qt}{V}\right)^{-\frac{3}{2}} + \left(\frac{Qt}{V}\right)^{-\frac{1}{2}}\right]$$

which at one mean residence time gives a gradient G of:

$$\frac{1}{2}\left[\frac{Lu}{\pi D}\right]^{\frac{1}{2}}$$

and a Péclét number of $4\pi G^2$.

7.5 Direct Application of the Residence Time Distribution Function to the Design of Reactors

It has been shown (Zwietering, 1959) that the residence time distribution does not uniquely define the reaction characteristics and in general further information about the nature of the molecular mixing process is required. However, first order reactions have been shown to yield the same conversion expression for both the cases of complete segregation and maximum mixedness. For a first order reaction the fractional conversion of material which has spent a time t is:

$$(1 - e^{-kt})$$

and since the fraction of the total throughput which spends time t in the system is $E(t)\,dt$, the overall fractional conversion from the system is:

$$f = \int_0^\infty (1 - e^{-kt})\, E(t)\, dt, \tag{7.90}$$

i.e.

$$f = \int_0^\infty E(t)\, dt - \int_0^\infty e^{-kt}\, E(t)\, dt. \tag{7.91}$$

The former integral is unity, and differentiation of equation (7.6) to give:

$$E(t)\, dt = dF(t) \tag{7.92}$$

enables the latter integral to be expressed as

$$\int_0^1 e^{-kt}\, dF(t)$$

using the fact that $F(t) = 1$ at $t = \infty$,

i.e.
$$f = 1 - \int_0^1 e^{-kt}\, dF(t), \tag{7.93}$$

and hence the area beneath a plot of e^{-kt} against experimental $F(t)$ values will enable the fractional conversion to be estimated. Alternatively an analytical expression for $E(t)\,dt$ could be utilised in the following development from equation (7.91):

$$f = 1 - \int_0^\infty e^{-kt}\, E(t)\, dt. \tag{7.94}$$

Example 4. Consider the application of equation (7.94) to the case of two equal well mixed reactors for which equation (7.44) gives:

$$E(t) = \left(\frac{Q}{v}\right)^2 t \exp\left(-\frac{Qt}{v}\right).$$

Hence
$$f = 1 - \int_0^\infty \left(\frac{Q}{v}\right)^2 t \exp\left\{-\left(k + \frac{Q}{v}\right)t\right\}$$

On repeated integration by parts this yields:

$$f = 1 - \left(\frac{Q}{v}\right)^2 \frac{1}{\left(k + \dfrac{Q}{v}\right)^2} = 1 - \frac{1}{\left(1 + \dfrac{kv}{Q}\right)^2}$$

or
$$\frac{c_2}{c_0} = \frac{1}{\left(1 + \dfrac{kv}{Q}\right)^2}$$

which is seen to be consistent with the result of equation (5.16).

Example 5. Establish the E- and F-diagrams for laminar flow in a tubular vessel and hence confirm the results obtained in section 6.6 for laminar flow reactors effecting first and second order reactions.

The fraction of the total flow passing through the elementary annulus bounded by r and $r + dr$ in a time between t and $t + dt$ is:

$$\frac{2\pi u_r\, dr}{\dfrac{\pi u_0 R^2}{2}} \quad \text{or} \quad \frac{4 r u_r\, dr}{u_0 R^2}.$$

Substituting $r\,dr$ and u_r from equations (6.33) and (6.28) gives:

$$E(t)\,dt = \frac{2L^2}{U_0^2}\frac{dt}{t^3} = 2t_0^2\frac{dt}{t^3}.$$

Since the minimum time of passage through the tube is t_0 the E-diagram is fully represented by:

$$t < t_0, \qquad E(t)\,dt = 0$$

$$t > t_0, \qquad E(t)\,dt = 2t_0^2\frac{dt}{t^3}$$

and on integration the corresponding portions of the F-diagram are:

$$t < t_0, \qquad F(t) = 0$$

$$t > t_0, \qquad F(t) = 1 - \frac{t_0^2}{t^2}.$$

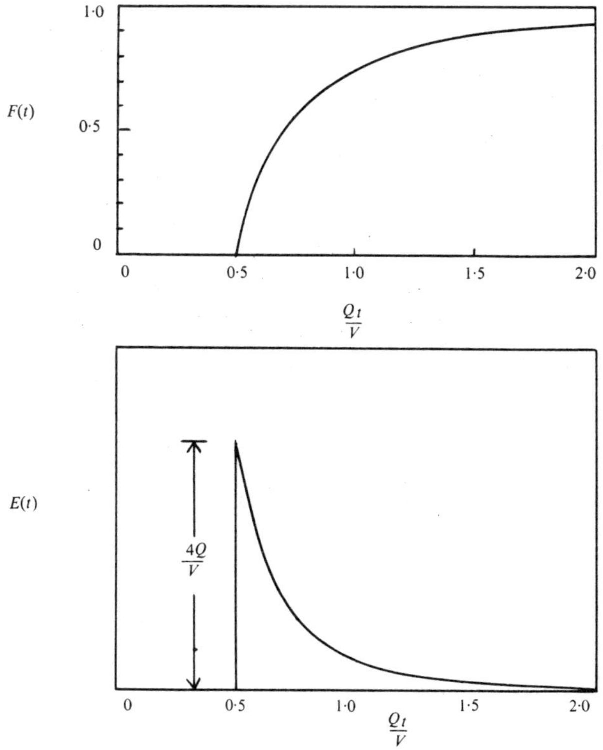

FIG. 7.16. $E(t)$ and $F(t)$ diagrams for laminar flow

These results are plotted in Fig. 7.16.

Application of (7.94) for a first order reaction now gives:

$$f = 1 - 2t_0^2 \int_{t_0}^{\infty} e^{-kt} \frac{dt}{t^3}$$

which is consistent with equation (6.42).

For a second order reaction:

$$f = \int_{t_0}^{\infty} \frac{kc_0 t}{1 + kc_0 t} 2t_0^2 \frac{dt}{t^3}$$

$$= 2kc_0 t_0^2 \int_{t_0}^{\infty} \frac{dt}{t^2(1 + kc_0 t)}$$

and was obtained earlier as equation (6.36).

Example 6. In this illustration, the data obtained from $E(t)$ in Example 1 are used to obtain a numerical result from equation (7.94). For a first order reaction whose rate constant is 0·20 min^{-1}:

t	kt	e^{-kt}	$E(t)$	$e^{-kt} E(t)$
0	0	1	0	0
5	1	0·368	0·03	0·01104
10	2	0·135	0·05	0·00675
15	3	0·050	0·05	0·00250
20	4	0·018	0·04	0·00072
25	5	0·007	0·02	0·00014
30	6	0·0025	0·01	0·000025
35	7		0	0

Using Simpson's rule as in Example 1, the final column of the table may be integrated to:

$$\tfrac{5}{3}[0 + 2(0{\cdot}00675 + 0{\cdot}00072 + 0{\cdot}000025) + 4(0{\cdot}01104 + 0{\cdot}00250 + 0{\cdot}00014 + 0)] = 0{\cdot}1162.$$

Hence $f = 1 - 0{\cdot}116 = 0{\cdot}884.$

From the table, it is seen that most of the unconverted material is in the part of the throughput which has a short residence time, and illustrates how a wide spread of residence-times can seriously affect reactor performance at high conversion.

The comparative performance of a plug flow reactor is given by:

$$f = 1 - \exp(0{\cdot}2 \times 15) = 1 - 0{\cdot}05 = 0{\cdot}95.$$

7.6 Use of Incompletely Stirred Models for Reactor Design

Step changes may be used as described in sections 7.3.3 to 7.3.9 to decide on the detailed type of model which represents a deviation from a well stirred vessel and at the same time deduce numerical values for the extent of the short-circuiting, stagnation, plug flow or circulation parameters. Having decided upon the model, the reactor performance may be developed and a few illustrations follow.

7.6.1 Short circuit with partial stagnation (Fig. 7.5)

For a reaction A → B, the performance of the well stirred region is given by:

$$f_1 Q c_0 = f_1 Q c + k f_2 V c \tag{7.95}$$

or

$$\frac{c}{c_0} = \frac{f_1 Q}{f_1 Q + k f_2 V}. \tag{7.96}$$

On mixing with the short circuiting material, the system outlet concentration c_1 is:

$$Q c_1 = f_1 Q c + (1 - f_1) Q c_0 \tag{7.97}$$

and

$$\frac{c_1}{c_0} = \frac{1 + \dfrac{k f_2 V (1 - f_1)}{f_1 Q}}{1 + \dfrac{k f_2 V}{f_1 Q}}. \tag{7.98}$$

Expressions are developed below for the comparison of the performance of such a reactor with that to be expected, first for a reactor without stagnation and secondly for a reactor with neither stagnation nor short circuiting.

Let the exit concentration from a well mixed vessel without

stagnation be c_M. Then the performance of such a reactor is given by putting $f_2 = 1$ in equation (7.98),

i.e.
$$\frac{c_M}{c_0} = \frac{1 + \dfrac{kV(1 - f_1)}{f_1 Q}}{1 + \dfrac{kV}{f_1 Q}}. \qquad (7.99)$$

On elimination of the group kV/f_1Q between equations (7.98) and (7.99):

$$\frac{c_1}{c_0} = \frac{\dfrac{c_M}{c_0}[1 - f_2(1 - f_1)] - [1 - f_1 - f_2 + f_1 f_2]}{\dfrac{c_M}{c_0}[1 - f_2] - [1 - f_1 - f_2]}. \qquad (7.100)$$

For a given extent of short circuiting f_1 this equation permits the deduction of the performance of a vessel with stagnation f_2 from the corresponding fully stirred result.

When there is no short circuiting, the corresponding result is obtained from equation (7.100) on setting $f_1 = 1$,

i.e.
$$\frac{c_1}{c_0} = \frac{\dfrac{c_M}{c_0}}{\dfrac{c_M}{c_0}[1 - f_2] + f_2}. \qquad (7.101)$$

7.6.2 Short circuit with series/parallel plug flow (Fig. 7.10)

The material balance on the well stirred vessel for a reaction A → B is:

$$f_1 Q c_0 + f_3 Q c_0 \exp\left\{-\frac{k(1 - f_2)V}{f_3 Q}\right\} = (f_1 + f_3) Qc + kf_2 Vc \qquad (7.102)$$

from which:

$$\frac{c}{c_0} = \frac{Q\left[f_1 + f_3 \exp\left\{-\dfrac{k(1 - f_2)V}{f_3 Q}\right\}\right]}{(f_1 + f_3) Q + kf_2 V}. \qquad (7.103)$$

Hence the concentration c_1 at the system outlet is found from:

$$Qc_1 = (1 - f_1 - f_3)Qc_0 + \frac{Q^2 c_0(f_1 + f_3)\left[f_1 + f_3 \exp\left\{-\frac{k(1-f_2)V}{f_3 Q}\right\}\right]}{(f_1 + f_3)Q + kf_2 V}, \tag{7.104}$$

i.e.

$$\frac{c_1}{c_0} = \frac{1 - f_3\left(1 - \exp\left\{-\frac{k(1-f_2)V}{f_3 Q}\right\}\right) + \frac{kf_2 V}{Q(f_1+f_3)}(1 - f_1 - f_3)}{1 + \frac{kf_2 V}{Q(f_1 + f_3)}}. \tag{7.105}$$

7.7 Tubular Reactor with Axial Diffusion

It has been explained elsewhere in the text that real tubular reactors will exhibit some degree of backmixing and that the reactor volume required for a given duty will exceed that for plug flow but will be appreciably less than that needed if complete mixing occurs. This backmixing effect may be treated by the addition of an axial diffusion term to the plug flow model of equation (6.1). Thus consider the mass balance over the element of volume shown in Fig. 7.17.

$$Qc - AD\frac{dc}{dl} = Qc + \frac{d}{dl}(Qc)\delta l - AD\left(\frac{dc}{dl} + \frac{d^2c}{dl^2}\delta l\right) - A\delta l r \tag{7.106}$$

or

$$AD\frac{d^2c}{dl^2} - \frac{d}{dl}(Qc) - Ar = 0. \tag{7.107}$$

Equation (7.107) is now developed for the case of constant Q. Such an assumption would be satisfied either if there is no change in the number of moles on reaction or if the reactants are dispersed in such a large quantity of diluent that change in the number of moles on reaction has a negligible effect on Q. Equation (7.107) then becomes:

$$\frac{AD}{Q}\frac{d^2c}{dl^2} - \frac{dc}{dl} - \frac{Ar}{Q} = 0 \tag{7.108}$$

and is frequently presented in terms of velocity u as:

$$\frac{D}{u}\frac{d^2c}{dl^2} - \frac{dc}{dl} - \frac{r}{u} = 0. \tag{7.109}$$

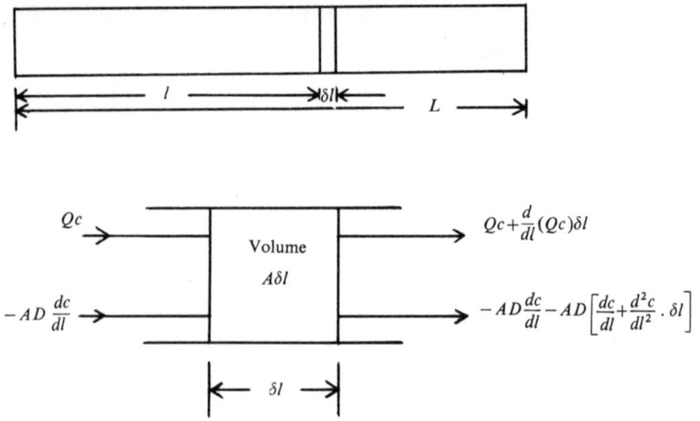

FIG. 7.17. Bulk flow and diffusion in a tubular reactor

7.7.1 FIRST ORDER IRREVERSIBLE REACTION

Analytical solutions may be obtained to equations (7.108) and (7.109) if the kinetics are either zero order or first order of the form:

$$r = kc \tag{7.110}$$

or

$$r = k_1 c_A - k_2 c_B. \tag{7.111}$$

For other orders there is no analytical solution of these equations, but a numerical solution has been presented by Fan and Bailie (1960) for orders ranging from 0·25 to 3.

Before proceeding with the solution for first order kinetics it is convenient to introduce a dimensionless position Z and a dimensionless concentration γ such that:

$$Z = \frac{l}{L} \tag{7.112}$$

and

$$\gamma = \frac{c}{c_0} \tag{7.113}$$

where l is a general point along the length of the reactor and L is the full reactor length, c is the concentration at point l and c_0 is the system inlet concentration.

With these substitutions and applying to a first order forward reaction, equation (7.109) may be rearranged to:

$$\frac{D}{uL}\frac{d^2\gamma}{dZ^2} - \frac{d\gamma}{dZ} - \frac{kL}{u}\gamma = 0. \tag{7.114}$$

The group uL/D is the Péclét number (Pe) and the term kL/u (the reaction rate number) will be denoted as R in subsequent manipulation. Equation (7.114) thus becomes:

$$\frac{1}{Pe}\frac{d^2\gamma}{dZ^2} - \frac{d\gamma}{dZ} - R\gamma = 0 \tag{7.115}$$

and the form of solution will depend upon the chosen boundary conditions. The general solution is:

$$\gamma = N_1\,e^{m_1 Z} + N_2\,e^{m_2 Z} \tag{7.116}$$

where N_1 and N_2 are constants to be determined from the boundary conditions and m_1 and m_2 are the roots of the auxiliary equation,

i.e.
$$m_1 \text{ and } m_2 = \frac{Pe}{2}\left[1 \pm \left(1 + \frac{4R}{Pe}\right)^{\frac{1}{2}}\right]. \tag{7.117}$$

Replacing the term $(1 + 4R/Pe)^{\frac{1}{2}}$ by a:

$$m_1 \text{ and } m_2 = \frac{Pe}{2}[1 \pm a] \tag{7.118}$$

and in subsequent working m_1 will be given the positive alternative.

One set of boundary conditions which has been applied is:

At inlet:
$$Z = 0, \gamma = 1 \tag{7.119}$$

After complete conversion:

$$Z = \infty, \gamma = 0 \tag{7.120}$$

Substitution into (7.116) gives:

$$N_1 = 0 \tag{7.121}$$

and
$$N_2 = 1 \tag{7.122}$$

so that
$$\gamma = \exp\frac{Pe}{2}(1 - a)\,Z. \tag{7.123}$$

This result is plotted in Fig. 7.18 as the curve BC for comparison with more extensive solutions of the differential equation. As diffusion coefficient tends to zero equation (7.123) may be shown to reduce to the plug flow expression:

$$\gamma = \exp(-RZ) \tag{7.124}$$

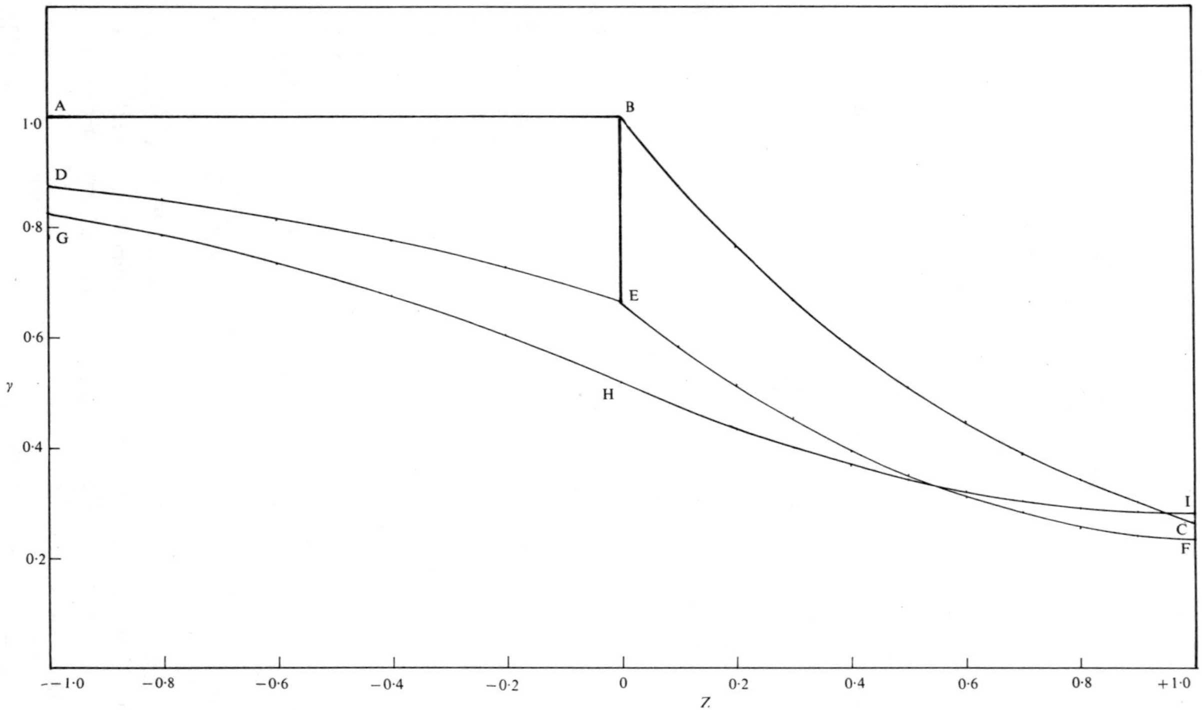

FIG. 7.18. Concentration profiles in pre-reaction and reaction zones

or for $Z = 1$:

$$c = c_0 \exp\left(-\frac{kL}{u}Z\right) = c_0 \exp\left(-\frac{kL}{u}\right) = c_0 \exp\left(-\frac{kV_R}{Q}\right). \tag{7.125}$$

However, the well-mixed reactor result is not obtained on allowing D to become infinite.

Two further sets of boundary conditions will be considered. These two sets differ in that the one considers diffusion to occur in the physical regions preceding the reactor entry and beyond the reactor outlet; the other set confines the diffusion to the reactor zone. It is conceivable for both of these situations to arise in practice. Thus a tubular reactor may be as shown in Fig. 7.19 with reaction starting at section AB where heating starts or catalyst exists and having unimpeded back diffusion into the pre-reaction zone. In contrast the distribution plate at entry to a fluidised bed reactor could be expected to impede diffusional mixing between the fluids on either side of it.

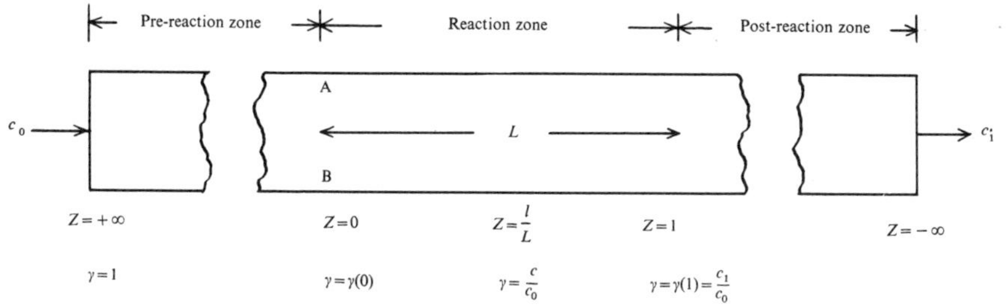

FIG. 7.19. Concentrations with diffusion in all three zones

The boundary concentrations appropriate to diffusion in the three zones are shown in Fig. 7.19 and were introduced by Wehner and Wilhelm (1956). At the extremities they are:

$$Z = -\infty, \qquad \gamma = 1; \tag{7.126}$$

$$Z = +\infty, \qquad \gamma \text{ is finite.} \tag{7.127}$$

Continuity of concentration at the two reaction zone boundaries requires:

$$\text{At } Z = 0: \gamma(0-) = \gamma(0+) \equiv \gamma(0), \tag{7.128}$$

$$\text{At } Z = 1: \gamma(1-) = \gamma(1+). \tag{7.129}$$

Conservation of mass by combined bulk flow and diffusion across the reactor inlet boundary requires that:

$$Q\gamma(0-) - \frac{AD_1}{L}\left(\frac{d\gamma}{dZ}\right)_{0-} = Q\gamma(0+) - \frac{AD_2}{L}\left(\frac{d\gamma}{dZ}\right)_{0+} \tag{7.130}$$

where D_1 and D_2 recognise the possibility of different diffusion coefficients in the two zones. Since

$$\gamma(0-) = \gamma(0+)$$

this equation may be simplified to:

$$\frac{1}{Pe_1}\left(\frac{d\gamma}{dZ}\right)_{0-} = \frac{1}{Pe_2}\left(\frac{d\gamma}{dZ}\right)_{0+} \tag{7.131}$$

with the Péclét number subscripts corresponding to the two diffusion coefficients. This condition implies a discontinuity of concentration gradient at the reactor inlet unless the diffusion coefficients are equal in the two zones. Similarly conservation across the reactor outlet boundary requires:

$$\frac{1}{Pe_2}\left(\frac{d\gamma}{dZ}\right)_{1-} = \frac{1}{Pe_3}\left(\frac{d\gamma}{dZ}\right)_{1+}. \tag{7.132}$$

The differential equation governing concentration in the pre-reaction zone will be equation (7.115) with the rate term set at zero,

i.e.
$$\frac{1}{Pe_1}\frac{d^2\gamma}{dZ^2} - \frac{d\gamma}{dZ} = 0 \tag{7.133}$$

which has a general solution:

$$\gamma = N_3 + N_4\,e^{Pe_1 Z}. \tag{7.134}$$

Equations (7.126) and (7.128) then give:

$$N_3 = 1 \tag{7.135}$$

and
$$N_4 = \gamma(0) - 1. \tag{7.136}$$

Hence for $Z < 0$:

$$\gamma = 1 + [\gamma(0) - 1]\,e^{Pe_1 Z}. \tag{7.137}$$

The value of $\gamma(0)$ at the boundary has still to be determined from the solution of the equations for the subsequent zones.

The reaction zone has the general solution:

$$\gamma = N_5 \exp\left\{\frac{Pe_2}{2}(1+a)Z\right\} + N_6 \exp\left\{\frac{Pe_2}{2}(1-a)Z\right\} \qquad (7.138)$$

Equation (7.128) yields:

$$\gamma(0) = N_5 + N_6. \qquad (7.139)$$

Differentiation of (7.137) and (7.138) and use of (7.131) leads to:

$$[\gamma(0) - 1] = \tfrac{1}{2}(1+a)N_5 + \tfrac{1}{2}(1-a)N_6. \qquad (7.140)$$

For the post reaction zone the general solution is:

$$\gamma = N_7 + N_8\,e^{Pe_3 Z}. \qquad (7.141)$$

Since the model under consideration requires γ to be finite at $Z = \infty$, it follows that:

$$N_8 = 0 \qquad (7.142)$$

and hence that the concentration for $Z > 1$ is constant at N_7.

Application of boundary conditions (7.129) and (7.132) yields the further two equations:

$$N_7 = N_5 \exp\left\{\frac{Pe_2}{2}(1+a)\right\} + N_6 \exp\left\{\frac{Pe_2}{2}(1-a)\right\} \qquad (7.143)$$

and

$$0 = N_5(1+a)\exp\left\{\frac{Pe_2}{2}(1+a)\right\} + N_6(1-a)\exp\left\{\frac{Pe_2}{2}(1-a)\right\}. \qquad (7.144)$$

Solution of the simultaneous equations (7.139), (7.140), (7.143) and (7.144) for N_5, N_6, N_7 and $\gamma(0)$ gives:

$$N_5 = -(1-a)N_9 \exp\left\{-a\frac{Pe_2}{2}\right\}$$

$$N_6 = (1+a)N_9 \exp\left\{a\frac{Pe_2}{2}\right\} \qquad (7.145)$$

$$N_7 = 2aN_9 \exp\left\{\frac{Pe_2}{2}\right\} \qquad (7.146)$$

and

$$\gamma(0) = \left[(1+a)\exp\left\{a\frac{Pe_2}{2}\right\} - (1-a)\exp\left\{-a\frac{Pe_2}{2}\right\}\right]N_9 \quad (7.147)$$

wherein

$$N_9 = \frac{2}{(1+a)^2\exp\left\{a\dfrac{Pe_2}{2}\right\} - (1-a)^2\exp\left\{-a\dfrac{Pe_2}{2}\right\}}. \quad (7.148)$$

Substitution of these constants into the general solutions of the differential equations gives:

For $Z \leqslant 0$:

$$\gamma = 1 + e^{Pe_1 Z} \times \left[2\frac{(1+a)\exp\left\{a\dfrac{Pe_2}{2}\right\} - (1-a)\exp\left\{-a\dfrac{Pe_2}{2}\right\}}{(1+a)^2\exp\left\{a\dfrac{Pe_2}{2}\right\} - (1-a)^2\exp\left\{-a\dfrac{Pe_2}{2}\right\}} - 1\right] \quad (7.149)$$

and is a function of both D_1 and D_2.
For $0 \leqslant Z \leqslant 1$:

$$\gamma = 2\exp\left\{\frac{Pe_2}{2}Z\right\} \times \left[\frac{(1+a)\exp\left\{a\dfrac{Pe_2}{2}(1-Z)\right\} - (1-a)\exp\left\{-a\dfrac{Pe_2}{2}(1-Z)\right\}}{(1+a)^2\exp\left\{a\dfrac{Pe_2}{2}\right\} - (1-a)^2\exp\left\{-a\dfrac{Pe_2}{2}\right\}}\right]. \quad (7.150)$$

This result and the following one are functions of Pe_2 only.
For $Z \geqslant 1$:

$$\gamma = \frac{4a\exp\left\{\dfrac{Pe_2}{2}\right\}}{(1+a)^2\exp\left\{a\dfrac{Pe_2}{2}\right\} - (1-a)^2\exp\left\{-a\dfrac{Pe_2}{2}\right\}}. \quad (7.151)$$

A plot of concentration against position is shown in Fig. 7.18 (DEF). The zero slope at $Z = 1$ can be confirmed analytically by differentiation of equation (7.150), and is of interest since it is one of the boundary conditions in the remaining set to be considered. Thus on re-arrangement and using the symbol N_9 again:

$$\gamma = N_9\left[(1+a)\exp\left\{a\frac{Pe_2}{2}\right\}\exp\left\{\frac{Pe_2}{2}(1-a)Z\right\} - (1-a)\exp\left\{-a\frac{Pe_2}{2}\right\}\exp\left\{-a\frac{Pe_2}{2}(1+a)Z\right\}\right] \quad (7.152)$$

and $$\frac{d\gamma}{dZ} = N_9(1+a)(1-a)\frac{Pe_2}{2}\exp\left\{\frac{Pe_2}{2}(Z-aZ+a)\right\} \times\left[1-\exp\{-aPe_2(Z-1)\}\right] \quad (7.153)$$

from which:

$$\left(\frac{d\gamma}{dZ}\right)_{Z=1} = 0. \quad (7.154)$$

The boundary conditions for the situation of no diffusion in the pre-reaction zone were applied by Danckwerts (1953) and in terms of the preceding notation are:

At $Z = 0-$: $$\gamma = 1. \quad (7.155)$$

With bulk flow only in the one zone and both bulk flow and diffusion in the other, conservation of mass across the reaction inlet boundary gives:

$$Q\gamma(0-) = Q\gamma(0+) - \frac{AD_2}{L}\left(\frac{d\gamma}{dZ}\right)_{0+}. \quad (7.156)$$

Hence from (7.155) and (7.156):

$$\gamma(0-) = \gamma(0+) - \frac{1}{Pe_2}\left(\frac{d\gamma}{dZ}\right)_{0+} = 1. \quad (7.157)$$

It should be noted for this model that:

$$\left(\frac{d\gamma}{dZ}\right)_{0-} = 0 \quad (7.158)$$

and $$\gamma(0-) \neq \gamma(0+), \quad (7.159)$$

i.e. there is a discontinuity of both composition and of composition gradient at the reactor inlet.
Finally at the reactor outlet $Z = 1$:

$$\left(\frac{d\gamma}{dZ}\right)_{1-} = 0. \tag{7.160}$$

Evaluation of the constants in the general solution using equations (7.157) and (7.160) yields a result identical with equation (7.150) for the variation of the concentration throughout the reaction zone.

In particular reactor design is concerned with the conversion achieved at the reactor exit, and on putting $Z = 1$ into equation (7.150):

$$\gamma(1) = \frac{2\exp\left\{\frac{Pe_2}{2}\right\}(2a)}{(1+a)^2\exp\left\{a\frac{Pe_2}{2}\right\} - (1-a)^2\exp\left\{-a\frac{Pe_2}{2}\right\}} \tag{7.161}$$

and hence in terms of inlet and exit concentrations, reaction zone diffusion coefficient and reactor length:

$$\frac{c_1}{c_0} = \frac{4a}{(1+a)^2\exp\left\{-\frac{uL}{2D}(1-a)\right\} - (1-a)^2\exp\left\{-\frac{uL}{2D}(1+a)\right\}}. \tag{7.162}$$

While it is a straightforward matter to substitute values of L into this equation and thereby evaluate the reactor performance, a numerical iteration is required to determine the necessary reactor length for a specified fractional conversion. Equation (7.162) may be applied to the full range of diffusion coefficients and may be reduced to both the well mixed and plug flow results.

For very large values of D after neglecting the higher terms of the exponential expansions:

$$\frac{c_1}{c_0} = \frac{4a}{(1+a)^2\left(1 - \frac{uL}{2D} + \frac{uLa}{2D}\right) - (1-a)^2\left(1 - \frac{uL}{2D} - \frac{uLa}{2D}\right)} \tag{7.163}$$

Hence:
$$\frac{c_1}{c_0} = \frac{4a}{\dfrac{uLa}{D} + 2a\left(2 - \dfrac{uL}{D}\right) + a^2\left(\dfrac{uLa}{D}\right)} \tag{7.164}$$

and on substitution for a^2 from (7.117):
$$\frac{c_1}{c_0} = \frac{1}{1 + \dfrac{kL}{u}}. \tag{7.165}$$

For very small values of D:
$$a \to 1 \tag{7.166}$$
$$1 + a \to 2 \tag{7.167}$$
$$1 - a \to 0 \tag{7.168}$$

and
$$\exp\left\{-\frac{uL}{2D}(1 - a)\right\} \to \exp\left\{+\frac{kL}{u}\right\} \tag{7.169}$$

Hence:
$$\frac{c_1}{c_0} = \frac{4}{4\exp\{kL/u\} - 0} = \exp\{-kL/u\} \tag{7.170}$$

Example 7. The concentration profiles will be developed for a tubular reactor with diffusion for the cases of:

(*i*) equal diffusion coefficients in the pre-reaction and reaction zones such that $Pe = 1$;
(*ii*) different diffusion coefficients in the two zones with $Pe_1 = 1$ and $Pe_2 = \frac{8}{3}$;
(*iii*) zero diffusion in the pre-reaction zone and $Pe_2 = \frac{8}{3}$.

In each case the parameter R is taken as 2.
Case (*i*) Using equations (7.118), (7.148) and (7.147)
$$a = [1 + 4 \times 2]^{\frac{1}{2}} = 3 \tag{I}$$

$$N_9 = \frac{2}{16\exp(1{\cdot}5) - 4\exp(-1{\cdot}5)}$$
$$= \frac{\exp(1{\cdot}5)}{2[4\exp(3) - 1]} = 0{\cdot}0283 \tag{II}$$

and
$$\gamma(0) = \frac{2\exp(3) + 1}{4\exp(3) - 1} = 0{\cdot}519. \tag{III}$$

Hence from equation (7.137) or (7.149) the concentration in the pre-reaction zone is given by:

$$\gamma = 1 - 0{\cdot}481 \exp Z \qquad \text{(IV)}$$

and for the reaction zone from equation (7.150) by:

$$\gamma = 0{\cdot}0283 \exp(Z/2)\{4 \exp[1{\cdot}5(1 - Z)] + 2 \exp - [1{\cdot}5(1 - Z)]\}. \qquad \text{(V)}$$

Evaluations from (IV) and (V) are summarised in the table below and plotted as the curve GHI in Fig. 7.18.

Case (*iii*) The values of *a*, N_9 and $\gamma(0)$ are the same as for case (*ii*) and

$$a = 2 \qquad \text{(VI)}$$

$$N_9 = \frac{2 \exp(2{\cdot}667)}{9 \exp(5{\cdot}333) - 1} = 0{\cdot}01545 \qquad \text{(VII)}$$

and
$$\gamma(0) = \frac{2[3 \exp(5.333) + 1]}{9 \exp(5{\cdot}333) - 1} = 0{\cdot}668. \qquad \text{(VIII)}$$

Hence for the pre-reaction zone:

$$\gamma = 1 - 0{\cdot}333 \exp Z \qquad \text{(IX)}$$

and for the reaction zone:

$$\gamma = 0{\cdot}01545 \exp(1{\cdot}333\, Z)\{3 \exp[2{\cdot}667(1 - Z)] + \exp[-2{\cdot}667(1 - Z)]\}. \qquad \text{(X)}$$

Results from equations (IX) and (X) are also tabulated and plotted as curve DEF, and the discontinuity of gradient at $Z = 0$ is demonstrated.

Case (*iii*) The values of *a*. Ng and $\gamma(0)$ are the same as for case (*ii*) and hence the profile in the reaction zone is also identical. However, in the pre-reaction zone the zero diffusion coefficient leads to a constant value of $\gamma = 1$. The plot in this case is ABEF.

The variation of concentration with position in the pre-reaction and reaction zones of a tubular reactor with diffusion for cases (*i*) and (*ii*) is tabulated below.

z	$Pe_1 = 1 = Pe_2$ γ	$Pe_1 = 1; Pe_2 = \frac{8}{3}$ γ
−1·0	0·823	0·877
−0·8	0·784	0·850
−0·6	0·736	0·817
−0·4	0·677	0·778
−0·2	0·606	0·727
0	0·519	0·667
0·1	0·474	0·585
0·2	0·435	0·516
0·3	0·400	0·454
0·4	0·368	0·396
0·5	0·342	0·352
0·6	0·320	0·312
0·7	0·304	0·282
0·8	0·291	0·256
0·9	0·283	0·240
1·0	0·280 (5)	0·234

Problems

1. It is desired to study the characteristics of a continuously operating reactor by measurement of the distribution of residence times. For this purpose 2000 mg of an inert tracer fluid are quickly mixed into the main stream of feed entering the reactor. All the fluids are completely miscible. The concentration in mg litre^{-1} of tracer in the fluid leaving the reactor at various instants after the moment of addition, are found to be as follows.

t (min)	0·1	0·2	1·0	2·0	5·0	10·0	20·0	30·0
concentration mg litre^{-1}	1·96	1·93	1·64	1·34	0·74	0·27	0·034	0·004

The effective volume of the reactor is 1000 litre and the feed rate is 200 litre min^{-1}. Does the flow characteristics of the reactor approximate at all closely to that of an idealised type? What percentage conversion would be expected if the reaction is first order with $k = 1{\cdot}8$ min^{-1} and the reactor operation is isothermal?

ANS. 89·5 %

2. For a first order reaction achieving 99% conversion with a velocity constant of 0·5 h^{-1} use the simple solution to the diffusion equation to find the throughput for a tubular reactor of cross-

sectional area 0·25 ft^2 and volume 500 litres when the longitudinal diffusion coefficient is (*a*) 0·8 ft^{2}h^{-1} and (*b*) 10 ft^{2}h^{-1}.

ANS. (*a*) 53·3 litre h^{-1}; (*b*) 49·8 litre h^{-1}.

Substitute these answers into the more exact solution of the diffusion equation to check that the conversion is of the order of 99 %.

3. It is proposed to design a pilot plant for the production of allyl chloride. The reactants will consist of 4 moles propylene per mole of chlorine and will enter the reactor at 200°C. The reactor will be 2·0 in I.D. and the feed rate will be 0·85 lb mole h^{-1}. Estimate the length of reactor to accomplish 80·0% conversion of the chlorine assuming isothermal operation at 200°C in:
(*i*) a plug flow reactor,
(*ii*) if the flow characteristics of the fluid are taken into consideration.
If the feed rate is increased to 5·25 lb mole per hour, how would this affect (*i*) and (*ii*)?

Data:

(a) Viscosity of fluid in reactor 0·13 c.p.
(b) Density of fluid in reactor 0·08 lb ft^{-3}.

Reactions taking place are:

(*c*) $C_3H_6 + Cl_2 \rightarrow CH_2 = CH - CH_2Cl + HCl$
$r_a = k_1 p_p p_{Cl_2}$; $k_1 = 0{\cdot}0215$ ft^3 (lb mole)$^{-1}$ h^{-1}

(*d*) $C_3H_6 + Cl_2 \rightarrow CH_2Cl - CHCl - CH_3$
$r_b = k_2 p_p p_{Cl_2}$; $k_2 = 0{\cdot}2$ ft^3 (lb mole)$^{-1}$ h^{-1}

ANS. Plug flow: 28 ft for 0·85 lb moles/h^{-1}

4. Use the equation of the *F*-diagram for two equal tanks to show that:

(*i*) The area below below the diagram between $t = 0$ and one mean residence time of the whole system (V/Q) is equal to the area bounded by the diagram and the line $F(t) = 1$ between $t = V/Q$ and infinity. [Each equal to $2V/e^2Q$.]

(*ii*) 59·4% of the input at any time will have emerged after one mean residence time.

(*iii*) 99% of the input will have emerged between three and four mean residence times.

REFERENCES

CHOLETTE, A. and CLOUTIER, L. 1959. Mixing efficiency determinations for continuous flow systems. *Can. J. Chem. Engng*, **37**, 105.

DANCKWERTS, P. V. 1953. Continuous flow systems—distribution of residence times. *Chem. Engng Sci.*, **2**, 1.

FAN, L. T., and BAILIE, R. C. 1960. Axial diffusion in isothermal tubular flow reactors. *Chem. Engng Sci.*, **13**, 63.

LEVENSPIEL, O. 1958. Longitudinal mixing of fluids flowing in circular pipes. *Ind. Engng Chem.*, **50**, 343.

MACDONALD, R. W., and PIRET, E. L. 1951. Continuous stirring tank reactor; agitation requirements. *Chem. Engng Prog.*, **47**, 363.

MANNING, F. S., and WILHELM, R. H. 1963. Concentration fluctuations in a stirred baffled vessel. *A.I.Ch.E.Jl.*, **9**, 12.

RICE, A. W., TOOR, H. L., and MANNING, F. S. 1964. Scale of mixing in a stirred vessel. *A. I. Ch. E. Jl.*, **10**, 125.

WEHNER, J. F., and WILHELM, R. H. 1956. Boundary conditions of a flow reactor. *Chem. Engng Sci.*, **6**, 89.

ZWIETERING, TH. N. 1959. The degree of mixing in continuous flow systems. *Chem. Engng Sci.*, **11**, 1.

8

HETEROGENEOUS REACTORS: SOLID–FLUID REACTOR ANALYSIS

8.1 Introduction

In a very large number of chemical reaction processes the chemical reaction is sustained at an adequate rate by means of a solid catalyst. These processes are continuous and generally the catalyst, in the form of pellets, is packed in the reactor to make a 'catalyst bed'. The feed of reactants is made to flow through the packed bed where chemical reaction occurs at the gas–solid surface, and the reaction products so formed are discharged continuously from the reactor. Therefore the problems encountered in the design and analysis of solid–fluid catalytic reactors are similar to those of other continuous reactors. Thus the design or analysis will include kinetic studies, heat transfer studies and the fluid mechanical studies associated with the flow of the mixture through the catalyst bed. However, because of the presence of the solid catalyst, solution of these problems will be somewhat different so that separate consideration must be given.

8.2 Solid–Fluid Reaction Kinetics

It is a well established fact that a catalyst alters the rate of a chemical reaction without itself undergoing chemical change. This does not mean that the catalyst does not take part in the reaction. It has been shown that the physical condition of most solid catalysts is greatly modified after reaction. The way in which the catalyst accelerates or retards a chemical reaction differs with each reaction and numerous theories of solid–fluid catalysis have been and are still being proposed. However, it is generally agreed that the rate steps of all solid–fluid catalytic reactions are as set out in section 3.13.4 and are:

(*i*) Diffusion of the reactants from the bulk of the reaction mixture to the active sites which may be on the external or internal surface.

(*ii*) Adsorption of one or more of the reactants on the solid surface.

(*iii*) Activation and chemical reaction on the surface.

(*iv*) Desorption of the reaction products off the surface.

(*v*) Diffusion of the reaction products from the catalyst surface back via the pores into the bulk of the reaction mixture.

These steps occur in series at different rates, and, therefore, if one step is very much slower than the remaining ones it will control the reaction process. Consequently each step deserves some consideration.

8.2.1 Diffusion processes

The diffusion of reactants from the bulk mixture and of products into the bulk mixture are very similar processes. They are mass transfer processes and include diffusion through the fluid to the vicinity of the external catalyst surface and thereafter, if the catalyst is porous through the pores in the catalyst itself, to the active sites. In a great many solid-fluid catalytic reaction processes the diffusion processes are much faster than the other steps, with the result that they need not be considered when the overall rate equation is being formulated for the reaction process. These conditions persist when the velocity of the fluids over the catalyst is high. Under these conditions it is justifiable to assume that the concentration of the reactant, or product, at the surface of the catalyst is the same as that in the bulk of the fluid in the reactor. When these conditions do not exist in the reactor it is necessary to obtain the concentration of the reactants at the surface of the catalyst in order to evaluate the reaction rate. This may be done by applying the conventional mass transfer rate equation. For gases this is:

$$r_A = k_g a (p_{Ag} - p_{Ai}) \tag{8.1}$$

while for liquids it is:

$$r_A = k_L a (c_{AL} - c_{Ai}). \tag{8.2}$$

In these equations:

k with the appropriate suffix is the gas phase or liquid phase mass transfer coefficient,

a is the surface area between the fluid and the solid and will be discussed in more detail later in the section on adsorption,

p_{Ag} is the partial pressure of a gas phase reactant A in the bulk mixture,
p_{Ai} is the partial pressure of a gas phase reactant A at the interface,
c_{AL} is the concentration of a liquid phase reactant A in the bulk mixture,
and c_{Ai} is the concentration of a liquid phase reactant A at the interface.

The mass transfer coefficient depends on the flow conditions in the vicinity of the catalyst surface, the physical properties of the fluid and on the species transported. For a gas, Wilke and Hougen (1945) showed that the mass transfer coefficient could be correlated by one of the following equations depending on the Reynolds number with respect to the catalyst particle:

For $Re > 350$:

$$J_D = \frac{k_g M_m p_f}{G}\left(\frac{\mu}{\rho D}\right)^{\frac{2}{3}} = 0{\cdot}989\left(\frac{d_p G}{\mu}\right)^{-0{\cdot}41}. \tag{8.3}$$

For $Re < 350$:

$$J_D = \frac{k_g M_m p_f}{G}\left(\frac{\mu}{\rho D}\right)^{\frac{2}{3}} = 1{\cdot}82\left(\frac{d_p G}{\mu}\right)^{-0{\cdot}51}. \tag{8.4}$$

In equations (8.3) and (8.4):

G is the mass rate of flow,
M_m is the mean molecular weight of the gaseous phase,
D is the molecular diffusivity,
p_f is the partial pressure film factor (Houghen and Watson, 1947),
d_p is the diameter of the sphere having the same area as the pellet of catalyst,
and the other symbols have their normal significance.

Equations (8.1) and either (8.3) or (8.4) can be used to estimate the significance of the partial pressure gradient to the overall rate process in the reactor, and from this the dependence of the overall rate on the diffusion steps can be assessed.

Example 1. Olson, Schuler and Smith (1950) obtained the following results for the rates of oxidation of sulphur dioxide at different mass flow rates in a fixed bed catalytic reactor packed with $\frac{1}{8}$ in $\times$ $\frac{1}{8}$ in cylindrical catalyst pellets operating at 753°K.

Mean conversion f	Rate of reaction g mole SO_2/hr^{-1} g of catalyst	Partial Pressure p_g atm SO_2	SO_3	O_2
0·1	0·0956	0·0603	0·0067	0·201
0·6	0·0189	0·0273	0·0409	0·187

The superficial mass rate of flow of the gases through the reactor was 147 lb h^{-1} ft^{-2}, and the feed to the reactor consisted of 6·42 % SO_2 and 93·58 % air by volume. Using the following data, estimates are made for $(p_i - p_g)$ and hence p_i at each conversion level for the backward diffusion of SO_3.
(*a*) Surface area of catalyst: 5·12 ft^2 lb^{-1} catalyst.
(*b*) Diffusivity of SO_3 in air: 1·6 ft^2 h^{-1} (neglecting the effect of SO_2 on the diffusion).
(*c*) Viscosity of air at 753°K: 0·09 lb ft^{-1} h^{-1}.
(*d*) Total pressure 790 mmHg.

Solution. Estimation of the Reynolds number:

$$\frac{d_p G}{\mu}$$

where for a cylinder of diameter d_C and equal height l:

$$\pi d_p^2 = \pi d_C l + 2\frac{\pi}{4} d_C^2 = \pi d_C^2 \left(1 + \frac{2}{4}\right). \tag{I}$$

Hence with $d_C = 0{\cdot}125$ in,

$$d_p = \sqrt{\left(\frac{1{\cdot}5 \times 0{\cdot}125^2}{144}\right)} = 0{\cdot}0128 \text{ ft}. \tag{II}$$

Then
$$Re = \frac{0{\cdot}0128 \times 147}{0{\cdot}09} = 21. \tag{III}$$

Estimation of the Schmidt number:

$$\rho_{AIR} = \frac{29 \times 790 \times 273}{359 \times 753 \times 760} = 0{\cdot}03 \text{ lb ft}^{-3}. \tag{IV}$$

Then
$$Sc = \frac{\mu}{\rho D} = \frac{0{\cdot}09}{0{\cdot}03 \times 1{\cdot}6} = 1{\cdot}85 \tag{V}$$

Therefore using equation (8.4):

$$\frac{k_g M_m p_f}{G}(1{\cdot}85)^{\frac{2}{3}} = 1{\cdot}82 \times 21^{-0{\cdot}51}. \tag{VI}$$

Estimation of M_m from the experimental results:
The pressure of 790 mm is equivalent to 1·0395 atm

	p_g (atm)	Mole fraction	Molecular weight	Weight
SO_2	0·0603	0·058	64	3·72
SO_3	0·0067	0·006	80	0·51
O_2	0·2010	0·193	32	6·17
N_2	0·7715	0·743	28	20·78
TOTAL	1·0395	1·000		31·18

i.e. when $f = 0{\cdot}1$, M_m is approximately 31·2.
Similarly when $f = 0{\cdot}6$, M_m is 31·7.
p_f is very close to 1·0.
From the stoichiometry of the reaction:

$$SO_2 + \tfrac{1}{2}O_2 \rightarrow SO_3$$

one mole SO_2 yields one mole SO_3, and so for the case $f = 0{\cdot}1$:

$r = 0{\cdot}0956$ g mole SO_3 h^{-1} g^{-1} of catalyst
or 0·0956 lb mole SO_3 h^{-1} lb^{-1} of catalyst. (VII)

Then $$0{\cdot}0956 = k_g a(p_i - p_g) \tag{VIII}$$

or $$(p_i - p_g) = \frac{0{\cdot}0956}{5{\cdot}12\,k_g}$$

$$= \frac{0{\cdot}0956 \times 31{\cdot}2 \times (21{\cdot}0)^{0{\cdot}51} \times (1{\cdot}85)^{0{\cdot}67}}{5{\cdot}12 \times 147 \times 1{\cdot}82} = 0{\cdot}016 \text{ atm}. \tag{IX}$$

Similarly when the conversion is 0·6:

$$p_i - p_g = 0{\cdot}0032 \text{ atm} = \Delta p_g. \tag{X}$$

Summarising:

f	p_g	$p_i = p_g + \Delta p_g$
0·1	0·0067	0·0227
0·6	0·0409	0·0441

The difference between p_g and p_i is significant in comparison with the value of p_i and therefore basing the overall rate equation on p_g would lead to appreciable error.

When diffusion processes control the overall rate process the basic design equation will take the form developed below for a first order reaction. Thus let the reaction

$$A \rightleftharpoons B \tag{8.5}$$

take place on the surface of the catalyst.

Since the reaction is rapid compared with the diffusion processes, chemical equilibrium will be established on the surface. Thus if K_s is the surface equilibrium constant

$$K_s = p_{Bi}/p_{Ai}. \tag{8.6}$$

Also the mass transfer rate of reactant A is:

$$r_A = k_{gA}(p_{Ag} - p_{Ai}) \tag{8.7}$$

and of product B:

$$r_B = k_{gB}a(p_{Bi} - p_{Bg}). \tag{8.8}$$

From equations (8·6), (8.7) and (8.8):

$$\frac{r_A}{k_{gA}a} = p_{Ag} - p_{Ai} \tag{8.9}$$

and

$$\frac{r_B}{k_{gB}K_s a} = p_{Ai} - \frac{p_{Bg}}{K_s}. \tag{8.10}$$

The rates r_A and r_B are equal and the interfacial partial pressure p_{Ai} is eliminated by addition of equations (8.9) and (8.10):

$$r = \frac{1}{\dfrac{1}{k_{gA}} + \dfrac{1}{k_{gB}K_s}}\left(p_{Ag} - \frac{p_{Bg}}{K_s}\right) \tag{8.11}$$

where r is now the reaction rate per unit surface area of catalyst.

If r is expressed by a conventional specific reaction rate for a first order reversible reaction:

$$r = k_1\left(p_{Ag} - \frac{p_{Bg}}{K_s}\right) \tag{8.12}$$

where, by comparison of equations (8.11) and (8.12) and use of either (8.3) or (8.4):

$$\frac{1}{k_1} = \frac{1}{k_{gA}} + \frac{1}{k_{gB}K_s} = \frac{1}{k_{gA}}\left[1 + \frac{1}{K_s}\left(\frac{D_A}{D_B}\right)^{\frac{2}{3}}\right]. \qquad (8.13)$$

8.2.2 ADSORPTION STEPS

The adsorption of the reactants on the surface and desorption of the products involve surface chemistry studies. Thus it is now well established that adsorption processes can be divided into

(*a*) Physical Adsorption
and (*b*) Chemical Adsorption.

(*a*) *Physical Adsorption* is similar to condensation. Thus the heat of adsorption is of the same order of magnitude as heat of condensation. Furthermore equilibrium between the adsorbed gaseous molecule and the surface is established very rapidly. Since the adsorption activation energy is small (about 1000 cal (g mole)$^{-1}$) desorption is easily achieved and this suggests that the physical adsorption phenomenon is likely to have only a very small effect in catalysis.
(*b*) *Chemical Adsorption* is a much stronger type of adsorption. Langmuir stated that chemisorbed molecules are attached to the surface with a tenacity similar to valency forces. The heat of chemisorption may be as large as 100000 calories (g mole)$^{-1}$, similar to heats of chemical reaction. Thus the energy possessed by chemisorbed molecules can be very different from that of the molecules themselves. Consequently the activation energy of reactions involving chemisorbed molecules will be very much less than that for the same molecules in a homogeneous reaction. Hence in contrast to physical adsorption, chemisorption is likely to be a significant factor during catalytic reactions on solid surfaces.

Chemisorption is also called activated adsorption because the chemisorbed molecules are activated by a transfer of energy to them. The free energy of activation of such molecules is far greater than for those which are physically adsorbed. Hence the chemisorption process must be slower at low temperatures than the physical adsorption process.

Thus the mass of adsorbed gas per unit area of solid surface is high at low temperatures because of the relatively rapid rate of physical adsorption. As the temperature is increased the fall off of physical adsorption rate is more than compensated by the rise of the activated adsorption rate, with the result that the mass adsorbed passes through a minimum and then rises again. As the temperature is increased further the physical rate becomes negligible and the mass adsorbed chemically becomes limited by the equilibrium value and so passes through a maximum before finally diminishing with even further rise of temperature. There is thus a temperature band, enclosing the maximum, in which the amount of activated adsorbed gas is significant for chemical reaction. The situation is illustrated in Fig. 8.1 which shows the likely useful temperature band

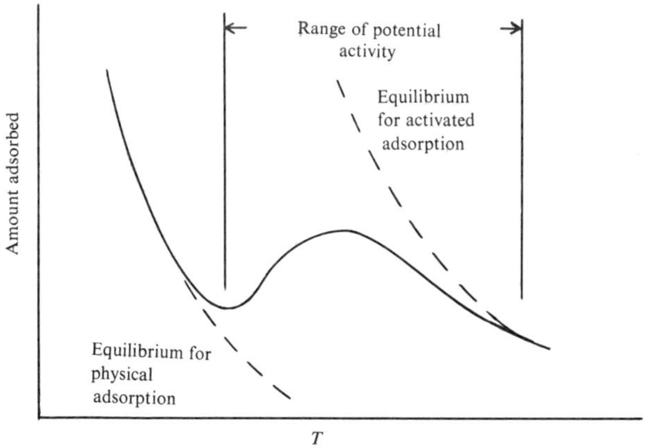

FIG. 8.1. Temperature range of potential catalyst activity

together with the tendency of the actual adsorption curve to coincide with the physical and chemical equilibrium curves at the temperature extremes. Such variation of adsorption with temperature has been confirmed experimentally and in addition it has been shown that the maximum amount of material that can be chemisorbed corresponds to a monomolecular layer. This would be expected from the fact that the forces involved are valency forces which diminish rapidly with distance.

The quantitative analysis of the extent of physical adsorption is based on the proposals of Freundlich, Polyani and Langmuir. Freundlich presented his adsorption isotherm as:

$$v = \phi p^{1/n} \tag{8.14}$$

where v is volume of gas adsorbed,
p is the gas pressure,
and n and ϕ are constants depending on the gas and the surface.

The Langmuir adsorption isotherm applies specifically to chemisorption and forms a basis for the analysis of chemical reactions as already considered in Chapter 3. If θ is the fraction of a solid surface covered by adsorbed molecules, then $(1 - \theta)$ is the fraction of surface available for adsorption. Thus the rate of adsorption r_a will be

$$r_a = kp(1 - \theta) \tag{8.15}$$

where k is an adsorption rate coefficient, and p is the gas pressure.

The rate of desorption will be proportional to the surface already covered. That is

$$r_d = k'\theta. \tag{8.16}$$

When equilibrium is established

$$\theta = \frac{kp}{k' + kp} = \frac{K_A p}{1 + K_A p} \tag{8.17}$$

where K_A is the adsorption equilibrium constant.
The development of equation (8.17) is based on the assumption that the whole surface behaves similarly towards adsorption. If only certain parts of the surface are active, these are called 'active centres' and then θ would be the fraction of the active centres already occupied.

Equation (8.17) forms the basis of the derivation of catalytic rate equations when equilibrium adsorption–desorption exists, while equations (8.15) and (8.16) may be utilised to obtain the net rate of adsorption under non-equilibrium conditions. This latter situation will next be considered. Thus let the total number of active centres per unit area of catalyst be S, and let S_v be the number of vacant active centres per unit area. Then the rate of adsorption of the gas, on the assumption that one molecule of a gas A occupies one site, will be:

$$r'_a = kps_v \tag{8.18}$$

where r'_a is expressed in molecules adsorbed per unit area of surface. If the surface area per unit mass of catalyst is S_g then the rate in gram moles per unit time, per unit mass of catalyst will be related to r'_a by

$$r_a = \frac{r'_a S_g}{N_0} = \frac{kpS_v S_g}{N_0} = kp\,c_v \qquad (8.19)$$

where N_0 is Avogadro's number, and the parameter c_v (equivalent to $S_v . S_g/N_0$) represents the molal concentration of adsorbed sites per unit mass of catalyst. Similarly the rate of desorption r_d will depend on the concentration of adsorbed molecules. Let S_A be the number of active centres occupied by A per unit area of surface. Then the molal surface concentration of adsorbed A is:

$$c_A = \frac{S_A S_g}{N_0} \qquad (8.20)$$

and the rate of desorption is:

$$r_d = k'\,c_A. \qquad (8.21)$$

Then the net rate of adsorption of A will be

$$r_A = r_a - r_d = k_A p_{Ai} c_v - k'_A\,c_A = k_A\left(p_{Ai}\,c_v - \frac{c_A}{K_A}\right) \qquad (8.22)$$

where p_{Ai} is the pressure of A at the surface of the catalyst.

8.2.3 Surface reaction

When the reaction process is controlled by the rate of the surface reaction it is necessary to establish whether chemical reaction occurs

(*a*) between an adsorbed molecule and an adjacent molecule in the gas film at the surface, or

(*b*) between two adsorbed molecules on adjacent active centres. Thus for a reaction of the type:

$$A + B \rightarrow C \qquad (8.23)$$

the rate equation for the case of A adsorbed and B not adsorbed would be developed as follows. The process could be written stoichiometrically as:

$$A + S \rightarrow A.S$$

followed by

$$A.S + B \rightarrow C.S \qquad (8.24)$$

where A . S and C . S represent the combination of A or C with a single active site. Then the rate equation becomes:

$$r_S = k_S c_A p_{Bi} - k_S' c_C = k_S \left(c_A p_{Bi} - \frac{c_C}{K_S} \right) \tag{8.25}$$

where k_S is the second order rate constant for the forward surface reaction,

and K_S is the surface reaction equilibrium constant.

The stoichiometric equation for the case of reaction between A and B both adsorbed on adjacent active sites in terms of the above notation would be:

$$A.S + B.S \rightarrow C.S + S \tag{8.26}$$

However, for reaction to occur, molecules of A must be adsorbed on sites adjacent to those of B. Hence the rate of the forward reaction will depend directly on the concentration of pairs of adjacent sites occupied by A and B. This concentration may be expressed as the product of the actual concentration of A, and the fraction of the adjacent centres occupied by B molecules. Thus if C_t is defined as the molal concentration of total centres, the fraction of total surface occupied by B molecules will be C_B/C_t and this fraction will be proportional to the fraction of adjacent centres occupied by B molecules. The rate of the forward reaction will then be:

$$r_f = \frac{k_S c_A c_B}{c_t}. \tag{8.27}$$

The reverse reaction rate is similarly dependent on the juxtaposition of an adsorbed C molecule and a vacant site. For the concentration of vacant sites of c_V, the reverse reaction rate is:

$$r_b = k_S' \frac{c_C c_V}{c_t}. \tag{8.28}$$

Combining equations (8.27) and (8.28) gives the net rate of the surface reaction

$$r_S = \frac{k_S c_A c_B - k_S' c_C c_V}{c_t} = \frac{k_S}{c_t} \left(c_A c_B - \frac{c_d c_V}{K_S} \right). \tag{8.29}$$

Equations (8.25) and (8.29) are two rate equations for different mechanisms of the surface chemical reaction represented by the stoichiometric equation (8.23). Different second order reactions would result in different mechanisms and these two rate equations illustrate how the form of rate equation may be developed for different models of the detailed mechanism.

8.2.4 OVERALL RATE EQUATIONS

In the above sections, rate equations for each of the steps in the overall process have been considered. In an actual reaction, more than one step may be slow and as a consequence all steps except the fast ones must be included in the analysis. The reaction of equation (8.24) is considered with adsorbed A reacting with adsorbed A and it is assumed that all steps are significant in determining the overall rate. The rates of each step are equal and are denoted by r.

The equimolar diffusion of A, B and C demanded by the stoichiometry give:

$$r = k_{gA}a(p_{Ag} - p_{Ai}) \tag{8.30}$$

$$r = k_{gB}a(p_{Bg} - p_{Bi}) \tag{8.31}$$

$$r = -k_{gC}a(p_{Cg} - p_{Ci}) \tag{8.32}$$

The net rates of adsorption and desorption of reactants and products are given by:

$$r = k_A\left(p_{Ai}c_V - \frac{c_A}{K_A}\right) \tag{8.33}$$

$$r = k_B\left(p_{Bi}c_V - \frac{c_B}{K_B}\right) \tag{8.34}$$

$$r = -k_C\left(p_{Ci}c_V - \frac{c_C}{K_C}\right) \tag{8.35}$$

The rate of reaction on the solid surface is:

$$r = \frac{k_S}{c_t}\left(c_A c_B - \frac{c_C c_V}{K_S}\right) \tag{8.36}$$

It is possible to eliminate from these seven equations the interfacial partial pressures and the surface concentrations c_A, c_B and c_C to obtain an expression for r in terms of the readily measurable bulk partial pressures and c_V. Thus from equations (8.30) and (8.34), elimination of p_{Ai} by re-arrangement and addition gives:

$$\frac{r}{k_A c_V} + \frac{r}{k_{gA}a} = p_{Ag} - \frac{c_A}{c_V K_A} \tag{8.37}$$

or

$$c_A = K_A \cdot \left[c_V\left(p_{Ag} - \frac{r}{k_{gA}a}\right) - \frac{r}{k_A}\right]. \tag{8.38}$$

Similar expressions for c_B and c_C may be developed from the appropriate pairs of equations and then substituted into equation (8.36) to give:

$$r = \frac{k_S}{c_t}\left\{K_A K_B\left[c_V\left(p_{Ag} - \frac{r}{k_{gA}a}\right) - \frac{r}{k_A}\right]\left[c_V\left(p_{Bg} - \frac{r}{k_{gB}a}\right) - \frac{r}{k_B}\right] - \frac{K_C c_v}{K_S}\left[c_V\left(p_{Cg} + \frac{r}{k_{gC}a}\right) + \frac{r}{k_C}\right]\right\}. \tag{8.39}$$

An expression for the concentration of vacant active sites is formed by substituting equations of the type:

$$c_t = \frac{Sa}{N_0} \tag{8.40}$$

into

$$S = S_A + S_B + S_C + S_V \tag{8.41}$$

to give:

$$c_t = c_A + c_B + c_C + c_V. \tag{8.42}$$

Equation (3.38) and the similar expressions for c_B and c_C may again be used to eliminate them from equation (8.42) yielding an expression for c_V in terms of r and the constant c_t which is a basic physical property for a given catalyst:

$$c_V = \frac{c_t + r\left[\dfrac{K_A}{k_A} + \dfrac{K_B}{k_B} - \dfrac{K_C}{k_C}\right]}{1 + K_A\left[p_{Ag} - \dfrac{r}{k_{gA}a}\right] + K_B\left[p_{Bg} - \dfrac{r}{k_{gA}a}\right] + K_C\left[p_{Cg} + \dfrac{r}{k_{gC}a}\right]} \tag{8.43}$$

Equations (8.39) and (8.43) yield a quartic polynomial in 'r' if c_V is eliminated. These equations may be readily degenerated to give simpler results when some of the steps are relatively rapid.

Thus when the surface reaction is controlling, k_{Ag}, k_{Bg}, K_{gC}, k_A, k_B and k_C are all large so that equations (8.39) and (8.43) on letting these six constants tend to infinity give respectively:

$$r = \frac{k_S}{c_t}\left[K_A K_B c_V^2 p_{Ag} p_{Bg} - \frac{K_C c_V^2}{K_S} p_{Cg}\right] \tag{8.44}$$

and
$$c_V = \frac{c_t}{1 + K_A p_{Ag} + K_B p_{Bg} + K_C p_{Cg}}. \tag{8.45}$$

Elimination of c_V then gives:

$$r = \frac{k_S\left[K_A K_B p_{Ag} p_{Bg} - \dfrac{K_C}{K_S} p_{Cg}\right]}{c_t\,[1 + K_A p_{Ag} + K_B p_{Bg} + K_C p_{Cg}]^2}. \tag{8.46}$$

As another example, if adsorption of A is slow while k_S, k_{Ag}, k_{Bg}, k_{Cg}, k_B, k_C all tend to infinity, then:

$$K_A K_B c_V\left[c_V p_{Ag} - \frac{r}{k_A}\right] - \frac{K_C c_V^2}{K_S} p_{Cg} = 0 \tag{8.47}$$

and
$$c_V = \frac{c_t + \dfrac{r K_A}{k_A}}{1 + K_A p_{Ag} + K_B p_{Bg} + K_C p_{Cg}} \tag{8.48}$$

which yield r as:

$$r = \frac{k_A c_t\left[p_{Ag} - \dfrac{K_C p_{Cg}}{K_A K_B K_S p_{Bg}}\right]}{1 + K_B p_{Bg} + K_C p_{Cg} + \dfrac{K_C}{K_B K_S}\dfrac{p_{Cg}}{p_{Bg}}}. \tag{8.49}$$

The combination of equilibrium constants:

$$\frac{K_A K_B K_S}{K_C}$$

is the equilibrium constant for the corresponding homogeneous reaction predicted from the standard free energy change of the reaction.

In this analysis, four equilibrium constants, seven rate constants and the quantity c_t for the number of active sites are needed to represent the model of the overall reaction process. In fact very many more rate equations such as (8.46) and (8.49) could be developed from the numerous combinations of finite values of the seven rate constants. While attempts might be made to fit experimental rate data to these equations, the likelihood of selecting the correct one is small. Since it is impracticable to develop an equation for the actual mechanism, when surface factors control it is possible to use simpler forms of the rate equation for design purposes. Thus it has been found that usually the rate data is represented satisfactorily either by:

(*i*) a first order irreversible or reversible equation:

$$r = kp_{Ag} \tag{8.50}$$

or

$$r = k(p_{Ag} - p_{Ae}) \tag{8.51}$$

in which p_{Ae} is the equilibrium partial pressure of the reactant A on the surface.

(*ii*) an *n*th order irreversible rate equation:

$$r = kp_{Ag}^{n} \tag{8.52}$$

or (*iii*) simplified expressions based on the active site theory such as:

$$r_A = \frac{kp_{Ag}}{1 + k_1 p_{Ag}} \tag{8.53}$$

or

$$r_A = \frac{k(p_{Ag} - p_{Ae})}{1 + k_1 c_A} \tag{8.54}$$

or

$$r_A = \frac{kp_{Ag}}{(1 + k_1 p_{Ag})^2} \tag{8.55}$$

or

$$r_A = \frac{k(p_{Ag} - p_{Ae})}{(1 + k_1 p_{Ag})^2}. \tag{8.56}$$

8.2.5 Qualitative Predictions from the Active Site Theory

The real value of the active site theory for purposes of analysis and design is that the rate expressions give a qualitative picture of what may happen on extrapolation to new operating conditions. Thus, the effects of a change in the operating conditions can be deduced as follows:

8.2.5.1 *Pressure*

In general for a catalytic reaction process whose rate is controlled by adsorption, at low pressure, the rate equation approximates to a first order reaction. As the pressure is increased a larger number of molecules will be adsorbed and eventually the active centres will become saturated and if the reaction rate is slow the surface reaction rate will control. The increase in rate, with increase in pressure, will attain a maximum and may decrease on further increase of pressure (Fig. 8.2). If the reaction rate is rapid but the products are not desorbed sufficiently rapidly the surface will become saturated and the rate will be controlled by the desorption process. In this case also the rate will level off.

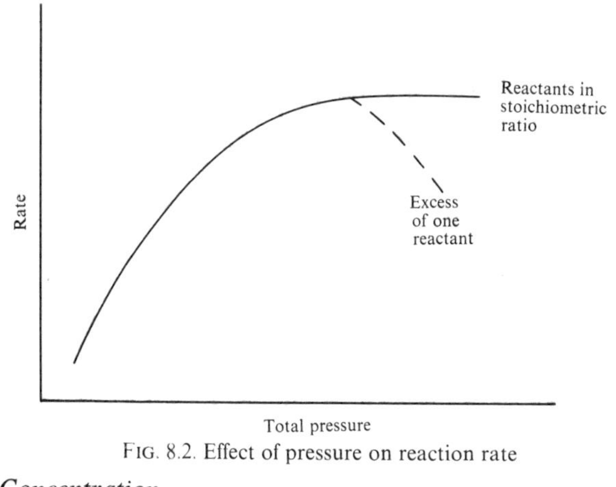

FIG. 8.2. Effect of pressure on reaction rate

8.2.5.2 *Concentration*

If the rate of reaction is controlled by the desorption step, increase in reactant concentration will have little affect on the rate because the desorption process is an equilibrium process between the product molecules attached to an active site and the free product molecules.

If the process is controlled by the surface reaction rate and the reaction involves two reacting species A and B adsorbed with, say, A in excess on the surface, then an increase in concentration of B will effect an increase in the rate. Increase in A will provide an even larger excess of this reactant, thereby reducing the rate.

8.3 Height of Reactor Unit

In the design of mass transfer equipment such as absorbers, extractors and packed distillation equipment the height of the mass transfer column is based on the 'Height of a Transfer Unit'. This concept can be applied to the mass transfer process occurring in the vicinity of the catalyst particle.

Thus the H.T.U. for a gas A is:

$$\text{H.T.U.} = \frac{h}{\int_{p_2}^{p_1} \frac{dp_{\text{Ag}}}{(p_{\text{Ag}} - p_{\text{Ai}})}} = \frac{h}{\text{N.T.U.}} \tag{8.57}$$

where $(p_{\text{Ag}} - p_{\text{Ai}})$ is the driving force and p_1 and p_2 the limits on p_{Ag} at the ends of the catalytic reactor. The integral is a measure of the difficulty of the transfer process and is assigned the Number of Transfer Units for the duty. Hurt (1943) extended this concept to gas solid catalytic reactors by introducing the following two additional units:

(*a*) The height of a catalytic unit (H.C.U.):

$$\text{H.C.U.} = \frac{h}{\int \frac{dp_{\text{Ag}}}{(p_{\text{Ai}} - p^*_{\text{Ag}})}} \tag{8.58}$$

where p^*_{Ag} is the pressure of A in equilibrium with the reaction products, and

(*b*) The height of a reactor unit (H.R.U.):

$$\text{H.R.U.} = \frac{h}{\int \frac{dp_{\text{Ag}}}{(p_{\text{Ag}} - p^*_{\text{Ag}})}}. \tag{8.59}$$

The logic of Hurt's proposals is that:

$$p_{\text{Ag}} - p^*_{\text{Ag}} = (p_{\text{Ag}} - p_{\text{Ai}}) + (p_{\text{Ai}} - p^*_{\text{Ag}}) \tag{8.60}$$

and substitution for these differences from the differential forms of the above equations, gives:

$$\text{H.R.U.} = \text{H.T.U.} + \text{H.C.U.} \tag{8.61}$$

From equation (8.57) H.T.U. is dependent on the mass transfer of the reactant to the catalyst surface and is hence a measure of bulk diffusional resistance. In equation (8.58) the difference in pressure between p_{Ai} and the partial pressure of A in equilibrium with the products on completion of the reaction, is a measure of the surface rate processes. Their sum is the height of a reactor unit H.R.U.

Example 2. The following data was obtained by Olson, Schuler and Smith (1950) for the oxidation of SO_2 in a differential reactor by $\frac{1}{8}$ in $\times$ $\frac{1}{8}$ in cylindrical platinum catalyst pellets:

(*i*) Temperature 673°K
(*ii*) Total pressure 1·04 atm.
(*iii*) Flow rate 350 lb h^{-1} ft^{-2}
(*iv*) Total Moles 62·45
(*v*) Moles SO_2 entering 4·074
(*vi*) Height of bed 0·0313 ft
(*vii*) Mass of catalyst 10·5 g
(*viii*) Conversion entering reactor $f = 0{\cdot}5937$
(*ix*) Conversion leaving reactor $f = 0{\cdot}6063$
(*x*) Equilibrium Constant 620 for reaction $SO_2 + \frac{1}{2}O_2 \rightarrow SO_3$
(*xi*) The Schmidt number for SO_2 in air is 1·28 at 673°K
(*xii*) Viscosity of air 0·077 lb hr^{-1} ft^{-1} at 673°K

Calculation of H.R.U.:

The Reynolds number $\dfrac{Gd_p}{\mu} = \dfrac{1}{8 \times 12} \times \dfrac{350}{0{\cdot}077} = 47.$

For $\frac{1}{8}$ in cylinders and this *Re*, the plot by Hurt (1943) gives:

$$\frac{\text{H.T.U.}}{(Sc)^{2/3}} = 0{\cdot}35 \text{ in.}$$

Hence H.T.U. $= 0{\cdot}35 \times 1{\cdot}28^{2/3} = 0{\cdot}35 \times 1{\cdot}18 = 0{\cdot}41$ in $= 0{\cdot}035$ ft.

The equilibrium value of the partial pressure of SO_2 in the reactor $(p_g^*)_{SO_2}$ is obtained from the equilibrium conversion by the procedure given in Chapter 2 on Chemical Equilibria. Application to this case gives:

$$\text{Mole per cent of } SO_2 = \frac{4{\cdot}074}{62{\cdot}45} = 6{\cdot}5\%$$

$$\text{Moles } O_2 + N_2 = 62{\cdot}45 - 4{\cdot}074 = 58{\cdot}376$$

$$O_2 = \frac{0{\cdot}21 \times 58{\cdot}376}{62{\cdot}45} = 19{\cdot}5 \text{ mole } \%$$

$$N_2 = 74{\cdot}0 \text{ mole } \%.$$

Then based on the 100 moles of mixture:

$$620 = \frac{6{\cdot}5f(100 - 3{\cdot}25f)^{\frac{1}{2}}}{6{\cdot}5(1 - f)(19{\cdot}5 - 3{\cdot}25f)^{\frac{1}{2}}(1{\cdot}04)^{\frac{1}{2}}} \qquad \text{(I)}$$

or

$$f = 0{\cdot}995$$

and

$$(p_g^*)_{SO_2} = \frac{6{\cdot}5(1 - f)\,1{\cdot}04}{100 - 3{\cdot}25f} = 0{\cdot}0003 \simeq 0.$$

For this zero value of equilibrium partial pressure, the evaluation of the integral of equation (8.59) gives:

$$\text{H.R.U.} = \frac{h}{\displaystyle\int_{p_2}^{p_1} \frac{dp_{Ag}}{p_{Ag}}} = \frac{h}{\ln \dfrac{p_1}{p_2}}. \qquad \text{(II)}$$

The numerical values of the limits are:

$$p_1 = \frac{1{\cdot}04\,(\text{Moles } SO_2 \text{ entering})}{\text{Total moles entering}}$$

Moles SO_2 in $= 6{\cdot}5(1 - f_1)$
Total Moles in $= 100 - 3{\cdot}25f_1$

$$p_1 = \frac{1{\cdot}04 \times 6{\cdot}5(1 - 0{\cdot}5937)}{100 - 3{\cdot}25 \times 0{\cdot}5937} = 0{\cdot}0280.$$

In the same way $p_2 = 0{\cdot}02715$.
Then

$$\text{H.R.U.} = \frac{0{\cdot}0313}{\ln\left(\dfrac{0{\cdot}0280}{0{\cdot}02715}\right)} \simeq 1{\cdot}02 \text{ ft.}$$

8.4 Effectiveness of Catalysts

The effectiveness of a catalyst for accelerating a chemical reaction is dependent on the number of active sites on the catalyst surface. These must be accessible to the reactants and therefore the production of the most efficient catalyst is not merely the production of a solid with the largest possible surface area; although the larger the surface area, the greater the possibility of increasing the number of active centres. Thus a porous material has a far larger surface area than a non-porous solid of the same volume. Yet the interior surface of a porous catalyst will not be as effective as the outer surface unless the active sites within the pores are readily accessible. This is due to the resistance to diffusion in the pores and since in many catalysts a very large fraction of the surface area is within the pores, the effectiveness of the catalyst is closely related to the resistance to diffusion within the pores.

The analysis of the effects of internal diffusion is complicated because of:

(*i*) the complex pore paths and variable cross-sectional area of the pores,

(*ii*) The relative effects of surface reaction and diffusion.

However, some idea of the importance of diffusion in the catalyst pores may be obtained by the following analysis.

8.4.1 Surface area, voidage, mean pore radius and mean pore length

In order to be able to assess the contribution the pore surface area makes to a catalytic reaction process it is necessary to know the pore radius and the way in which pores may be interconnected. This can be done as follows.

The surface area per gram of catalyst S_g can be estimated by the 'Brunauer–Emmett–Teller' method which consists of adsorbing adsorbed against the pressure. The temperature is controlled to somewhere near the boiling point of the gas, e.g. for CO_2 the temperature would be near to $-78°C$. The type of curve obtained (Fig. 8.3) has a steep gradient at low pressure, then flattens out to a linear section at intermediate pressures after which there is an increase in slope again at higher pressures. Emmett and Teller showed that the part of the curve up to the flattening corresponded to the formation of a complete monomolecular layer, and that the subsequent portion described the build-up of further layers of gas. They showed that the multi-layer system was represented by the equation:

$$\frac{p}{v(p_0 - p)} = \left(\frac{C - 1}{v_m C}\right)\frac{p}{p_0} + \frac{1}{v_m C} \qquad (8.61)$$

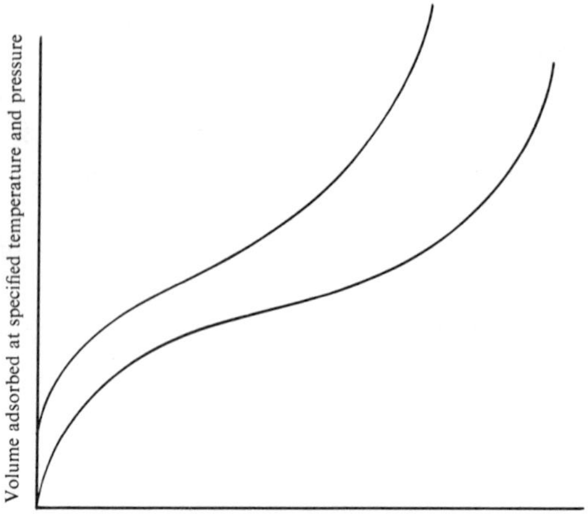

FIG. 8.3. Adsorption isotherms

where: v_m is the volume adsorbed in forming a complete monolayer.
v is the volume adsorbed at pressure p.
p_0 is the saturated vapour pressure of adsorbent at the temperature of the experiment.

Thus by plotting $[p/v(p_0 - p)]$ against $[p/p_0]$ a straight line results, the slope of which is equal to $[(C - 1)/v_m C]$ and whose intercept is $[1/v_m C]$. From these two parameters the values of v_m and the constant C can be obtained. When v_m has been evaluated the value of S_g is obtained from the relation

$$S_g = \left[\frac{v_m N_0}{V_m}\right]\phi \tag{8.62}$$

where ϕ is the 'parking area' per molecule,
V_m is the molar volume at the pressure and temperature of the experiment,
N_0 is Avogadro's number.

Typical values of S_g are:

Activated carbon 1500 m^2g^{-1}
SiO_2 or Al_2O_3 200 to 500 m^2g^{-1}.

In order to evaluate the pore radius it is necessary to estimate the pore volume per unit mass, V_g. This is accomplished by the helium-mercury displacement method. The volume of helium displaced by a known mass of catalyst is measured. This gives the actual volume occupied by the solids within the catalyst. Then the volume of mercury displaced by a known mass of catalyst is estimated at normal pressures. Since mercury will not enter the pores at such pressures the displaced volume represents the external volume of the catalyst. The displacements per unit mass of catalyst are then evaluated and the difference between these values gives the volume occupied by the pores within a unit mass of the catalyst.

Let V_H be the helium displaced per unit mass. Then the solid density ρ_C is:

$$\rho_C = \frac{1}{V_H}. \tag{8.63}$$

Similarly if V_M is the volume of mercury displaced per unit mass, the particle density ρ_P is:

$$\rho_P = \frac{1}{V_M}. \tag{8.64}$$

If the difference between these volumes is denoted V_g, the porosity δ within the particles is:

$$\delta = \frac{V_g}{V_M} = \frac{V_M - V_H}{V_M} = 1 - \frac{V_H}{V_M}, \tag{8.65}$$

i.e.

$$\delta = 1 - \frac{\rho_P}{\rho_C}. \tag{8.66}$$

The porosity may alternatively be expressed in terms of the pore volume, on elimination of V_M from equation (8.65), as:

$$\delta = \frac{V_g \rho_C}{1 + V_g \rho_C}. \tag{8.67}$$

Then, assuming that a catalyst particle of mass m_p having surface area S_g and void volume V_g g^{-1} contains n pores of mean length $\bar{L}$ and mean radius $\bar{r}$:

$$m_p S_g = 2\pi\bar{r}\,(n\bar{L}) \tag{8.68}$$

and

$$m_p V_g = \pi\bar{r}^2\,(n\bar{L}) \tag{8.69}$$

which on division give:

$$\bar{r} = \frac{2V_g}{S_g}. \tag{8.70}$$

In order to assess the diffusion it is also necessary to estimate the mean pore length. Thus if the area of the total external surface of the catalyst particle is S_x, the area of the pore holes in the surface is δS_x on the assumption that the void fraction of the external surface is the same as that for the internal solid. Then the number of pores per particle is:

$$n = \frac{S_x \delta}{\pi \bar{r}^2 \sqrt{2}}. \qquad (8.71)$$

The factor $\sqrt{2}$ is introduced because the pore holes, on average, reach the surface at 45° and the intersections are therefore ellipses rather than circles. By means of equations (8.69) and (8.71) an expression for $\bar{L}$ is obtained:

$$\bar{L} = \frac{m_p V_g}{S_x \delta} \sqrt{2}. \qquad (8.72)$$

8.4.2 Reaction on internal pore surface

Assuming the pore geometry to be known, consider a reaction in a pore of length $\bar{L}$ and radius $\bar{r}$. Let the concentration at the catalyst surface at the mouth of the pore be $C_{A.0}$ and let a first order reaction take place on the surface of the pores.

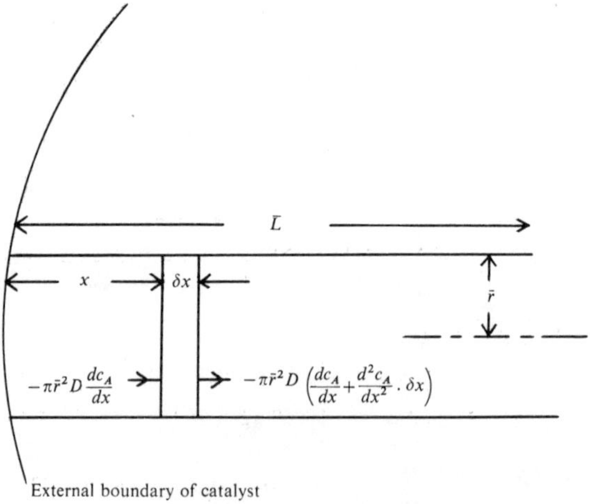

FIG. 8.4. Diffusion and reaction in a catalyst pore

A mass balance on an element (Fig. 8.4) of the pore, of thickness δx at distance x inside is

$$-\pi\bar{r}^2 D\frac{dc_A}{dx}$$

$$= -\pi\bar{r}^2\left\{D\frac{dc_A}{dx} + \frac{d}{dx}\left(D\frac{dc_A}{dx}\right)\delta x\right\} + 2\pi\bar{r}\delta x k_S c_A \qquad (8.73)$$

or

$$D\frac{d^2c_A}{dx^2} - \frac{2k_S c_A}{\bar{r}} = 0. \qquad (8.74)$$

This equation has a general solution:

$$c_A = A\,e^{\alpha x} + B\,e^{-\alpha x} \qquad (8.75)$$

where

$$\alpha = \left[\frac{2k_S}{D\bar{r}}\right]^{\frac{1}{2}} \qquad (8.75a)$$

with the boundary conditions that at $x = 0$, $c_A = c_{A,0}$ and at $x = \bar{L}$, $dc_A/dx = 0$ equation (8.75) yields:

$$c_{A0} = A + B \qquad (8.76)$$

and

$$A\,e^{\alpha\bar{L}} = B\,e^{-\alpha\bar{L}}. \qquad (8.77)$$

Hence the solution of equation (8.74) is in this case:

$$c_A = c_{A.0}\frac{\cosh m(1 - x/\bar{L})}{\cosh m} \qquad (8.78)$$

where

$$m = \bar{L}\left[\frac{2k_S}{D\bar{r}}\right]^{\frac{1}{2}}. \qquad (8.79)$$

The rate of reaction in the pore r_p will be the rate of diffusion into the mouth of the pore, which is:

$$r_p = -\pi\bar{r}^2 D\left(\frac{dc_A}{dx}\right)_{x=0} = \frac{\pi\bar{r}^2 D}{\bar{L}}\,c_{A.0}\,m\,\mathrm{Tanh}\,m. \qquad (8.80)$$

If there were no diffusional resistance the concentration of A would be $c_{A.0}$ at all points in the pore of the catalyst and then the reaction rate r_{max} for the pore would be greater than r_p and have the value:

$$r_{max} = 2\pi\bar{r}\bar{L}k_S c_{A,0}. \qquad (8.81)$$

Then the effectiveness of the catalyst E may be defined by comparing these two rates:

i.e.
$$E = \frac{r_p}{r_{max}} = \left(\frac{\bar{r}D}{2k_S\bar{L}^2}\right) m \operatorname{Tanh} m = \frac{\operatorname{Tanh} m}{m}. \qquad (8.82)$$

Equation (8.82) gives the 'effectiveness factor' E of the catalyst in terms of the modulus m.

In the above derivation the rate constant k_S has been based on unit surface area of catalyst. It is more convenient to base the rate on one gram of catalyst. Thus, the rate constant per gram of catalyst k_1 and k_S are connected by:

$$k_1 = k_S S_g. \qquad (8.83)$$

This rate constant is obtained from the experimental value k_{expt} through the relationship

$$r = k_{expt}\, c_{A.0} = k_1 E c_{A.0}. \qquad (8.84)$$

Finally using equation (8.55) the modulus m is related to the properties of the catalyst and an effective pore diffusion coefficient D_S is introduced instead of D to give:

$$m = \frac{2V_p}{S_x} \sqrt{\left(\frac{k_1}{2V_g D_S}\right)} = \frac{m_p}{S_x \delta} \sqrt{\left(\frac{2k_1 V_g}{D_S}\right)}. \qquad (8.85)$$

The Effective Diffusion Coefficient D_S will depend on the pore size. If the pore diameter is greater than the 'mean free path' of the molecules, diffusion will be similar to that in the gas phase adjacent to the catalyst. Satterfield and Sherwood (1963), propose that under these conditions

$$D_S = \frac{\delta}{\tau} D_M \qquad (8.86)$$

where δ is the porosity given by equations (8.66) or (8.67), τ is the 'Tortuosity', and D_M is the molecular diffusivity.

If the pore size is less than the mean free path the rate of diffusion along the pores will be determined by the number of collisions with the walls of the pore. Under these conditions Knudsen diffusion occurs and the diffusion coefficient is:

$$D_k = 9{\cdot}7 \times 10^3 \bar{r} \sqrt{\left(\frac{T}{M}\right)} \text{cm}^2\,\text{sec}^{-1}. \qquad (8.87)$$

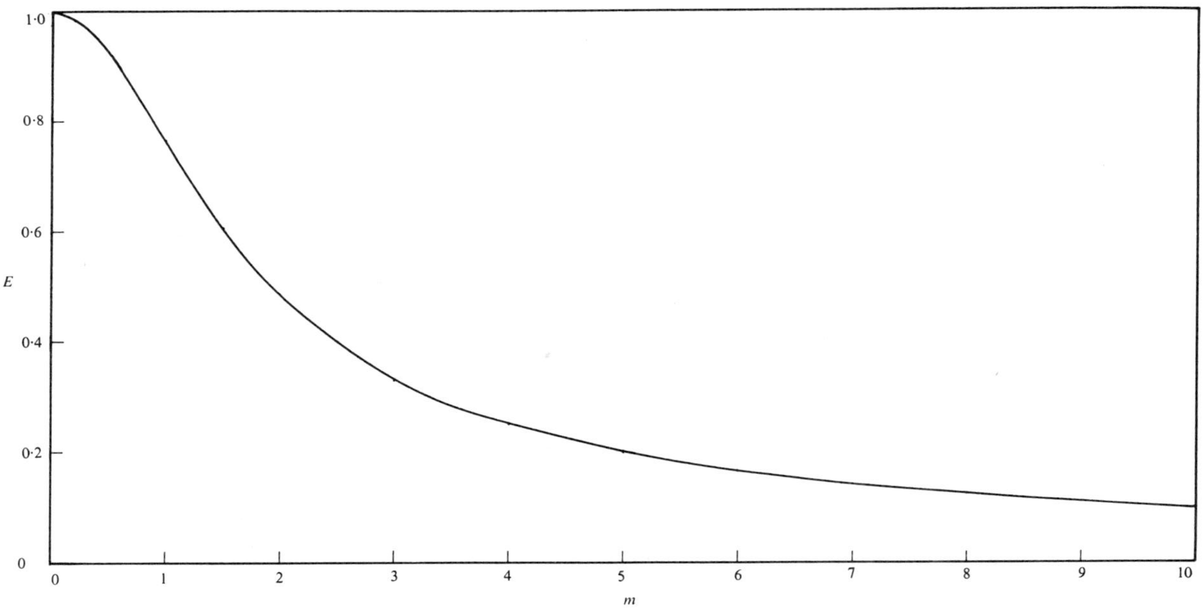

FIG. 8.5. Catalyst effectiveness as a function of the modulus m for a first order reaction

Again allowances must be introduced for the porosity and tortuosity so that under these conditions:

$$D_S = \frac{\delta}{\tau} D_k. \tag{8.88}$$

In the pores of a number of catalysts both types of diffusion may be significant. When this is the case Pollard and Present (1948) found that a good approximation for D_S was:

$$\frac{1}{D_S} = \frac{\tau}{\delta}\left(\frac{1}{D_M} + \frac{1}{D_k}\right). \tag{8.89}$$

The relationship between E and m as given by equation (8.82) is plotted in Fig. 8.5. From this figure, or equation (8.82) it can be seen that

(*a*) When $m < 0{\cdot}5, E \rightarrow 1{\cdot}0$

suggesting that in this range the reactant concentration in the pores is the same as on the external surface. These conditions arise if the pore depth is small or the diffusivity large compared with the rate constant and in all such cases diffusion in the pores is insignificant.

(*b*) When $m > 5{\cdot}0, E \simeq 1/\mathrm{m}.$

Under these conditions the internal diffusion in the pores dominates the reaction process.

The above analysis has been developed for a first order reaction. It can be extended to higher order reactions by considering these to be pseudo first order reactions. Thus for the reaction:

$$nA \rightarrow \mathrm{B} \tag{8.90}$$

the rate equation can be written:

$$r = kc_A^n = (kc_A^{n-1})c_A. \tag{8.91}$$

The term in brackets is inserted into equation (8.85) in place of k.

The difficulty in applying the above analysis is the evaluation of the tortuosity factor. It is not readily predicted although many attempts have been made to estimate its value from first principles. Hoogschafen (1955) and Scott and Dullien (1962) give values of τ

ranging from 1·0 to 12·5. However, for catalysts of moderate porosity it appears that the tortuosity is approximately equal to:

$$\tau = \delta^{-\frac{1}{2}}. \tag{8.92}$$

However, these correction factors should be calculated only when experimental values of $[\delta/\tau]$ are not available.

8.5 Non-isothermal Fixed Bed Reactors

The above discussion has been confined to isothermal analysis of the behaviour of solid–fluid reaction processes. However, since catalytic reactions also proceed with absorption or evolution of heat, the heat effects must be taken into consideration in the design and analysis of these reactors. These effects will be influenced by the rate of heat transfer through the bed of catalyst containing the reaction mixture. The pressure of the solids will impede the flow of heat and therefore, in this type of reactor, it would be expected that there would be a radial temperature gradient as well as a longitudinal temperature gradient. In the analysis of tubular homogeneous reactors it was assumed that the rate of heat transfer was sufficiently rapid within the fluid to enable the temperature in any radial plane to be assumed to be constant. These conditions cannot be accepted in fixed bed reactors, although when the reactor is made up of small diameter tubes packed with catalyst and the rate of flow through the tubes is high the assumption of uniform temperatures over each cross section is reasonable. However, for low flow rates through tubes of large diameter some consideration must be given to the assessment of the effects of a radial temperature profile and the effect of the catalyst particles on the rate of heat transfer to or from the bed. These will be considered in the following sections.

8.5.1 Heat transfer in packed beds

The rate of heat transfer through a bed of catalyst particles, in which a chemical reaction occurs, is affected by the mechanism by which heat is transferred through the fluid to the walls, and by the mechanism by which heat flows through particles. In addition the effects of the interactions between the fluid and the particles are also significant. Therefore, for convenience of analysis, these interrelated heat transfer processes within the packed bed have been replaced by a single heat transfer process. For this purpose, the packed bed containing the fluid mixture is assumed to be replaced by a solid of uniform texture and the heat transfer rate through this hypothetical solid will be the same as through the reactor. Since heat transfer through solids is a conduction process, it is necessary to define 'a thermal conductivity' in order to apply the hypothesis to predict radial and longitudinal temperature profiles for the reaction process. This thermal conductivity is called the 'effective thermal conductivity' k_e. It is usually determined by experiment for the solid–fluid system to be analysed in the reaction process, and the way in which it is utilised as illustrated in the following example.

Example 3. Using the data given below obtain a preliminary estimate of the diameter of the tubes to be installed in a fixed bed catalytic reactor which is to be used for the synthesis of vinyl chloride from acetylene and hydrogen chloride. The tubes are to contain mercuric chloride catalyst deposited on 0·1 inch particles of carbon and the heat of reaction is to be used to generate steam at 250°F. It is estimated that the temperature drop through the tube wall and surface films will be 50 deg F and in order to ensure a satisfactory life of the catalyst, its temperature should not exceed 485°F.

The following data may be used:

(*a*) The rate of reaction can be approximated by the expression: $r(T) = r_0(1 + AT)$ lb mole lb^{-1} catalyst where $r_0 = 0{\cdot}12$: $A = 0{\cdot}024$ and T is the temperature in °F above 200°F.

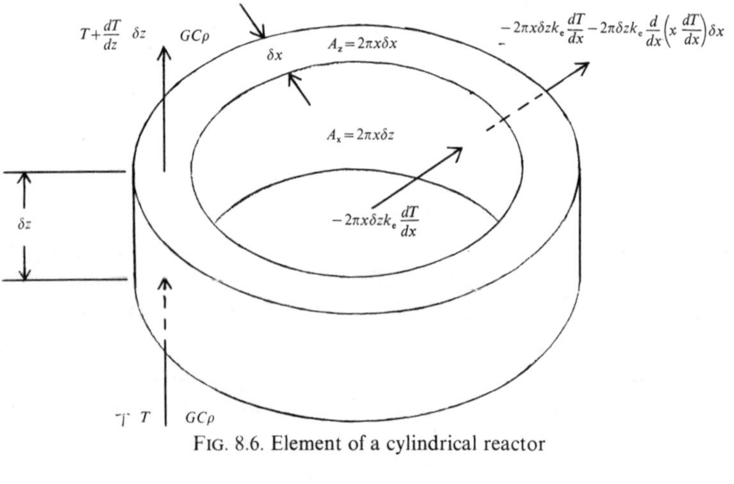

FIG. 8.6. Element of a cylindrical reactor

(b) Effective Thermal Conductivity: $k_e = 4{\cdot}0$ Btu h^{-1} ft^{-2} $degF^{-1}$ ft.

(c) Heat of Reaction ΔH: − 46200 Btu (lb mole)$^{-1}$.

(d) Bulk density of catalyst: 18 lb ft^{-3}

In order to make a preliminary estimate of the reactor tube diameter it will be assumed that the temperature of the bed is constant at any radius at all distances from the bed entrance.

Then a steady state heat balance over the element shown in Fig. 8·6 is:

input and reaction:

$$2\pi x\delta x G \,.\, C_p\rho T + 2\pi x\delta x\delta z\rho\Delta H r - 2\pi x k_e \frac{\partial T}{\partial x}\ \delta z \qquad \text{(I)}$$

output:

$$2\pi x\delta x G C_p\rho\left(T + \frac{\partial T}{\partial z}\,.\,\delta z\right)$$
$$- \left[2\pi x k_e \frac{\partial T}{\partial x}\delta z + \frac{\partial}{\partial x}\left(2\pi x k_e \frac{\partial T}{\partial x}\,.\,\delta z\right)\delta x\right]. \qquad \text{(II)}$$

Hence

$$\frac{\partial^2 T}{\partial x^2} + \frac{1}{x}\frac{\partial T}{\partial x} - \frac{G C_p \rho}{k_e}\,.\,\frac{\partial T}{\partial z} + \frac{\rho r \Delta H}{k_e} = 0. \qquad \text{(III)}$$

Since it is assumed that the bed temperature is constant at any radius at all values of z, at this point $\partial T/\partial z = 0$ and equation (III) becomes:

$$\frac{d^2T}{dx^2} + \frac{1}{x}\frac{dT}{dx} + \frac{\rho\Delta H r_0}{k_e}(1 + AT) = 0. \qquad \text{(IV)}$$

Equation (IV) can be converted into a Bessel equation. Thus let

$$Ay = 1 + AT$$

Then

$$\frac{dy}{dx} = \frac{dT}{dx} \quad \text{and} \quad \frac{d^2y}{dx^2} = \frac{d^2T}{dx^2} \qquad \text{(V)}$$

Substituting into (IV) gives

$$\frac{d^2y}{dx^2} + \frac{1}{x}\frac{dy}{dx} + Qy = 0 \qquad \text{(VI)}$$

where

$$Q = \frac{\rho \Delta H r_0 A}{k_e}.$$

Equation (VI) is a zero order Bessel Equation, the solution of which is

$$y = C_1 J_0(xQ^{\frac{1}{2}}) + C_2 Y_0(xQ^{\frac{1}{2}}) \qquad \text{(VII)}$$

where C_1 and C_2 are arbitrary constants.
When $x = 0, J_0(xQ^{\frac{1}{2}}) = 1{\cdot}0$ and $Y_0(xQ^{\frac{1}{2}}) = -\infty$

$\therefore C_2$ must be zero.

Then $$T + \frac{1}{A} = C_1 J_0(xQ^{\frac{1}{2}})$$

or $$T = C_1 J_0(xQ^{\frac{1}{2}}) - \frac{1}{A}$$

also when $x = 0$; $T = 285$.
Then since $A = 0{\cdot}024$, $C_1 = 326{\cdot}7$

$$\sqrt{Q} = \sqrt{\frac{18 \times 46200 \times 0{\cdot}12 \times 0{\cdot}024}{4{\cdot}0}} = 24{\cdot}5$$

or $$T = 326{\cdot}7\, J_0\,(24{\cdot}5x) - 41{\cdot}7.$$

When $$x = R, T = 100$$

or $$100 = 326{\cdot}7\, J_0(24{\cdot}5R) - 41{\cdot}7$$

or $$J_0\,(24{\cdot}5R) = 0{\cdot}434$$

or $$R = 0{\cdot}8 \text{ in.}$$

For safety, tubes of diameter 1·5 inches would be recommended.

8.5.2 Temperature profiles

In the above example it was assumed that the temperature did not vary with bed depth, and while this is a reasonable assumption in many cases it cannot be assumed that this situation exists in all packed bed reactors. When this is not the case the radial temperature profile is predicted by solution of the partial differential equation: equation (III) of the above example.

$$\frac{\partial^2 T}{\partial x^2} + \frac{1}{x}\frac{\partial T}{\partial x} - \frac{GC_{p}\rho}{k_e}\cdot\frac{\partial T}{\partial z} + \frac{\rho r \Delta H}{k_e} = 0 \qquad (8.93)$$

for T in terms of x between the limits $x = 0$ and $x = R$. Also the longitudinal temperature profile can be obtained by solution of this equation for T in terms of z between the limits $z = 0$ and $z = L$.

A convenient way to solve equation (8.93) is to convert the equation to finite difference form and then obtain a numerical solution. This is accomplished by converting all derivatives into their finite difference forms as follows:

$$\left(\frac{\partial T}{\partial x}\right)_{m,n} = \frac{T_{m+1,n} - T_{m,n}}{\Delta x} \qquad (8.94)$$

$$\left(\frac{\partial^2 T}{\partial x^2}\right)_{m,n} = \frac{T_{m+1,n} - 2T_{m,n} + T_{m-1,n}}{(\Delta x)^2} \qquad (8.95$$

$$\left(\frac{\partial T}{\partial z}\right)_{m,n} = \frac{T_{m,n+1} - T_{m,n}}{\Delta z}. \qquad (8.96)$$

Substitution of equations (8.94), (8.95), and (8.96) into equation (8.93) and rationalising gives

$$\frac{T_{m,n+1} - T_{m,n}}{\Delta z} = \frac{k_e}{GC_{p}\rho} \times \left[\frac{T_{m+1,n} - 2T_{m,n} + T_{m-1,n}}{(\Delta x)^2} + \frac{T_{m+1,n} - T_{m,n}}{m(\Delta x)^2}\right] + \frac{\rho r \Delta H}{GC_{p}\rho} \qquad (8.97)$$

where $x = m\Delta x$.

Equation (8.97) gives a method of computing the temperature at any point in the bed. This equation can be simplified by letting the modulus

$$\left[\frac{k_e(\Delta z)}{GC_p\rho(\Delta x)^2}\right] = \tfrac{1}{2} \qquad (8.98)$$

by choosing appropriate values for Δx and Δz.

The final equation then becomes:

$$T_{m,n+1} = \tfrac{1}{2}\left[\frac{1}{m}(T_{m+1,n} - T_{m,n}) + (T_{m+1,n} + T_{m-1,n})\right]. \qquad (8.99)$$

Equation (8.99) can be used to estimate the radial and longitudinal temperature profiles.

8.5.3 PREDICTION OF THE DIMENSIONS OF FIXED BED REACTORS

The finite difference equation (8.99) can be combined with a mass balance equation of the same type in order to predict the conversion as a function of height of catalyst. The analysis is developed by Jenson and Jeffreys (1965).

Problems

1. Estimate the voidage and mean pore size of a catalyst which gave the following results on physical testing:

(*i*) A 50 g sample displaced 110 cm^3 of helium and 130 cm^3 of mercury

(*ii*) A 3 g sample had a surface area of 1680 m^2.

2. The following quantities of nitrogen were adsorbed on the surface of a 2·47 sample of solid at the pressures indicated.

P atm	v cm^3 (converted to N.T.P.)
0·05	51·3
0·10	58·8
0·15	64·0
0·20	68·9
0·25	74·2

Find the specific surface area of the solid given that the area occupied by a nitrogen molecule is 16·2 $Å^2$

Avogadro's Number: $6{\cdot}023 \times 10^{23}$

1Å : 10^{-8} cm.

3. In order to assess the effectiveness of a catalyst for the removal of sulphur from petroleum Naphthas the catalyst, 0·35 cm diameter, was tested by reducing thiophene with hydrogen at 660°K and 30 atm pressure. The experimental rate constant was found to be 0·3 ml g^{-1}. If the surface area of the catalyst is 180 m^2 g^{-1} as obtained in a B.E.T. test and the pore volume is 0·25 ml what is the effectiveness factor? The specific gravity of the material in the catalyst is 2·65 g ml^{-1}.

ANS. $\delta = 0{\cdot}4$; $D_m = 0{\cdot}54$; $D_R = 0{\cdot}022$; $E = 0{\cdot}3$

4. An exothermic chemical reaction is carried out in a catalytic tubular reactor 2·0 ft diameter. If the effective thermal conductivity of the catalyst is 200 Btu h^{-1} $ft^{-2}degF^{-1}$, the tube wall temperature is 200°F and the heat release per ft^3 of catalyst bed is equal to $\Delta Hr = (400 + 600T)$ where T is in °F estimate the maximum bed temperature

ANS. $T = 528$°F

REFERENCES

HOOGSCHAGEN, J. 1955. Diffusion in porous catalysts and adsorbents. *Ind. Engng Chem.*, 906.

HOUGEN, O. A., and WATSON, K. M. 1947. *Chemical process principles. Part* 3. *Kinetics and catalysis.* Wiley, New York.

HÜRT, D. M. 1943. Principles of reactor design. Gas–solid interface reactions. *Ind. Engng Chem.* **35**, 522.

JENSON, V. G., and JEFFREYS, G. V. 1965. *Mathematical methods for chemical engineers.* Academic Press, New York.

OLSON, R. W., SCHULER, R. W., and SMITH, J. M. 1950. Catalytic oxidation of sulphur dioxide. Effect of diffusion. *Chem. Engng Prog.*, **46**, 614.

POLLARD, W. G., and PRESENT, R. D. 1948. On gaseous self-diffusion in long capillary tubes. *Phys. Rev.*, **73**, 762.

SATTERFIELD, C. N., and SHERWOOD, T. K. 1963. *Role of diffusion in catalysis.* Addison–Wesley, New York.

SCOTT, D. S., and DULLIEN, F. A. L. 1962. Diffusion of ideal gases in capillaries and porous solids. *A.I.Ch.E.Jl*, **8**, 113.

WILKE, C. R., and HOUGEN, O. A., 1945. Mass transfer in the flow of gases through granular solids extended to low modified Reynolds number. *Trans. Am. Inst. chem. Engng*, **41**, 445.

9

GAS–LIQUID REACTORS

9.1 Introduction

In recent years numerous industrial applications of gas–liquid reactions have arisen. These are to be found in many processes ranging from paper manufacture to organic chemical manufacture. In all of these processes the gas–liquid reactors are operated continuously; usually the gas flow is countercurrent to the liquid although as recent research has shown, this is not necessarily the best method of operation. In such reactors, the solute in the gas phase diffuses to the gas–liquid interface, dissolves at the interface and thereafter reacts with some component in the liquid. Therefore, in these reactors simultaneous mass transfer and chemical reaction occurs, and the overall rate equations must include the transport process and the chemical reaction process. This can be done in terms of either the two-film theory or the penetration theory of mass transfer and hence both must be considered.

9.2 Infinitely Rapid Chemical Reactions and the Two Film Theory

This analysis was first proposed by Hatta (1928) to explain the rate of absorption of carbon dioxide in caustic alkali. The analysis is, however, applicable to any rapid irreversible second order chemical reaction. Thus consider a gas containing a soluble component A that dissolves in a solution containing a reactive constituent B. A and B react in the liquid phase according to the equation:

$$A + B \rightarrow AB \tag{9.1}$$

to give a product AB which is non-volatile but soluble in the liquid. Hence, when the gas and liquid phases are brought together, A at the interface dissolves in the liquid and immediately reacts with B according to equation (9.1). The compound AB so formed diffuses from the interface into the bulk of the liquid. Since the interface zone is soon depleted of B, more diffuses from the bulk solution into the surface zone. Generally the rate of diffusion of the dissolved gas is greater than the rate of diffusion of a non-volatile electrolyte. Hence B will be completely depleted from the surface layers and A, after dissolving in the interface, will diffuse into the surface reaction zone more rapidly than B diffuses from the bulk of the liquid with the result that the chemical reaction plane recedes into the bulk of the liquid until equilibrium between the rates of diffusion of A and B is attained. The equilibrium condition is established within a few seconds of exposure of the gas and liquid phases to one another. When this condition exists, the concentration gradients of the different species in the system will be as shown in Fig. 9.1 where RZ represents the equilibrium position of the reaction zone to which A and B diffuse and react. The zone from the interface to the plane RZ is called the reaction zone. In this zone the concentration of the product AB is constant. Let this concentration be m.

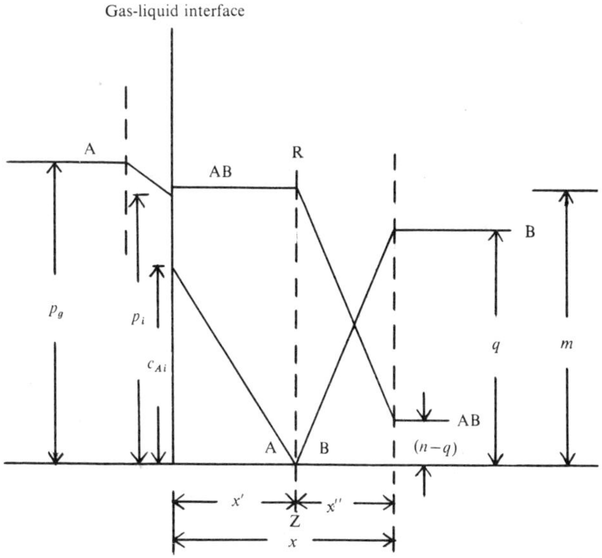

FIG. 9.1. Reactant and product concentrations with infinitely rapid second order irreversible reaction

When equilibrium has been established let the total concentration of B, both free and combined, in the main body of the liquid be n. Let the uncombined B be q. Then the concentration of the reaction product in the bulk of the liquid is $(n - q)$.

Then the rate of diffusion of AB from reaction zone to the main body of the liquid per unit area is:

$$N_{AB} = \frac{D_{AB}}{x''}[m - (n - q)]. \tag{9.2}$$

The rate of diffusion of A through the reaction zone is:

$$N_A = \frac{D_A}{x'}(c_{Ai} - 0) = k_g(p_g - p_i). \tag{9.3}$$

Finally the rate of diffusion of B from the bulk of the liquid to the reaction zone is:

$$N_B = -\frac{D_B}{x''}(q - 0). \tag{9.4}$$

When steady state conditions have been established

$$N_A = N_{AB} = -N_B \tag{9.5}$$

Let it be assumed that the equilibrium of A between gas and liquid can be represented by a form of Henry's Law. That is

$$p = H'c. \tag{9.6}$$

Then $$N_A = k_g H'(c_{Ag} - c_{Ai}) = k_g H'\left(c_{Ag} - \frac{N_A x'}{D_A}\right) \tag{9.7}$$

or $$\frac{N_A D_A}{k_g H'} + N_A x' + N_A x'' = D_A c_{Ag} + D_B q = \frac{D_A p_g}{H'} + D_B q. \tag{9.8}$$

That is $$N_A = \frac{D_A p/H' + D_B q}{\dfrac{D_A}{k_g H'} + x} = \frac{p/H + (D_B/D_A)q}{\dfrac{1}{k_g H'} + \dfrac{x}{D_A}}. \tag{9.9}$$

Equation (9.9) shows that the rate of mass transfer with simultaneous chemical reaction is:

(*a*) Directly proportional to the overall gas driving force plus an equivalent driving force due to the reacting solute B in the liquid;

(*b*) Inversely proportional to the overall resistance to transport of A on the assumption that this solute diffuses through the whole of the liquid film.

If the above analysis is carried out on the basis of transport through the liquid the resulting rate equation is:

$$N_A = \frac{c_{Ai} + (D_B/D_A)q}{\dfrac{x}{D_A}}. \tag{9.10}$$

Also in terms of an experimental mass transfer coefficient k_l, the rate is

$$N_A = k_l c_{Ai} \tag{9.11}$$

and therefore, from equations (9.10) and (9.11):

$$\frac{k_l}{\dfrac{D_A}{x}} = 1 + \frac{D_B}{D_A}\frac{q}{c_{Ai}} = \frac{k_l^\circ}{k_l}. \tag{9.12}$$

Equation (9.12) gives the ratio of the experimental mass transfer coefficient k_l to the mass transfer coefficient D_A/x or k_l° for physical absorption.

9.3 Moderately Rapid Chemical Reaction and the Two Film Theory

Quite frequently the chemical reaction rate is comparable with that of the transport process. Thus consider the situation in which the solute A diffuses through the gas phase, thereafter undergoing a first order chemical reaction in the liquid. The concentration profiles in the gas and liquid phases will be as shown in Fig. 9.2. The mass balance on an element of unit area and depth dx in the liquid at a distance x from the interface will lead to:

$$\frac{d^2 c_A}{dx^2} - \frac{k_l}{D_A} c_A = 0. \tag{9.13}$$

The derivation of equation (9.13) is identical with that of equation (8.74) and therefore need not be repeated. In addition since the boundary conditions are the same, the solution is:

$$c_A = \frac{c_{AX} \sinh(\alpha x) + c_{Ai} \sinh\ [\alpha(X - x)]}{\sinh(\alpha X)} \tag{9.14}$$

where $$\alpha = \sqrt{\frac{k_1}{D_A}}.$$

Equation (9.14) expresses the form of the concentration profile through the liquid film. The rate of the reaction process is equal to

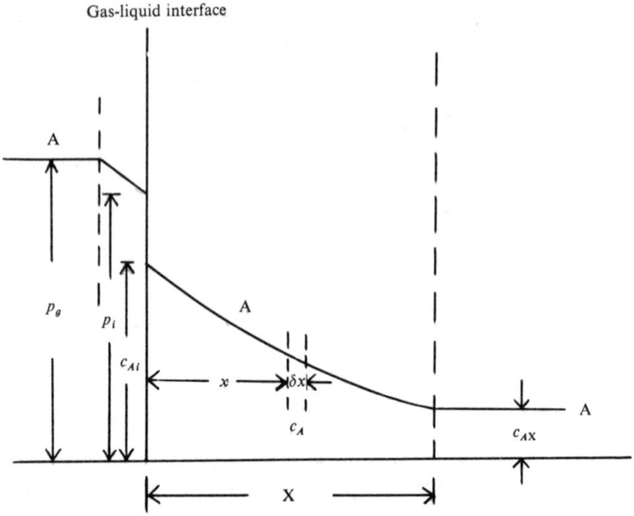

FIG. 9.2. Reactant concentration with moderately rapid first order reaction

the rate of mass transfer through the interface, which is:

$$(N_A)_{x=0} = -D_A\left(\frac{dc_A}{dx}\right)_{x=0}$$

$$= D_A c_{Ai}\left(\frac{\alpha\cosh\alpha X - \alpha c_{AX}/c_{Ai}}{\sinh\alpha X}\right). \tag{9.15}$$

Lightfoot (1958) used equation (9.15) to compare the rate of mass transfer accompanied by chemical reaction with the rate of mass transfer without chemical reaction and was able to show that over a fairly wide range of 'αX' the rate of absorption was independent of the rate of chemical reaction.

9.4 Absorption Rate in Terms of the Penetration Theory of Mass Transfer

Consider a thin film of liquid to flow down a flat plate. Let the thickness of the film be δ and let there be a soluble constituent A in the gas phase which on absorption in the liquid reacts chemically by a first order mechanism. Finally the laminar velocity profile in the film is represented by the equation

$$u = V\left(1 - \frac{x^2}{\delta^2}\right) \tag{9.16}$$

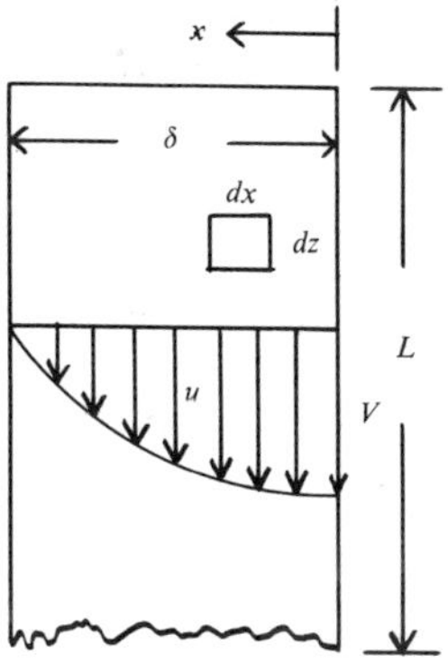

FIG. 9.3. Reaction in a falling film

where u is the velocity at x, and V is the surface velocity. A mass balance in terms of A on an element of liquid (Fig. 9.3) in the film of unit width, height dZ and breadth dx results in the equation:

$$-V\left(1-\frac{x^2}{\delta^2}\right)\frac{\partial c_A}{\partial z}+D_A\frac{\partial^2 c_A}{\partial x^2}=k_1 c_A. \qquad (9.17)$$

However, since x is very small compared with δ when C_A is reduced to zero, this equation can be simplified to:

$$-V\frac{\partial c_A}{\partial z}+D_A\frac{\partial^2 c_A}{\partial x^2}=k_1 c_A. \qquad (9.18)$$

The solution of this equation is given by Jenson and Jeffreys (1965) and is

$$\frac{c_A}{c_{Ai}}=\tfrac{1}{2}\exp\left[-x\sqrt{\frac{k_1}{D_A}}\right]\operatorname{erfc}\left[\frac{x}{2}\sqrt{\frac{V}{D_A L}}-\sqrt{\frac{k_1 L}{V}}\right]$$

$$+\tfrac{1}{2}\exp\left[x\sqrt{\frac{k_1}{D_A}}\right]\operatorname{erfc}\left[\frac{x}{2}\sqrt{\frac{V}{D_A L}}+\sqrt{\frac{k_1 L}{V}}\right]. \qquad (9.19)$$

The rate of absorption through the surface is:

$$N_A=-D\left(\frac{dc_A}{dx}\right)_{x=0}\simeq c_{Ai}\sqrt{(k_1 D_A)}\left(t+\frac{1}{2k_1}\right) \qquad (9.20)$$

where $t=\sigma L/V$ and when kt is large.

Equation (9.20) is used extensively in laboratory determination of the reaction rate constant for fairly fast reactions in gas–liquid reactors.

9.5 Equilibrium Relations in Gas–Liquid Reactive Systems

In the derivation of all the above rate equations the concentration of the solute A in the liquid at the interface is required. Since A reacts in the liquid its equilibrium value cannot be determined by direct experimentation. However, in gas absorption operations the equilibrium relationship of the solute gas is usually expressed by Henry's Law:

$$c = Hp \tag{9.21}$$

where c is the concentration of the dissolved solute in the liquid,
p is the partial pressure of the solute in the gas phase,
and H is the Henry Law constant.

Equation (9.21) applies to the dissolution of a gas in a pure solvent. It does not apply when the liquid contains dissolved electrolyte because of the 'salting out' effect of the dissolved salts. However, van Krevelen and Hoftijzer (1948) found experimentally that an expression of the type of equation (9.21) still applied if the value of H is modified to allow for the presence of the dissolved salt. That is, the equilibrium relation is written as:

$$c = H'p \tag{9.22}$$

where the modified constant H' depends on

(*a*) The nature of the gas.
(*b*) The nature of the dissolved electrolyte.
(*c*) The ions forming the electrolyte.
(*d*) The temperature; the salting out effect diminishes with rise of temperature.

Van Krevelen and Hoftijzer showed that the modified Henry Law constant H' was related to the thermodynamic chemical potential difference between the gas and the liquid as follows:

$$\mu_g^* - \mu_S^* = RT\,\mathrm{Ln}\frac{\gamma C}{p} \tag{9.23}$$

or

$$\exp\frac{(\mu_g^* - \mu_S^*)}{RT} = H' = \frac{\gamma C}{p} \tag{9.24}$$

where γ is the activity coefficient of the solute in the liquid.

The above authors found that $\mathrm{Ln}\,\gamma$ was proportional to the ionic strength of the dissolved electrolyte,

i.e.
$$\gamma = e^{k_S I} \tag{9.25}$$

where I = Ionic strength which can be estimated from:

$$I = \tfrac{1}{2}\Sigma c_n Z_n^2 \tag{9.26}$$

where c_n is the ionic concentration of ion n,
and Z_n is the valency of ion n.

For equation (9.25), the constant k_S to be used was determined experimentally, its value being dependent upon
(*a*) the solute gas,
and (*b*) the type of electrolyte dissolved in the liquid.

However, for any gas and a particular parent liquid, and the same concentration of different electrolytes having a common cation, the values of k_S for the electrolytes are related by

$$k_A = k_B + \Delta k_S: \tag{9.27}$$

van Krevelen found that Δk_S was independent of the solute gas. Values of k_S abstracted from the results of these workers are presented in Table 9.1.

Since Δk_S is independent of the absorbent gas and the temperature it is possible to estimate the value of H' for a gas between the gaseous phase and a reactive liquid. That is k is determined for a non-reactive gas and the solution of the reactive electrolyte, to give a value of, say k_A. Then k is determined for the same gas and a second electrolyte having a common cation to give a value of k_B. Thereafter Δk_S, for the non-reactive gas and the two electrolytes, is obtained from equation (9.27). Finally k is determined for the reactive gas and the second electrolyte to give, say k_C. From these results k for the reactive system is obtained as

$$k = k_C + \Delta k_S. \tag{9.28}$$

When k is known, I can be estimated from the normality of the electrolyte solution and thereafter H' calculated. The procedure is illustrated in the following example.

Example 1. Estimate the value of the modified Henry law constant for the equilibrium of carbon dioxide in a 1·0 N sodium hydroxide solution. The Henry Law constant for the equilibrium of carbon dioxide in water is $1{\cdot}5 \times 10^{-2}$ litre atm mole^{-1}.

Solution. The ionic strength I of 1·0 N NaOH is:

$$I = \tfrac{1}{2}\Sigma c_n Z_n^2 = \tfrac{1}{2}[(1{\cdot}0 \times 1{\cdot}0^2) + (1{\cdot}0 + 1{\cdot}0^2)] = 1{\cdot}0.$$

TABLE 9.1 Values of the constant k_S at 15°C

	H_2	O_2	N_2O	CO_2
HCl		0·054	0·027	0·011
HNO_3		0·033	0·006	−0·01
H_2SO_4			0·027	0·011
NaOH		0·188		
NaCl	0·117	0·148		0·105
$NaNO_3$	0·096			
KOH			0·138	
KCl	0·094		0·098	0·082
KNO_3	0·073			0·061
KI			0·082	0·066
KBr			0·088	0·072
K_2SO_4		0·125		
K_2CO_3	0·094			
$MgSO_4$	0·069			
$ZnSO_4$	0·069			
$CaCl_2$	0·074			
NH_4Cl			0·058	0·042

Let oxygen be the reference gas and sodium chloride the reference electrolyte. Then from Table 9.1:

k_A for oxygen in sodium hydroxide is 0·188

k_B for oxygen in sodium chloride is 0·148

and
$$\Delta k_S = 0{\cdot}040.$$

k_C for carbon dioxide in sodium chloride is 0·105.

Hence k for carbon dioxide in sodium hydroxide is

$$0{\cdot}105 + 0{\cdot}04 = 0{\cdot}145.$$

Then $H' = H \exp(-Ik) = 1{\cdot}5 \times 10^{-2} \times e^{-0{\cdot}145}$

or $H' = 1{\cdot}29 \times 10^{-2}$ litre atm mole^{-1}.

Van Krevelen and Hoftijzer showed that k could be predicted directly since in addition to the gas, the cation and anion of the electrolyte each contributed to its value. The values of these contributions are given in Table 9.2 from which the value of k for the system carbon dioxide–sodium hydroxide is:

(*i*) Contribution of sodium ion:	0·094
(*ii*) Contribution of hydroxyl ion:	0·061
Total electrolyte contribution	0·155
(*iii*) Contribution for CO_2 at 15°C:	− 0·01
Hence as before, k:	0·145

TABLE 9.2 Cation, anion and gas contributions to k

Cations		Anions		Gases	
H^+	0	NO_3^-	0	H_2	0·002
NH_4^+	0·031	I^-	0·005	O_2	0·033
K^+	0·071	Br^-	0·011	N_2O	0·006
Na^+	0·094	Cl^-	0·021	CO_2	−0·010
Mg^{2+}	0·046	OH^-	0·061		
Zn^{2+}	0·046	SO_4^{2-}	0·021		
Ca^{2+}	0·051	CO_3^{2-}	0·021		

It will be noticed from Table 9.2 that the contribution to k of the solute gas varies with temperature although the electrolytic effect is independent of temperature.

9.6 Analysis of Gas–Liquid Reactors

Gas–liquid reactors are generally continuously operated reactors, although many well-known processes have been operated batchwise in the past by sparging the gas into a kettle containing the liquid phase. Notable examples of this type of operation are the chlorination of hydrocarbons, the manufacture of hexamine and the oxidation of cumene to produce phenol. However, most of these processes have now been converted to continuous operation and therefore the following discussion will be confined to continuously operated gas–liquid reactors.

Continuous gas–liquid reactors are generally tall columns containing packing or plates and are similar to conventional absorption equipment. When the gas–liquid reaction rate is very high and is the controlling step the reactor may take the form of a simple wetted wall column. This type of equipment has the advantage that the temperature of the reaction process can be accurately controlled and the residence time of the liquid phase restricted. The distribution of residence times is also small in this type of reactor.

These reactors may be operated under countercurrent or co-current flow of the gas and liquid phases. In conventional gas absorption, countercurrent flow is preferred, although recent work on gas liquid reactions has shown that there are some advantages in co-current operation. However, irrespective of the type of equipment or the method of continuous operation, the design and analysis of these reactors is carried out as described below. The procedure will be applied to packed and wetted wall columns initially and later developed for stagewise operated reactors. Thus, consider an infinitely rapid chemical reaction to occur between a solute A absorbed from a gas into a liquid containing a reactive constituent B. The chemical reaction can be represented by the equation:

$$A + B \rightarrow AB \tag{9.29}$$

and takes place in a packed tower operated continuously under countercurrent flow conditions (Fig. 9.4). Let the upward gas flow rate be G and let the concentration of A in the gas at inlet be y_1. Let the liquid flow rate be L and the concentration of B in the inlet liquid be q_4. The liquid enters the top of the tower and flows down over the packing where it comes into contact with the ascending gas. Let the interfacial area between the gas and liquid be a ft^2 for each cubic foot of tower. Under these operating conditions the gas entering the base of the column may come into contact with liquid which contains no reactant B and in these circumstances there will be simple physical absorption. Therefore, near the bottom of the column there may be a 'physical absorption zone', denoted by h_1 in Fig. 9.4. Similarly, near the top of the column the gas phase, containing a low concentration of A, will come into contact with fresh absorbent rich in B.

Consequently near the top of the tower A and B will react in the surface of the liquid because there will be excess B present and the rate of the reaction process will be controlled by the rate of mass transfer through the gas film. Hence near the top of the column there will be a 'surface reaction zone', h_3. In between these two zones the concentrations of A and B in the gas and liquid will be comparable so that A would be expected to dissolve and penetrate into the liquid before being consumed by chemical reaction. This middle zone will be the 'interior reaction zone', h_2.

If there is an overall excess of B present there will not be a physical absorption zone but in the most general case the reactor/absorber will contain three zones. The estimation of the height of packing in each of these zones will now be undertaken, starting from the top of the column.

Consider an element of the surface reaction zone of height δh. A mass balance over this element is

$$N_A a\delta h = G\frac{dy}{dh}\delta h = k_g a P y \delta h \tag{9.30}$$

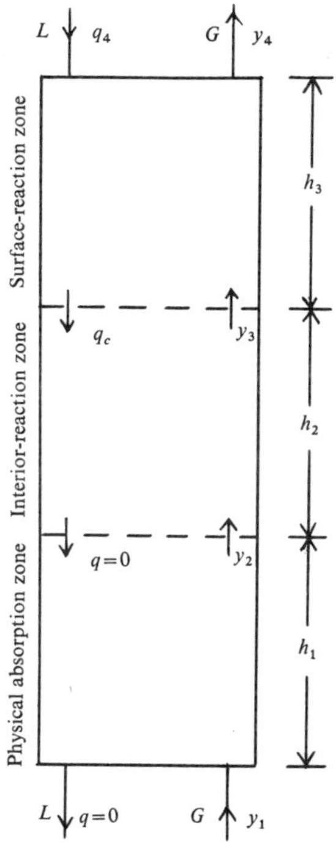

FIG. 9.4. Reactor zones for infinitely rapid reaction between A in gas feed and reactive B in liquid

where P is the total pressure
and y is the mole fraction of A in the gas phase (its concentration at the interface will be zero).

Integration of the second and third terms in equation (9.30) gives:

$$h_3 = \frac{G}{k_g aP} \ln \frac{y_3}{y_4}. \tag{9.31}$$

Equation (9.31) is valid as long as the chemical reaction remains on the surface. This depends on the concentration of B in the liquid, since when the concentration of B falls below a critical value the reaction zone recedes into the liquid as described in Hatta's model discussed above. Secor and Southworth (1961) showed that the critical concentration of B, q_C, could be estimated as follows. Consider Fig. 9.5 which illustrates the concentration profiles of the gas and liquid corresponding to the critical concentration q_C.

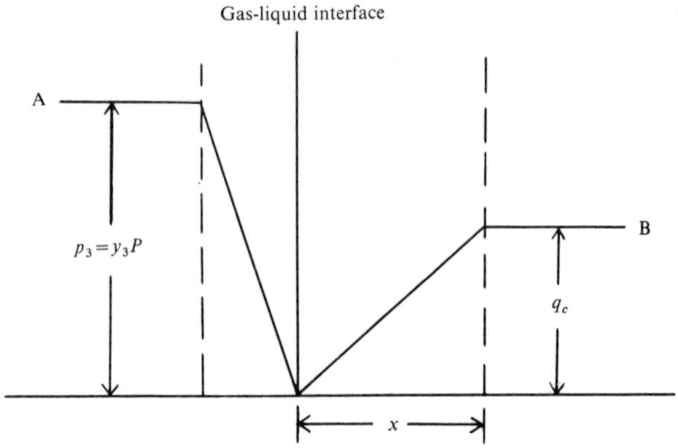

FIG. 9.5. Critical concentration in profiles

The concentration of A and B at the interface must be zero because at an infinitesimal concentration difference below q_C the rate of transport B will be insufficient to combine with A at the interface, so that A will enter the liquid in order to reach B. Under these conditions the rate of mass transfer of A and B are seen from the figure to be:

$$N_A a = k_g a P y_3 = -N_B a = \frac{D_B}{x}\frac{\rho}{z} a q_C \tag{9.32}$$

where z is the number of moles of B that react with one mole of A. and ρ is the molar density of the liquid.
Rearranging equation (9.32) gives:

$$q_C = \frac{k_g P y_3 z x}{\rho D_B} = \left(\frac{k_g}{k_l^0}\right)\left(\frac{D_A}{D_B}\right)\left(\frac{Pz}{\rho}\right) y_3 \tag{9.33}$$

where from equation (9.12) $k_l^0 = D_A/x$.
The value of y_3 is obtained in terms of the boundary conditions (y_4, q_4) by means of a material balance between the top of the column and the critical zone:

$$G(y_3 - y_4) = \frac{L}{z}(q_4 - q_C). \tag{9.34}$$

Substitution of the value of y_3 obtained from equation (9.34) into equation (9.33) gives:

$$q_C = \frac{\left(\frac{L}{G}\right)\left(\frac{k_g}{k_l^0}\right)\left(\frac{D_A}{D_B}\right)\left(\frac{P}{\rho}\right)q_4 + \left(\frac{k_g}{k_l^0}\right)\left(\frac{D_A}{D_B}\right)\left(\frac{Pz}{\rho}\right)y_4}{1 + \left(\frac{k_g}{k_l^0}\right)\left(\frac{D_A}{D_B}\right)\left(\frac{PL}{\rho G}\right)} \tag{9.35}$$

When the concentration of B in the liquid falls below q_C and the concentration of A in the gas increases above y_3 the conditions are those of the interior reaction zone and the Hatta model of the rate process is applicable. A mass balance on an element of height δh in this zone will be:

$$N_A a\delta h = G\frac{dy}{dh}\delta h = \left(\frac{1}{k_g a} + \frac{H'}{k_l^0 a}\right)^{-1}\left(p_g + \frac{D_B}{D_A}\frac{\rho H'}{Z}q\right). \tag{9.36}$$

Converting the modified Henry law constant H' so that concentrations can be expressed in mole fractions gives:

$$m = \frac{\rho H'}{P}. \tag{9.37}$$

Let the overall rate constant for the interior reaction zone be K_{gR} where:

$$\frac{1}{K_{gR}} = \frac{1}{k_g} + \frac{mP}{\rho k_l^0} \tag{9.38}$$

Finally a material balance over the interior reaction zone gives:

$$q = q_C - \frac{GZ}{L}(y - y_3). \tag{9.39}$$

Substitution of equations (9.37), (9.38) and (9.39) into equation (9.36) gives:

$$\begin{aligned} N_A a &= G\frac{dy}{dh} \\ &= K_{gR}aP\left[\left(1 - \frac{mGD_B}{LD_A}\right)y + \frac{mq_C D_B}{zD_A} + \frac{mGD_B}{LD_A}y_3\right]. \end{aligned} \tag{9.40}$$

Let
$$1 - \frac{mGD_B}{LD_A} = \alpha$$

and
$$\frac{mD_B q_C}{zD_A} + \frac{mGD_B}{LD_A} y_3 = \beta,$$

Equation (9.40) on integration becomes:

$$h_2 - \frac{G}{K_{gR} aP\alpha} \Big[\ln(\alpha y + \beta) \Big]_{y_3}^{y_2} \tag{9.41}$$

and on substitution of the limits:

$$h_2 = \frac{G}{K_{gR} aP \left(\dfrac{mGD_B}{D_A L} - 1 \right)} \ln \left[\frac{\dfrac{mD_B q_C}{D_A z} + y_3}{\dfrac{mGD_B}{LD_A}(y_3 - y_2) + \dfrac{mD_B q_C}{D_A z} + y_2} \right] \tag{9.42}$$

For application of equation (9.42) the value of y_2 can be obtained from a mass balance between the top of the tower and the plane containing y_2 since q_2 must be zero corresponding to the complete reaction of B. That is:

$$y_2 = y_4 + \frac{L}{Gz} q_4. \tag{9.43}$$

Finally the height of the physical absorption zone can be estimated from the following equation to be found in many chemical engineering texts:

$$h_1 = \frac{G}{k_g aP \left(1 - \dfrac{mG}{L} \right)} \ln \left[\left(1 - \frac{mG}{L} \right) \frac{y_1}{y_2} + \frac{mG}{L} \right]. \tag{9.44}$$

The overall height h of the column reactor is finally obtained by summing the values for the various zones:

$$h = h_1 + h_2 + h_3. \tag{9.45}$$

The above procedure is illustrated by the following example.

Example 2. 22 500 lb h^{-1} of a gas mixture containing 6·6% CO_2 by volume in carbon monoxide as carrier gas is to be treated with

20000 lb h^{-1} of a 20% sodium hydroxide solution in order to remove 98% by weight of the CO_2. The operation is to be carried out at 1 atm pressure in a tower of cross-sectional area 4 ft^2 containing ring packing of interfacial area 100 ft^2 for each cubic foot of tower. Using the following data estimate the height of packing for this duty.

Data: $k_g = 0{\cdot}78$ lb mole h^{-1} ft^{-2} atm^{-1}
$k_l^0 = 0{\cdot}915$ ft h^{-1}
$H' = 575$ ft^3 atm(lb mole)$^{-1}$
D_{CO_2} in solution $= 4{\cdot}64 \times 10^{-5}$ ft^2 h^{-1}
D_{NaOH} in water $= 11{\cdot}6 \times 10^{-5}$ ft^2 h^{-1}
Reaction rate constant $= 4{\cdot}5 \times 10^8$ ft^3 (lb mole)$^{-1}$ h^{-1}
Specific gravity of liquid $= 1{\cdot}22$

Solution:

Weight per cent CO_2:

$$\frac{0{\cdot}066 \times 44}{(0{\cdot}066 \times 44) + (0{\cdot}934 \times 28)} = \frac{2{\cdot}91}{29{\cdot}11} = 10\%\ CO_2$$

$$CO_2 \text{ in feed gas}: 2250\,\frac{\text{lb}}{\text{h}} \text{ or } \frac{2250}{44}\,\frac{\text{lb moles}}{\text{h}} = 51{\cdot}1$$

CO_2 to be absorbed: $2250 \times 0{\cdot}98 = 2206$ lb

∴ CO_2 in effluent: 44 lb or 1·0 lb mole

Moles carbon monoxide passing through tower

$$= \frac{20250}{28} = 723 \text{ lb moles}$$

$$\therefore \text{ mole fraction } CO_2 \text{ in effluent gas} = \frac{1}{724} = 0{\cdot}0014.$$

The chemical reaction taking place is:

$$2\,NaOH + CO_2 \rightarrow Na_2\,CO_3 + H_2O$$

∴ 100·2 lb moles of NaOH are required

NaOH available is $0{\cdot}2 \times 20000 = 4000$ lb $= 100$ lb mole

∴ a small zone at the base of the tower will function as a physical absorber.

Now feed liquor contains 100 lb moles NaOH

$$+ \left(\frac{16000}{18} = 890\right) \text{lb moles } H_2O$$

∴ mole fraction NaOH in feed liquor $q_4 = 0{\cdot}101$.

From the above, the boundary condition of the surface reaction zone is obtained as follows:

$$\text{Molar density of liquor} = \frac{990 \times 62{\cdot}4 \times 1{\cdot}22}{20000} = 3{\cdot}77\,\frac{\text{lb mole}}{\text{ft}^3}$$

$$q_C = \frac{\left(\dfrac{0{\cdot}78}{0{\cdot}915}\right)\left(\dfrac{4{\cdot}64 \times 10^{-5}}{11{\cdot}6 \times 10^{-5}}\right)\left[\left(\dfrac{990 \times 1}{774{\cdot}1 \times 3{\cdot}77}\right)0{\cdot}101 + \left(\dfrac{2}{3{\cdot}77}\right)0{\cdot}0014\right]}{1 + \left(\dfrac{0{\cdot}78}{0{\cdot}915}\right)\left(\dfrac{4{\cdot}64}{11{\cdot}6}\right)\left(\dfrac{990}{774{\cdot}1 \times 3{\cdot}74}\right)}$$

i.e. $q_c = 0{\cdot}0115$
and from equation (9.34):

$$y_3 = 0{\cdot}0014 + \frac{990}{2 \times 774}(0{\cdot}101 - 0{\cdot}0115) = 0{\cdot}058.$$

∴ height of packing in surface reaction zone from equation (9.31) is:

$$h_3 = \frac{774{\cdot}1 \times 2{\cdot}303}{0{\cdot}78 \times 100 \times 4}\log\frac{0{\cdot}058}{0{\cdot}0014} = 9{\cdot}4\text{ ft.}$$

Consider the interior reaction zone:

$$\text{The Henry Law constant } m = \frac{\rho H}{P} = 3{\cdot}77 \times 575 = 2170.$$

The lower gas concentration from equation (9.43) is:

$$y_2 = 0{\cdot}0014 + \frac{990 \times 0{\cdot}101}{774 \times 2} = 0{\cdot}065.$$

The overall mass transfer coefficient for the interior reaction zone is:

$$\frac{1}{K_{gR}} = \frac{1}{0{\cdot}78} + \frac{2170}{3{\cdot}77 \times 0{\cdot}915} = 1{\cdot}3 + 629 = 630{\cdot}3.$$

Hence the height of packing from equation (9.42) is:

$$h_2 = \frac{774 \times 2{\cdot}303 \log\left[\dfrac{\dfrac{11{\cdot}6}{4{\cdot}64} \times \dfrac{2170}{2} \times 0{\cdot}0115 \quad + 0{\cdot}058}{4270(0{\cdot}058 - 0{\cdot}065) + 31{\cdot}3 + 0{\cdot}065}\right]}{\dfrac{1}{630{\cdot}3} \times 100 \times \left[\dfrac{11{\cdot}6}{4{\cdot}64} \times \dfrac{774}{990} \times \dfrac{2170}{1} - 1\right] \times 4}$$

$$= 0{\cdot}66\log\frac{31{\cdot}36}{1{\cdot}48} = 0{\cdot}9\text{ ft.}$$

Finally the height of packing required for the physical absorption section by equation (9.44) is estimated to be 0·5 ft.

∴ total height of packing required is:

$$h = 9{\cdot}4 + 0{\cdot}9 + 0{\cdot}5 = 10{\cdot}8 \text{ ft.}$$

Recommend say 11·0 ft of packing.

9.7 Analysis of Gas–Liquid Reactors with Moderate Reaction Rate

Most gas absorption and simultaneous chemical reaction processes that are industrially important follow the stoichiometric mechanism

$$\mathrm{A} + \mathrm{B} \rightarrow \mathrm{R} + \mathrm{S} \tag{9.46}$$

by moderately fast second order kinetics. Frequently this reaction is the chemical rate determining step and is followed by a second reaction which is much faster, such as:

$$\mathrm{R} + (m - 1)\,\mathrm{B} \rightarrow \mathrm{Q} \tag{9.47}$$

For example the absorption of carbon dioxide in alkaline solutions is represented by

$$CO_2 + OH^- \rightarrow HCO_3^-$$

which is rate controlling and followed by:

$$HCO_3^- + OH^- \rightarrow H_2O + CO_3^{2-}$$

If the chemical reaction mechanism represented by equations (9.46) and (9.47) is accepted a material balance on an element of liquid in the liquid film will result in the equation:

$$D_{\mathrm{A}} \frac{d^2 c_{\mathrm{A}}}{dx^2} = \frac{D_{\mathrm{A}}}{m} \frac{d^2 c_{\mathrm{B}}}{dx^2} = k_2\, c_{\mathrm{A}}\, c_{\mathrm{B}} \tag{9.48}$$

where the symbols have their usual significance and the diffusion coefficients of A and B are assumed equal at D_{A}.

There is no general analytical solution to equation (9.48) to which different boundary conditions can be attached. For example, if c_{B} is much greater than c_{A} the reaction becomes pseudo-first order, and the solution will be that given by equation (9.15). However, when the concentrations of c_{A} and c_{B} are comparable equation (9.48) becomes

$$D_{\mathrm{A}} \frac{d^2 c_{\mathrm{A}}}{dx^2} - k_2\, c_{\mathrm{A}}\, c_{\mathrm{B}} = 0 \tag{9.49}$$

and van Krevelen and Hoftijzer (1948a) converted equation (9.49) into dimensionless form as follows:

$$\frac{d^2 c_A}{d\left(\frac{x}{x_L}\right)^2} = \frac{k_2 c_{BL} c_{Ai}{}^2 \cdot D_A c_A}{c_{Ai}{}^2 \left(\frac{D_A}{x_L}\right)^2} = \left(\frac{R_1}{M_1}\right)^2 c_A \tag{9.50}$$

where $R_1 = \sqrt{(k_2 c_{BL} D_A \cdot c_{Ai})}$ and $M_1 = (D_A/x_L)c_{Ai} = k_l^\circ c_{Ai}$. Integration of equation (9.50) between the limits: when $(x/x_L) = 0$, $c_A = c_{Ai}$ and when $(x/x_L) = 1{\cdot}0$, $c_A = c_{AL}$ with R_1 and M_1 as parameters and $D_A = D_B$ leads to the graphical solution shown in Fig. 9.6.

Inspection of this figure shows that when R_1 is small the ratio (k_l/k_l°) approaches unity and implies that a slow reaction has no effect on the rate of physical absorption. Alternatively for large values of R_1 it can be seen that (k_l/k_l°) approaches an asymptote for each value of the ratio (c_{BL}/mc_{Ai}). This asymptotic region corresponds to a region of very fast reaction rate where increase in the value of the reaction rate constant will have no effect on the overall absorption rate. Under these conditions the ratio of the mass transfer coefficients is related to the concentrations by equation (9.12). That is

$$\frac{k_l}{k_l^\circ} = 1 + \frac{D_B}{D_A}\frac{c_{BL}}{c_{Ai}}. \tag{9.51}$$

Finally for any values of R_1 the ratio (k_l/k_l°) approaches its asymptote at very large values of (c_{BL}/mc_{Ai}). The curve having an infinite values of (c_{BL}/mc_{Ai}) is the locus of all the asymptotic values and for this curve to be approached c_B must be very large compared with c_{Ai}, i.e. the reaction will become pseudo-first order. For $(k_l/k_l^\circ) > 2{\cdot}0$ the pseudo-first order curve is represented by the equation

$$k_l = \sqrt{(k_2 D_A c_{BL})}. \tag{9.52}$$

Under these conditions the acutal mass transfer coefficient is independent of the physical absorption coefficient and the liquid flow rate.

To sum up it is apparent that Fig. 9.6 enables a chemical gas–liquid reaction to be assessed qualitatively. In addition this chart can be applied to the design and analysis of packed continuous gas–liquid reactors. Essentially the procedure recommended by van Krevelen and Hoftijzer (1953) consists of drawing the operating line for the

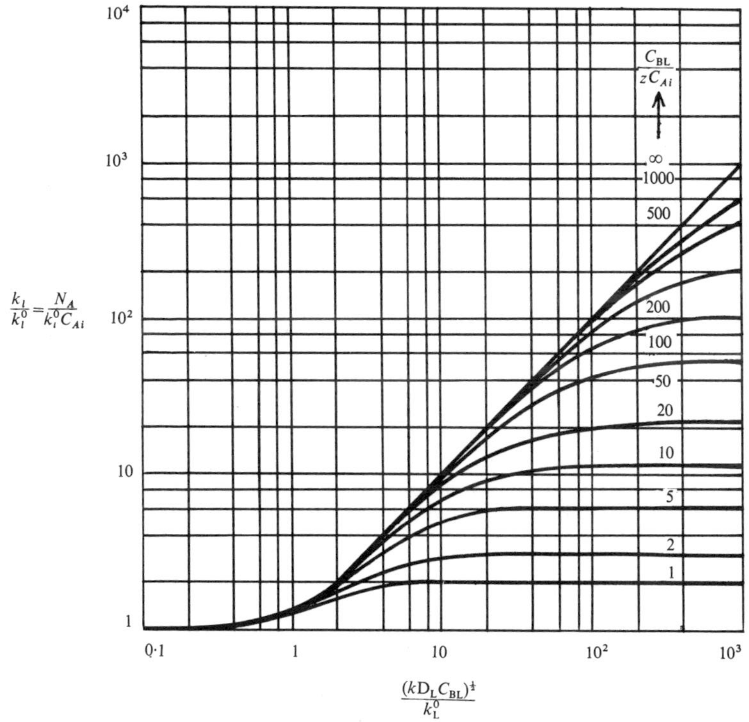

FIG. 9.6. $\frac{N_{A'}}{k_l^\circ c_{Ai}}$ versus $\frac{(k_{Di} c_{BL})^{\frac{1}{2}}}{k_l^{\frac{1}{2}}}$ with $\frac{c_{BL}}{Zc_A}$ as parameter

conditions at the top and bottom of the reaction tower in terms of partial pressure of the solute A in the gas and concentration of B in the liquid In addition the equilibrium line relating c_{Ai} to p_{Ai} is constructed from the modified Henry law constant. Then a value of p_{Ai} is assumed and the point rate of mass transfer is calculated from

$$N_A = k_g(p_{Ag} - p_{Ai}). \tag{9.53}$$

From this value of p_{Ai} the corresponding value of c_{Ai} is estimated and (c_{BL}/mc_{Ai}) calculated. The appropriate value of c_{BL} corresponding to p_{Ag} is obtained from a material balance between the point under consideration and the top or bottom of the tower. Thereafter the parameter R_1 is evaluated for the chosen p_{Ai} and from Fig. 9.6 the value of the ordinate $(N_A/k_l^\circ c_{Ai})$ is estimated. If the value of N_A obtained graphically is equal to that obtained from equation (9.53) the value of p_{Ai} is accepted. If not the procedure is repeated. A number of calculations of this type are made in order to integrate equation (9.53) graphically to obtain the height of packing for the given duty.

$$h = \frac{G}{k_g a} \int_{p_1}^{p_2} \frac{dp}{(p_g - p_i)}. \qquad (9.54)$$

The procedure is illustrated in the following example.

Example 3. Determine the height of packing required for the absorption of carbon dioxide in sodium hydroxide under the following operating conditions.

Operating pressure: 1·0 atm
Superficial gas velocity: 0·216 m sec^{-1}
Inlet CO_2 content: 4·2% by volume
Effluent CO_2 desired maximum: 1·4% by volume
Superficial liquid velocity: $9{\cdot}4 \times 10^{-4}$ m sec^{-1} (free flow through column)
The irrigating liquid contains 0·794 moles sodium hydroxide per litre and 0·0105 moles sodium carbonate per litre
The mass transfer coefficients are:

$$k_g = 1{\cdot}05 \times 10^{-3} \text{ kg mole sec}^{-1} m^{-2} \text{ atm}^{-1}$$
$$k_l^\circ = 7{\cdot}75 \times 10^{-5} \text{ metre sec}^{-1}$$

The Henry law constant $H = 35{\cdot}8 \dfrac{m^3 \text{ atm}}{\text{kg mole}}$
Diffusivity of CO_2 in liquid $D_A = 1{\cdot}2 \times 10^{-5}$ cm^2 sec^{-1}
Diffusivity of NaOH in water $D_B = 2{\cdot}5 \times 10^{-5}$ cm^2 sec^{-1}

Specific reaction rate constant $k = 7{\cdot}75 \times 10^6 \dfrac{cm^3}{\text{g mole.sec}}$

Interfacial area of packing: $430 \dfrac{m^2}{m^3}$

Solution. Consider one metre square of the column cross section. Then the carbon dioxide absorbed per second is:

$$0{\cdot}216 \times \left(0{\cdot}042 - \frac{0{\cdot}014 \times 95{\cdot}8}{98{\cdot}6}\right) \times \frac{1000}{22{\cdot}4} = 0{\cdot}274 \text{ g mole } CO_2$$

Sodium hydroxide in feed liquor is:

$$(9{\cdot}4 \times 10^{-4} \times 0{\cdot}794 \times 1000) = 0{\cdot}745 \text{ mole NaOH}$$

The overall reaction is:

$$2NaOH + CO_2 \rightarrow Na_2CO_3 + H_2O$$

∴ 1 mole CO_2 reacts with 2 moles NaOH.
That is moles NaOH required is: $0{\cdot}274 \times 2 = 0{\cdot}548$
and moles NaOH leaving tower per second:

$$0{\cdot}745 - 0{\cdot}548 = 0{\cdot}197 \text{ moles NaOH.}$$

The concentration of NaOH in effluent liquor is:

$$\frac{0{\cdot}197}{0{\cdot}94} = 0{\cdot}21 \frac{\text{moles}}{\text{litre}}$$

Therefore the points at the extreme of the operating line are:
Tower top
Gas: $p_g = 1{\cdot}4 \times 10^{-2}$ atm
Liquid: $c_{BL} = 0{\cdot}794$ mole litre^{-1}
Tower base
Gas: $p_g = 4{\cdot}2 \times 10^{-2}$ atm
Liquid: $c_{BL} = 0{\cdot}21$ mole litre^{-1}.
From the above, the operating line is drawn on Fig. 9.7.
At tower top: $p_g = 1{\cdot}4 \times 10^{-2}$
As a first trial let $p_i = 1{\cdot}3 \times 10^{-2}$
Then $N_A = k_g(p_g - p_i) = 1{\cdot}05 \times 10^{-3}(1{\cdot}4 \times 10^{-2} - 1{\cdot}3 \times 10^{-2})$

$$= 1{\cdot}05 \times 10^{-6} \frac{\text{kg mole}}{\text{m}^2 \text{ sec}}$$

$$\frac{\sqrt{(kc_{BL}D_A)}}{k_l^\circ} = \frac{\sqrt{(7{\cdot}75 \times 10^6 \times 0{\cdot}794 \times 1{\cdot}2 \times 10^{-5} \times 10^{-7})}}{7{\cdot}75 \times 10^{-5}} = 35.$$

The factor 10^{-7} within the square root is necessary to make the group dimensionally consistent.
Now

$$c_{Ai} = \frac{p_i}{H} = \frac{1{\cdot}3 \times 10^{-2}}{35{\cdot}8} = 3{\cdot}63 \times 10^{-4}$$

and since $z = 2$.

$$\frac{c_{BL}}{zc_{Ai}} = \frac{0{\cdot}794}{2 \times 3{\cdot}63 \times 10^{-4}} = 1093.$$

Therefore from fig. 9.6:

$$\frac{N_A}{M_1} = \frac{N_A}{k_l^{\circ} c_{Ai}} = 34{\cdot}8.$$

Hence:

$$N_A = 34{\cdot}8 \times 3{\cdot}63 \times 10^{-4} \times 7{\cdot}75 \times 10^{-5} = 1{\cdot}0 \times 10^{-6} \frac{\text{kg mole}}{\text{m}^2 \text{ sec}}$$

from which it is seen that the mass transfer rate based on the gas film agrees with the mass transfer rate in the chemical reactive liquid. Hence the trial value of p_i is accepted.

The procedure given above for checking estimated values of p_i corresponding to arbitrary values of p_g on the operating line was repeated between the top and bottom conditions in the tower, and the results obtained are given below.

p_g	p_i	N_A gas film	N_A liquid film	$(p_g - p_i)^{-1}$	$\frac{\sqrt{(kDc_{BL})}}{k_L^0}$	$\frac{c_{BL}}{zc_{Ai}}$
$1{\cdot}4 \times 10^{-2}$	$1{\cdot}3 \times 10^{-2}$	$1{\cdot}05 \times 10^{-6}$	$1{\cdot}0 \times 10^{-6}$	1000	35	1093
$2{\cdot}2 \times 10^{-2}$	$2{\cdot}08 \times 10^{-2}$	$1{\cdot}24 \times 10^{-6}$	$1{\cdot}26 \times 10^{-6}$	834	31	547
$2{\cdot}8 \times 10^{-2}$	$2{\cdot}65 \times 10^{-2}$	$1{\cdot}59 \times 10^{-6}$	$1{\cdot}59 \times 10^{-6}$	658	28·0	339
$3{\cdot}27 \times 10^{-2}$	$3{\cdot}13 \times 10^{-2}$	$1{\cdot}73 \times 10^{-6}$	$1{\cdot}75 \times 10^{-6}$	599	25·0	232
$3{\cdot}70 \times 10^{-2}$	$3{\cdot}54 \times 10^{-2}$	$1{\cdot}64 \times 10^{-6}$	$1{\cdot}65 \times 10^{-6}$	637	21·5	152
$4{\cdot}2 \times 10^{-2}$	$4{\cdot}05 \times 10^{-2}$	$1{\cdot}58 \times 10^{-6}$	$1{\cdot}58 \times 10^{-6}$	667	18	93

From the values in the above table p_g was plotted (Fig. 9.7) against $(p_g - p_i)^{-1}$ and the area under the curve between $p_g = 4{\cdot}2 \times 10^{-2}$ and $p_g = 1{\cdot}4 \times 10^{-2}$ is 20 and represents the integral of equation (9.54).

$$\therefore \text{ height of packing} = \frac{0{\cdot}216}{22{\cdot}4 \times 1{\cdot}05 \times 10^{-3} \times 430} \times 20 = 0{\cdot}411 \text{ m.}$$

Van Krevelen and Hoftijzer's absorption and simultaneous chemical reaction model is based on the Whitman–Lewis two film theory of mass transfer and Peaceman (1951) showed that solution of equation (9.50) by Fig. 9.6 deviated from a computer solution by about 8 %. However, Brian, Hurley and Hasseltine (1961) developed a mathematical model of the simultaneous gas absorption and second order chemical reaction process, described above, in terms of the penetration theory of mass transfer. They found that the maximum deviation of their solutions from those presented graphically by van Krevelen was 9 % thus showing that there would be very

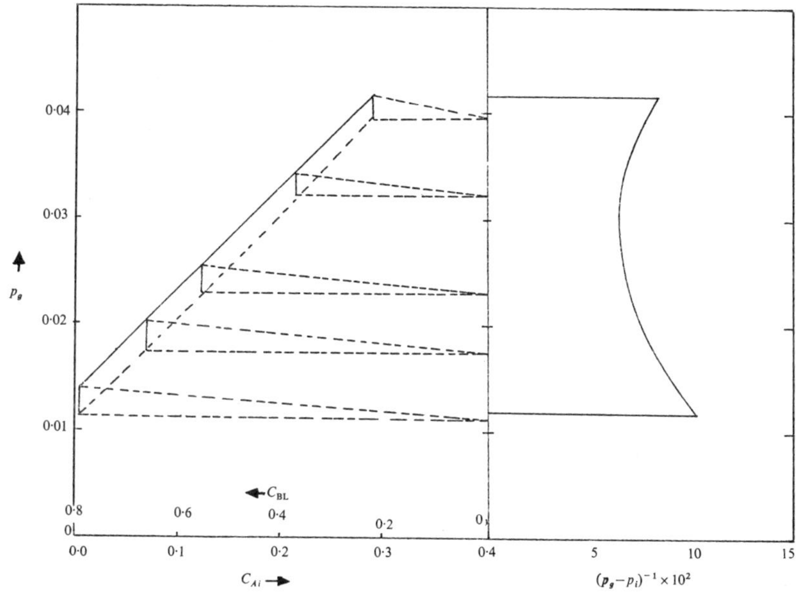

FIG. 9.7. Solution to Example 3

little difference between the results of analysing or designing gas–liquid reactors on the basis of the film model of mass transfer or by use of the penetration model.

9.8 Simultaneous Gas Absorption and Chemical Reaction in Plate Towers

In the past, chemical absorption processes have been carried out almost exclusively in packed towers; this is probably because these contactors are cheap to construct and economical to operate. Furthermore, these towers were packed with ceramic pieces which are very corrosion resistant to the caustic liquors used for the absorption duties. Finally, the concentration of the reactive solute was low and therefore the heats of reaction and of solution of the solute gas small so that the temperature rise of the liquid would not be detrimental to the process. However, chemical absorption processes are being extended to concentrated solutions so that the absorber is in fact a chemical reactor. When this is so the heats of reaction and solution are considerable, and could arrest or completely reverse the reaction through adverse physical or chemical equilibrium conditions at the higher temperatures. Hence some form of cooling system must be installed in the equipment in order to maintain favourable and safe operating conditions. Since cooling coils inside a packed tower are practically ineffective, plate towers tend to be introduced when the temperature rise must be controlled. It is comparatively simple to insert cooling coils on sieve, bubble cap or other type of plate and the removal of generated heat is effective. Therefore it is necessary to be able to predict the number of stages for a given chemical absorption duty taking into account the chemical reaction on the plate. This is conveniently done as follows.

9.8.1 Estimation of the number of stages in a plate column performing a chemical absorption duty

Consider the plate tower illustrated diagramatically in Fig. 9.8 to operate continuously under isothermal conditions. G lb mole h^{-1} inert gas containing Y_0 lb mole of a soluble constituent A per mole

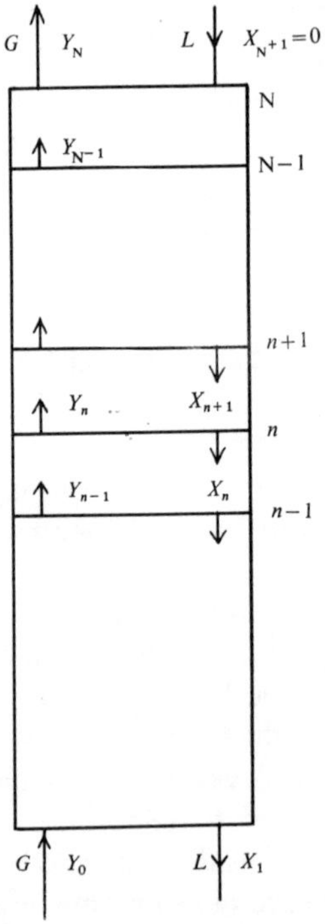

FIG. 9.8. Plate absorption tower

of inert gas enters the base of the tower and Llb mole h^{-1} of solute free liquid absorbent containing X_{n+1} lb mole of solute A per mole of carrier liquid enters the top of the tower. The solute A is absorbed by the liquid and reacts chemically to form a soluble constituent B by a pseudo-first order mechanism with respect to the solute A. Each plate is assumed to contain a cooling system so that isothermal conditions prevail throughout the tower.

The rate of chemical reaction can be expressed as follows

$$-\frac{dc}{dt} = k_1 c_A = \frac{1}{V}\frac{dn'_A}{dt} = \frac{kn'_A}{V} = \frac{kX_n H}{V} \tag{9.55}$$

where n'_A = total moles of A
and H = lb mole of carrier liquor held-up on each plate.
Let the equilibrium relationship between the solute A at the interface Y^*_n and A in the liquid X_n be given by:

$$Y^*_n = mX_n. \tag{9.56}$$

Then using the symbols of Fig. 9.8 a material balance over plate n will be:

$$G(Y_{n-1} - Y_n) - L(X_n - X_{n+1}) = kX_n H. \tag{9.57}$$

Now the Murphree stage efficiency E° is:

$$E^\circ = \frac{Y_{n-1} - Y_n}{Y_{n-1} - Y^*_n}. \tag{9.58}$$

Substitution of equations (9.56) and (9.57) into (9.58) and rearranging gives the second order finite difference equation:

$$Y_{n+1} + PY_n + QY_{n-1} = 0 \tag{9.59}$$

where

$$P = [(E^\circ - 2) - \alpha(\beta + 1)]$$

and

$$Q = [(1 - E^\circ)(1 + \alpha\beta) + \alpha]$$

and where

$$\alpha = \frac{GmE^\circ}{L} \quad \text{and} \quad \beta = \frac{Hk}{GmE^\circ}.$$

The solution of equation (9.59) is:

$$Y_n = Rz_1^n + Sz_2^n \tag{9.60}$$

where z_1 and z_2 are the roots of the auxiliary equation:

$$z^2 + Pz + Q = 0. \tag{9.61}$$

Inserting the boundary conditions at the top and bottom of the tower that when $n = 0$; $Y = Y_0$ and when $n = N + 1$; $X = 0$ gives the number of stages required as

$$N = \frac{\log\left\{\dfrac{Y_0[(1/z_1) - (1/z_2)]}{Y_{N-1} - Y_N/z_2}\right\}}{\log\left(\dfrac{1}{z_1}\right)}. \tag{9.62}$$

The above analysis is given in greater detail in *A problem in Chemical Engineering Design* (Jeffreys, 1961) and the following example has been taken from that text by kind permission of the Institution of Chemical Engineers, London.

Example 4. Calculate the number of plates required to absorb into glacial acetic acid 99 mole% of the ketene from 135 lb mole h^{-1} of a gas containing 4·6% by volume of ketene. 138 lb mole h^{-1} of glacial acetic acid is to be used and the tower hydraulics indicate that the plate dimensions will be such that the acid hold up will be 2·09 lb mole per plate. The equilibrium relation is $Y^* = 2{\cdot}0\ X$, the specific reaction rate constant is 0·075 sec^{-1}, and the plate efficiency is estimated to be 0·4.

Solution. Carrier gas rate $G = 0{\cdot}954 \times 135 = 129{\cdot}0$ mole

$$\therefore\ \alpha = \frac{129{\cdot}0 \times 2{\cdot}0 \times 0{\cdot}4}{138{\cdot}0} = 0{\cdot}75$$

$$\beta = \frac{2{\cdot}09 \times 0{\cdot}075 \times 3600}{129{\cdot}0 \times 0{\cdot}4 \times 2{\cdot}0} = 5{\cdot}46$$

$$P = [0{\cdot}4 - 2{\cdot}0 - 0{\cdot}75\,(5{\cdot}46 + 1)] = -\ 6{\cdot}43$$

$$Q = [(1{\cdot}0 - 0{\cdot}4)\,(1{\cdot}0 + 0{\cdot}75 \times 5{\cdot}46) + 0{\cdot}75)] = 3{\cdot}8$$

$$z_1 = 0{\cdot}54 \text{ and } z_2 = 5{\cdot}88$$

From a material balance on the top plate:

$$Y_{N-1} = \frac{0{\cdot}00046\,[(0{\cdot}075 \times 3600 \times 2{\cdot}09) + 138{\cdot}0 + 103{\cdot}2]}{0{\cdot}6\,(270 \times 2{\cdot}09) + 103{\cdot}2}$$

$$= 0{\cdot}00063$$

and hence using equation (9.62):

$$N = \frac{\log 0{\cdot}046\left[\left(\frac{1}{0{\cdot}54} - \frac{1}{5{\cdot}88}\right)\Big/\left(0{\cdot}00063 - \frac{0{\cdot}00046}{5{\cdot}88}\right)\right]}{\log\left(\frac{1}{0{\cdot}54}\right)}$$

and $N = 11{\cdot}3$ (say 12 plates).

REFERENCES

BRIAN, P. L. T., HURLEY, J. F., and HASSELTINE, E. H. 1961. Penetration theory for gas absorption accompanied by a second order chemical reaction. *A.I.Ch.E.Jl*, 7, 226.

HATTA, S. 1928/9. On the absorption velocity of gases by liquids. *Tôhoku Imperial University Technical Reports*, **8**, 1.

JEFFREYS, G. V. 1961. *A problem in chemical engineering design*. London, Institution of Chemical Engineers.

JENSON, V. G., and JEFFREYS, G. V. 1965. *Mathematical methods in chemical engineering*. New York, Academic Press.

LIGHTFOOT, E. N. 1958. Steady state absorption of a sparingly soluble gas in an agitated tank with simultaneous irreversible first-order reaction. *A.I.Ch.E.Jl*, **4**, 499.

PEACEMAN, D. W. 1951. Liquid-side resistance in gas absorption with and without chemical reaction. *D.Sc. Thesis M.I.T.*

SECOR, R. M., and SOUTHWORTH, R. W. 1961. Absorption with an infinitely rapid chemical reaction in packed towers. *A.I.Ch.E.Jl*, 7, 705.

VAN KREVELEN, D. W., and HOFTIJZER, P. J. 1948. Kinetics of simultaneous absorption and chemical reaction. *Chem. Engng Prog.*, **44**, 529.

VAN KREVELEN, D. W., and HOFTIJZER, P. J. 1948a. Kinetics of gas–liquid reactions. Part I. General theory. *Rec. trav. chem.*, **67**, 563.

VAN KREVELEN, D. W., and HOFTIJZER, P. J. 1953. Graphical design of gas–liquid reactors. *Chem. Engng Sci.*, **2**, 145.

10

LIQUID–LIQUID REACTORS

10.1 Introduction

In a number of chemical reaction processes, such as nitration or sulphonation, it is advantageous to distribute the reactants between two liquid phases. In others, the liquid reactants are only partially miscible in one another so that the reaction process must, of necessity, be a two phase reaction. Processes involving reactions of this kind are well-known and have been practiced for many years in the explosives industry. Some of these reactions are extremely exothermic and therefore the concentrations of the reactants must be kept low, thereby limiting the reaction rate and the generation of heat. This is conveniently accomplished by a two phase reaction process where one reactant is stored in a solvent phase and diffuses into the reaction phase to replace that consumed by chemical reaction. The concentration of the distributed reactant in the reactive phase is limited by the distribution coefficient and by the rate of transport across the interface, and in this way the rate of heat generation may be controlled.

Two phase reaction processes are generally operated by introducing the reactants to the reactor in separate streams, and mass transfer and chemical reaction occurs on mixing in the reaction vessel. The reaction may occur in one phase, or in both phases simultaneously, but generally the solubility relationships are such that the extent of chemical reaction in one of the phases is so small that it can be neglected in the analysis of this kind of reactor. Hence in the following discussion it will be assumed that the chemical reaction occurs in one phase only. This will be called the reactive phase and the other will be called the extractive phase.

Finally two phase reactions can be carried out batchwise or continuously and both of these methods have been practiced commercially. Two phase batch reactions are generally carried on in reaction kettles or autoclaves while two phase continuous reactions are carried out in mixer-settler extractors or conventional extraction columns. Tubular reactors are unsuitable for two phase reactions. Whatever type of continuously operated equipment is used for these reaction processes, the two phases may be contacted in countercurrent or crosscurrent flow in a similar manner to conventional extraction processes not involving a chemical reaction. These different types of operation will now be considered.

10.2 Batch Operation

Consider a first order reversible reaction:

$$aA \underset{k_2}{\overset{k_1}{\rightleftharpoons}} bB \tag{10.1}$$

to take place in a two phase system under isothermal conditions. Let the reaction occur in one phase only, with the reactive phase designated by R and the extractive phase by E. Let the distribution ratios between the phases of A and B be constant at m_A and m_B, so that the equilibrium relationships are:

$$y_A = m_A x_A \tag{10.2}$$

and

$$y_B = m_B x_B \tag{10.3}$$

where y_A and y_B are the mole fractions of A and B in the extractive phase, and x_A and x_B are their mole fractions in the reactive phase.

Letting the total moles of A at any time be N_A, the distribution of A between the phases is given by the following two equations:

$$N_A = N_{AE} + N_{AR} \tag{10.4}$$

where N_{AE} and N_{AR} are the amounts of A in the two phases,

and

$$m_A = \frac{y_A}{x_A} = \frac{N_{AE}}{N_{AR}} \cdot \frac{N_{TR}}{N_{TE}} = \frac{1}{L} \cdot \frac{N_{AE}}{N_{AR}} \tag{10.5}$$

where N_{TE} and N_{TR} are the total moles in the two phases, having a ratio:

$$L = \frac{N_{TE}}{N_{TR}}. \tag{10.6}$$

From equations (10.4) and (10.5):

$$dN_A = dN_{AE} + dN_{AR} = (m_A L + 1)\, dN_{AR}. \tag{10.7}$$

The rate of conversion of A in the reactive phase of volume V_R is:

$$-\frac{dN_A}{dt} = V_R[k_1 C_A - k_2 C_B] \tag{10.8}$$

or
$$-\frac{dN_A}{dt} = k_1 x_A N_{TR} - k_2 x_B N_{TR}. \qquad (10.9)$$

Eliminating dN_A between equations (10.7) and (10.9) gives:

$$-(m_A L + 1)\frac{dN_{AR}}{dt} = (k_1 x_A - k_2 x_B)N_{TR}. \qquad (10.10)$$

From the stoichiometry of the reaction, one mole of A yields b/a moles of B, and hence N_B the moles of B at any time is given by:

$$N_B = N_{BO} + \frac{b}{a}(N_{AO} - N_A) \qquad (10.11)$$

where N_{AO} and N_{BO} are the moles of A and B charged to the batch reactor at zero time.

Applying the above analysis to component B:

$$N_B = N_{BE} + N_{BR} = (m_B L + 1)N_{BR} \qquad (10.12)$$

and
$$x_B = \frac{N_{BR}}{N_{TR}} = \frac{N_{BO} + \frac{b}{a}(N_{AO} - N_A)}{(m_B L + 1)N_{TR}} \qquad (10.13)$$

Since
$$N_A = x_A N_{TR}(m_A L + 1) \qquad (10.14)$$

equation (10.13) becomes:

$$x_B = \frac{N_{BO} + \frac{b}{a}N_{AO} - \frac{b}{a}x_A(m_A L + 1)N_{TR}}{(m_B L + 1)N_{TR}}. \qquad (10.15)$$

Substituting this value of x_B into equation (10.10) establishes the required differential equation for x_A:

$$-\frac{dx_A}{dt} = \frac{k_1 x_A}{m_A L + 1} + \frac{k_2 b x_A}{a(m_B L + 1)} - \frac{k_2(aN_{BO} + bN_{AO})}{aN_{TR}(m_A L + 1)(m_B L + 1)}. \qquad (10.16)$$

Putting
$$\frac{k_1}{m_A L + 1} + \frac{k_2 b}{a(m_B L + 1)} = \beta \qquad (10.17)$$

and
$$\frac{k_2(aN_{BO} + bN_{AO})}{aN_{TR}(m_A L + 1)(m_B L + 1)} = \lambda \qquad (10.18)$$

equation (10.16) takes the form:

$$-\frac{dx_A}{dt} = \beta x_A - \lambda \tag{10.19}$$

or

$$t = \frac{1}{\beta}\left[\text{Ln}\,(\beta x_A - \lambda)\right]_{x_A}^{x_{A0}} \tag{10.20}$$

which on substitution of the limits and solving for x_A gives the following expression for variation of A's mole fraction with time in the reactive phase:

$$x_A = \frac{[bk_2(m_AL + 1) + ak_1(m_BL + 1)]N_{TR}x_{A0} - k_2(bN_{A0} + aN_{B0})(1 - e^{\beta t})}{[bk_2(m_AL + 1) + ak_1(m_BL + 1)]N_{TR}\,e^{\beta t}}. \tag{10.21}$$

Substituting $t = 0$, confirms the initial condition that $x_A = x_{A0}$ while putting $t = \infty$ and using Rolle's theorem to obtain a determinate result gives the equilibrium value x_{Ae} as:

$$x_{Ae} = \frac{k_2[bN_{A0} + aN_{B0}]}{[bk_2(m_AL + 1) + ak_1(m_BL + 1)]N_{TR}}. \tag{10.22}$$

Example 1. Consider the batch hydrolysis of beef tallow (M.W. = 840) to give glycerol (M.W. = 92) and fatty acid (M.W. = 802) at 240°C by the reaction:

$$\underset{\text{(tallow)}}{A} + \underset{\text{(water)}}{3B} \leftrightharpoons \underset{\text{(glycerol)}}{C} + \underset{\text{(fatty acid)}}{D} \tag{I}$$

The experimental results of acid production versus time are shown in Fig. 10.1 and are taken from (Jenson and Jeffreys, 1967). In the analysis, allowance will be made for the variation with time of the total masses of the water and fat phases.

If the stepwise reaction is idealised to its slowest step, the mechanism can be described by a second order forward reaction rate constant, k_1, and an equilibrium constant, K, in the rate equation:

$$\frac{dN_A}{dt} = -\frac{k_1}{V_R}\left[N_AN_{BR} - \frac{N_{CR}N_D}{K}\right]. \tag{II}$$

The volume of the fat phase, V_R, assuming no volume change on mixing is given by:

$$V_R = \frac{N_A}{\rho_A} + \frac{N_{BR}}{\rho_B} + \frac{N_{CR}}{\rho_C} + \frac{N_D}{\rho_D}. \tag{III}$$

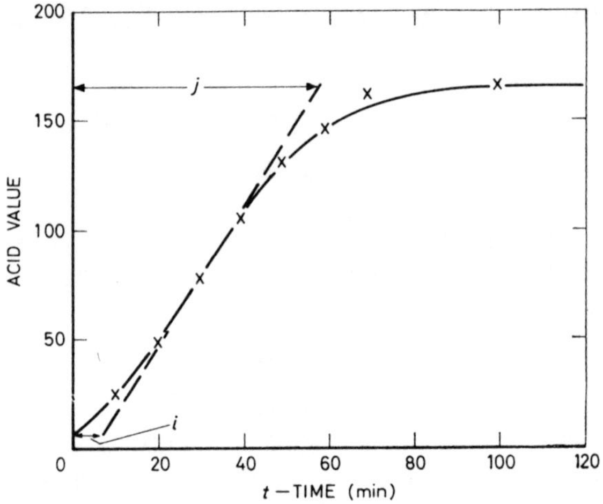

FIG. 10.1. The course of the reaction as determined by experiment using beef tallo[L] oil

Material balances for an initial change of N_{A0} and N_{B0} and L as the ratio of the total mass of the water phase to that of the fat phase give:

$$N_D = \frac{802}{840}(N_{A0} - N_A) \qquad \text{(IV)}$$

$$\left(\frac{m_B - 1}{m_B L + 1}\right)\cdot\left[N_{B0} - \frac{54}{840}(N_{A0} - N_A)\right] + \left(\frac{m_C - 1}{m_C L + 1}\right) \qquad \text{(V)}$$

$$\times\left[\frac{92}{840}(N_{A0} - N_A) - (N_A + N_D)\right)\right] = 0$$

$$N_{BR} = \frac{N_{B0} - \frac{54}{840}(N_{A0} - N_A)}{1 + m_B L} \qquad \text{(VI)}$$

$$N_{CR} = \frac{92(N_{A0} - N_A)}{840(1 + m_B L)}. \qquad \text{(VII)}$$

The value of m_C is known to be constant at 12·5. It is also known that water is less soluble in fat than in the derived fatty acid, but only the solubility in fatty acid can be determined experimentally. A digital computer was used, therefore, to infer the solubility of water in tallow oil along with the reaction rate and equilibrium constant. The value of m_B is determined from:

$$m_B = \frac{100}{S_1(1 - z) - S_0 z} \tag{VIII}$$

where z is the mass fraction of unreacted fat in the fat phase,
S_0 is the solubility of water in fat,
and S_1 is the solubility of water in fatty acid.

Equation (II) can be re-arranged with k_1 and t combined to form a new independent variable. Thus:

$$\frac{dN_A}{d(k_1 t)} = -\frac{1}{V_R}\left[N_A N_{BR} - \frac{N_{CR} N_D}{K}\right]. \tag{IX}$$

Equation (IX) may be integrated numerically using a step-by-step procedure provided its right-hand side can be evaluated at each step. This is achieved in the following sequence of calculations:

(*a*) A value of N_A is currently available at the stage of integration already reached; N_{A0} then enables N_D to be obtained from equation (IV).
(*b*) N_A and N_D determine z and hence m_B from equation (VIII).
(*c*) L is the only unknown remaining in equation (V).
(*d*) Equations (VI) and (VII) then yield N_{BR} and N_{CR}.
(*e*) V_R evaluated from equation (III) completes the evaluation of the terms in the right-hand side of equation (II).

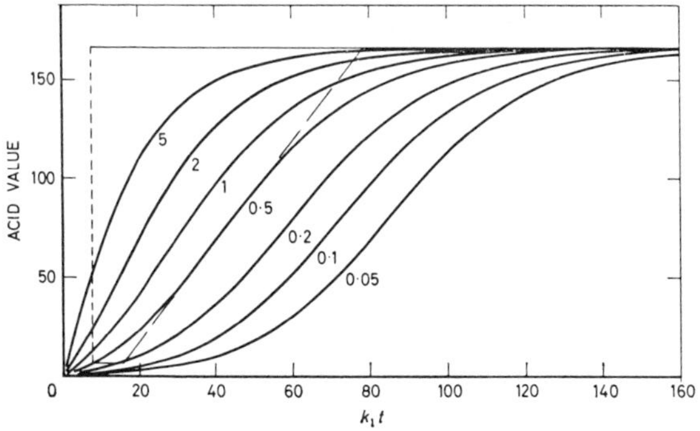

The figures along each curve represent the solubility of water in fat as a percentage

FIG. 10.2. The computed courses of the reaction

Initially, N_A and N_{A0} are set equal so that the calculation can be started. The three unknowns (k_1, K, S_0) are treated differently. The term k_1 combined with t, is evaluated by comparison of scales between the computed and experimental reaction profiles. Then K is determined by an iteration loop which forces the computer profile to reach the same equilibrium conversion as the experimental profile. A graphical trial and error method determines S_0 by generation of a family of profiles as shown in Fig. 10.2 for a series of values of S_0. Each computed and experimental curve is characterised by the ratio of lengths j/i shown in Figs 10.1 and 10.2 where the tangent to the curve has been drawn through the point of inflexion. This ratio eliminates k_1 and allows the appropriate solubility S_0 to be

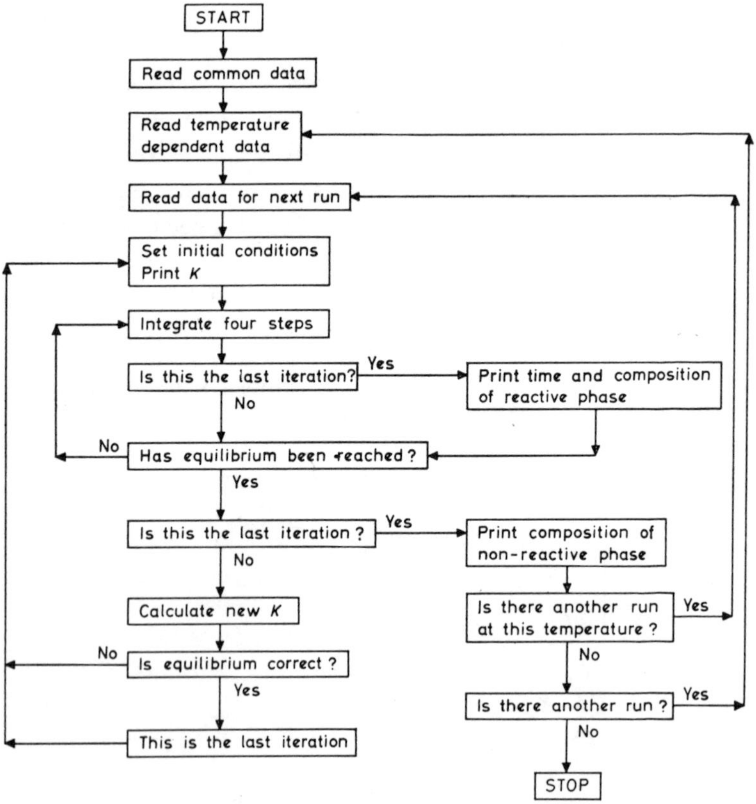

FIG. 10.3. Computer for the analysis of a batch reactor

determined. The value of k_1 then follows, as does the superimposition of the experimental points on the appropriate computed curve. The computer flow diagram used in this analysis is shown in Fig. 10.3.

For the temperature of 240°C, the deduced values of S_0 and k_1 are 0·5 and 1·23 respectively.

10.3 Cross Flow Operation

The flow diagram for this type of operation is illustrated in Fig. 10.4. It has been assumed, for convenience, that the reaction process is being carried out in a mixer–settler plant although the following analysis will apply to any stagewise equipment that can be operated in this way. In the flow diagram M represents a reaction mixer and S a settler. Then consider a first order reaction of the type:

$$A \underset{k_2}{\overset{k_1}{\rightleftharpoons}} B \tag{10.23}$$

to occur in a two phase liquid process in a mixer–settler unit operated as shown in the above flow diagram. The operation will be assumed to be isothermal, and the concentrations of the reactive species are such that the solubility of each component is independent of the components with which it coexists. The equilibrium relationships of A and B between the phases will be assumed to be represented as before by:

$$y_A = m_A x_A \tag{10.2}$$

and

$$y_B = m_B x_B. \tag{10.3}$$

Let R lb mole h^{-1} of a feed liquid be fed to stage 1 of an N stage process operating under cross flow conditions. Let x_{A0} be the mole fraction of reactant A in the feed. Also let E lb mole h^{-1} of solvent be fed to each stage of the process and let y_{A0} be the mole fraction of reactant A in the solvent. y_{A0} could, but need not necessarily be zero. However when the two phases are contacted in the mixers, chemical reaction occurs in the reactive phase in accordance with equation (10.23). The reaction product is considered to be more soluble in the extract phase so that it passes, via the settler, from the system to storage as shown in the diagram.

The rate of reaction is as given by equation (10.9),

i.e.

$$-\frac{dN_A}{dt} = k_1 x_A N_{TR} - k_2 x_B N_{TR} \tag{10.9}$$

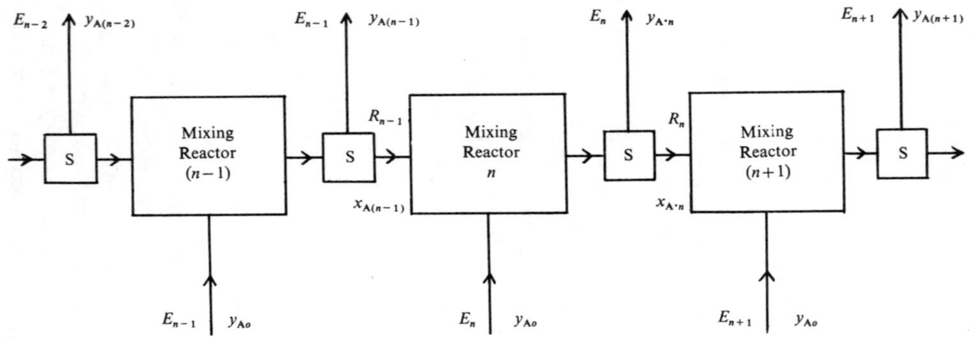

FIG. 10.4. Crosscurrent flow operation

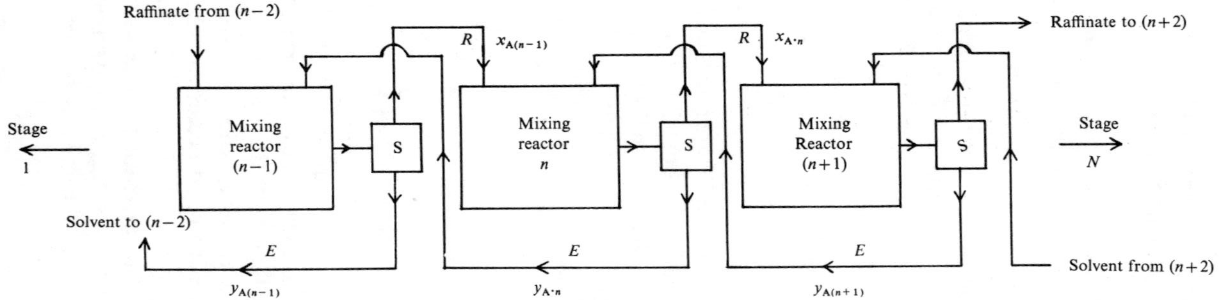

FIG. 10.5. Countercurrent flow operation

where N_{TR} is the total number of moles held-up in the reactive phase in any stage. Let the two phases be totally immiscible so that for stage n:

$$E_n + R_n = \phi, \text{ a constant for all } n. \tag{10.24}$$

However, the ratio H of the phases held-up in the reactor, may not be the same as the ratio of the flow rates of extractive phase and reactive phase. Hence let:

$$\frac{E_n}{R_n} = \alpha \frac{N_{TE.n}}{N_{TR.n}} = \alpha H \tag{10.25}$$

where α is the proportionality constant between these ratios. Then a steady state mass balance over stage n with respect to the reactant A is:

$$[R_{n-1}x_{A(n-1)} + E_n y_{A0}] - [(R_n x_{An} + E_n y_{A.n}) + V_R r_n] = 0. \tag{10.26}$$

Since the phases are completely immiscible:

$$R_1 = R_2 = R_3 = \ldots = R_n = \ldots = R_N = R \tag{10.27}$$

and if the solvent rate to each stage is the same, E_n is also constant. Then dividing equation (10.26) by R gives:

$$x_{A(n-1)} + \alpha H y_{A0} - x_{A.n} - \alpha H y_{A.n} = (k_1 x_{A.n} - k_2 x_{B.n})\frac{N_{TR}}{R}. \tag{10.28}$$

If pure solvent is fed to each stage $y_{A0} = 0$ and then using equation (10.2) to eliminate $y_{A.n}$ gives:

$$x_{A(n-1)} - \left(1 + \alpha m_A H + \frac{k_1 N_{TR}}{R}\right)x_{A.n} + \frac{k_2 N_{TR}}{R} x_{B.n} = 0. \tag{10.29}$$

Let

$$1 + \alpha m_A H = A_1 \tag{10.30}$$

$$1 + \alpha m_B H = A_2 \tag{10.31}$$

$$\frac{k_1 N_{TR}}{R} = B_1 \tag{10.32}$$

and

$$\frac{k_2 N_{TR}}{R} = B_2 \tag{10.33}$$

equation (10.29) then becomes:

$$x_{A(n-1)} - (A_1 + B_1)x_{A.n} + B_2 x_{B.n} = 0. \tag{10.34}$$

A similar analysis in terms of the product B yields:

$$x_{B(n-1)} - (A_2 + B_2)x_{B.n} + B_1 x_{A.n} = 0. \qquad (10.35)$$

Equations (10.34) and (10.35) are two simultaneous finite difference equations. These equations may be expressed in terms of the finite difference operator E as:

$$[1 - E(A_1 + B_1)]x_{A(n-1)} + B_2 E x_{B(n-1)} = 0 \qquad (10.36)$$

and

$$[1 - E(A_2 + B_2)]x_{B(n-1)} + B_1 E x_{A(n-1)} = 0. \qquad (10.37)$$

Elimination of $x_{B(n-1)}$ and rearrangement leads to:

$$[E^2(A_1A_2 + A_1B_2 + A_2B_1) - E(A_1 + A_2 + B_1 + B_2)E + 1]x_{A(n-1)} = 0. \qquad (10.38)$$

The solution of this will be similar to that given in section 5.7 and is:

$$x_{A.n} = G_1\rho_1^n + G_2\rho_2^n \qquad (10.39)$$

where G_1 and G_2 are arbitrary constants to be evaluated from the boundary conditions that when $n = 0$, the feed concentrations are $x_A = x_{A0}$ and $x_B = 0$.

ρ_1 and ρ_2 are the roots of the auxiliary equation and are hence:

$$\rho_1 \text{ and } \rho_2 = \frac{(A_1 + A_2 + B_1 + B_2) \pm \sqrt{\{(A_1 + B_1 - A_2 - B_2)^2 + 4B_1B_2\}}}{2[(A_1 + B_1)(A_2 + B_2) - B_1B_2]}. \qquad (10.40)$$

Also from equations (10.34), (10.35) and (10.39):

$$x_{B.n} = \frac{\rho_1^n}{B_2}\left[G_1\left(A_1 + B_1 - \frac{1}{\rho_1}\right)\right] + \frac{\rho_2^n}{B_2}\left[G_2\left(A_1 + B_1 - \frac{1}{\rho_2}\right)\right]. \qquad (10.41)$$

The constants G_1 and G_2 can be evaluated from equations (10.39) and (10.41) when the roots ρ_1 and ρ_2 have been evaluated by equation (10.40).

Cross flow reactive extraction processes of the type presented above have been analysed by Piret, Penny and Trambouze (1960). These reaction processes can be utilised advantageously for reversible reactions since addition of fresh solvent brings about the removal of the product B from the reaction phase thereby allowing the reaction to proceed to completion. If the process were operated under counter current flow conditions the concentration of the product B near the feed end of the stagewise process would be such that chemical equilibrium would exist and therefore there would be very little reaction in this section of the plant.

10.4 Countercurrent Flow Operation

10.4.1 Analysis of stagewise equipment

The flow diagram for a process operating under counterflow conditions is illustrated in Fig. 10.5 and is self-explanatory. Therefore let the reaction

$$\mathrm{A} \xrightarrow{k} \mathrm{B} \tag{10.42}$$

take place in this system, and let the equilibrium relations be as expressed by equations (10.2) and (10.3). Let the flows and concentions be represented by the same symbols as those given in section 10.3. However under the conditions considered here the solvent will be fed to stage N and will pass through each stage in turn emerging finally from stage 1. Therefore let the solvent flow rate E lb mole h^{-1} be fed to stage N. Then, under steady state conditions a mass balance on stage n in terms of the reactant A is:

$$[Rx_{\mathrm{A}(n-1)} + Ey_{\mathrm{A}(n+1)}] - [Rx_{\mathrm{A}.n} + Ey_{\mathrm{A}.n}] - kx_{\mathrm{A}.n}N_{\mathrm{TR}} = 0. \tag{10.43}$$

Similarly a mass balance for the product B is:

$$[Rx_{\mathrm{B}(n-1)} + Ey_{\mathrm{B}(n+1)}] + kx_{\mathrm{A}.n}N_{\mathrm{TR}} - [Rx_{\mathrm{B}.n} + Ey_{\mathrm{B}.n}] = 0. \tag{10.44}$$

Substituting for y in terms of x by the equilibrium relationship and then dividing equation (10.43) throughout by $m_{\mathrm{A}}E$ gives:

$$x_{\mathrm{A}(n+1)} + \frac{R}{m_{\mathrm{A}}E}x_{\mathrm{A}(n-1)} - \left(1 + \frac{R}{m_{\mathrm{A}}E} + \frac{kN_{\mathrm{TR}}}{m_{\mathrm{A}}E}\right)x_{\mathrm{A}.n} = 0. \tag{10.45}$$

Let

$$1 + \frac{R}{m_{\mathrm{A}}E} + \frac{kN_{\mathrm{TR}}}{m_{\mathrm{A}}E} = C \tag{10.46}$$

and

$$\frac{R}{m_{\mathrm{A}}E} = D \tag{10.47}$$

then equation (10.45) becomes:

$$x_{\mathrm{A}(n+1)} - Cx_{\mathrm{A}.n} + Dx_{\mathrm{A}(n-1)} = 0. \tag{10.48}$$

Equation (10.48) is a linear finite difference equation the solution of which is:

$$x_{A.n} = G_3\rho_3^n + G_4\rho_4^n \tag{10.49}$$

where G_3 and G_4 are arbitrary constants and

$$\rho_3 \text{ and } \rho_4 = \frac{C \pm \sqrt{(C^2 - 4D)}}{2}. \tag{10.50}$$

In a similar way it can be shown that equation (10.44) can be condensed to:

$$x_{B(n+1)} - Px_{B.n} + Qx_{B(n-1)} = \frac{kN_{TR}}{m_B E}(G_3\rho_3^n + G_4\rho_4^n). \tag{10.51}$$

The solution of equation (10.51) is:

$$x_{B,n} = G_5\rho_6^n + G_6\rho_6^n + \frac{FG_2\rho_2^n}{\rho_1^2 + P\rho_1 + Q} + \frac{FG_4\rho_4^n}{\rho_2^2 + P\rho_2 + Q} \tag{10.52}$$

G_5 and G_6 are also arbitrary constants and

$$\rho_5 \text{ and } \rho_6 = \frac{P \pm \sqrt{(P - 4Q)}}{2}. \tag{10.53}$$

All the arbitrary constants can be evaluated from the boundary conditions.

In the above analyses it has been assumed that the two phases are in equilibrium throughout the reaction process. That is, the assumption is made that the concentration of the reactant in the bulk of the reactive phase is the same as the concentration of the reactant at the interface. The extent to which this condition is satisfied can be checked by considering the equation developed in the previous chapter expressing the concentration profile through the laminar film. From equation (9.14) it can be shown that:

$$\frac{c_{AL}}{c_{Ai}} = \frac{c_{AL}\cosh L\sqrt{\frac{k}{D}} + c_{Ai}}{c_{AL} + c_{Ai}\cosh L\sqrt{\frac{k}{D}}}. \tag{10.54}$$

Equation (10.54) expresses the fraction of the reactant A that reaches the bulk of the reactive phase without undergoing chemical reaction This ratio will approach unity, i.e. c_{AL} will tend to c_{Ai}, if:

$$L\sqrt{\frac{k}{D}} \leqslant 0{\cdot}2. \tag{10.55}$$

For liquids D is of the order of 10^{-5} cm^2 sec^{-1} and if L is taken to be 10^{-2} cm, then for mass transfer rates to be neglected, k must be less than 2×10^{-3} sec^{-1}.

Consequently the above analysis applies only to very slow reactions. In all other cases the analysis must include the effect of the rate of mass transfer through the two films.

10.4.2 Continuous operation—analysis in terms of transfer units

Consider a hydrolysing column (Jeffreys, Jenson and Miles, 1961) which operates continuously under steady state conditions as shown in Fig. 10.6 where L lb h^{-1} of animal fat are fed into the base of the tower to ascend as the continuous phase. This fat is hydrolysed by G lb h^{-1} of water sprayed into the top to descend through the rising fat in the form of droplets. The ascending fat will be termed the raffinate, and the descending water, the extract.

Taking an element of column of height δh as shown in Fig. 10.6 where the rate of transfer of glycerine from the fat to the water phase is:

$$\text{Rate} = K_E a \,.\, S(y^* - y)\,\delta h \tag{10.56}$$

whilst the rate of removal of fat by chemical reaction by a first order mechanism is:

$$\text{Reaction in element} = kS\delta h\rho z \tag{10.57}$$

or if w is the weight of fat to produce 1 lb of glycerine, then the

$$\text{Rate of production of glycerine} = \frac{kS\delta h\rho z}{w}. \tag{10.58}$$

Now a glycerine balance over the element is:

$$Lx + G\left(y + \frac{dy}{dh}\,.\,\delta h\right) + \frac{kS\delta h\rho z}{w} = L\left(x + \frac{dx}{dh}\,.\,\delta h\right) + Gy \tag{10.59}$$

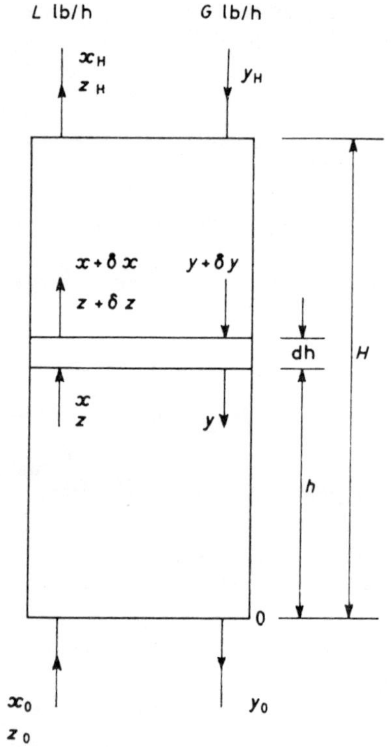

FIG. 10·6. Continuously operating fat-hydrolysing column under steady state conditions

which can be rearranged to express the rate of transfer thus:

$$Lx + \frac{kS\delta h\rho z}{w} - L\left(x + \frac{dx}{dh}\,.\,\delta h\right) = Gy - G\left(y + \frac{dy}{dh}\,.\,\delta h\right) \qquad (10.60)$$

$$= K_E a\,.\,S(y^* - y)\,\delta h.$$

An equivalent glycerine balance between the element and the base of the tower is:

$$\frac{Lz_0}{w} + Gy = L\left(x + \frac{z}{w}\right) + Gy_0. \qquad (10.61)$$

Substitution of the values of z/w obtained from equation (10.61) into equation (10.60) gives:

$$S\rho k\left[\frac{z_0}{w} + \frac{G}{L}(y - y_0) - x\right]\delta h - L\,.\,\frac{dx}{dh}\,.\,\delta h$$

$$= K_E a\,.\,S(y^* - y)\,\delta h. \qquad (10.62)$$

But the equilibrium is

$$y^* = mx \tag{10.63}$$

which, if combined with the right-hand side of equation (10.60), gives:

$$K_E a\, S\, mx = K_E a\, Sy - G\,.\frac{dy}{dh}. \tag{10.64}$$

Substitution of equation (10.64) into equation (10.62), rearranging, and multiplying throughout by $(K_E aSm)/LG$ gives:

$$\frac{d^2y}{dh^2} - \frac{dy}{dh}\left[\frac{K_E a\,.\,S}{G} - \frac{K_E aSm}{L} - \frac{S\rho k}{L}\right] + \left[\frac{K_E a\,.\,S^2 m\rho k}{L^2} - \frac{K_E aS^2\rho k}{LG}\right] y = \frac{K_E aS^2 m\rho k}{GL}\left[\frac{G}{L}y_0 - \frac{z_0}{w}\right]. \tag{10.65}$$

Equation (10.65) is a second order linear differential equation which is factorisable. Thus let

$$\alpha = \frac{mG}{L}; \qquad \beta = \frac{S\rho k}{L}; \qquad \gamma = \frac{K_E aS}{G}(\alpha - 1)$$

from which the solution of equation (10.65) is:

$$y = A\exp(-\beta h) + B\exp(-\gamma h) + \frac{\alpha y_0 - mz_0/w}{\alpha - 1} \tag{10.66}$$

where the boundary conditions to evaluate the arbitrary constants A and B are:

$$\text{at } h = 0, x = 0; \qquad \text{and at } h = H, y = 0.$$

Substitution of these conditions into equation (10.66) and letting

$$\lambda = \frac{\gamma - \beta + \alpha\beta}{\gamma} = 1 + \frac{k\rho G}{K_E aL}$$

the final equation is:

$$y_0 = \frac{mz_0}{w[\alpha - \exp(-\gamma H)]} \times \left[1 + \left(\frac{\lambda - 1}{\alpha - \lambda}\right)\exp(-\gamma H) - \left(\frac{\alpha - 1}{\alpha - \lambda}\right)\exp(-\beta H)\right]. \tag{10.67}$$

The glycerine fraction y_0, in the final extract from the hydrolyser, will be known from the experimental results, and therefore the height of the transfer unit on the overall extract basis can be calculated from the relation:

$$[\text{H.T.U.}]_{\text{T.O.E.}} = \frac{G}{K_E a S}. \tag{10.68}$$

Finally it can be shown by substituting equation (10.67) into equation (10.66) that the variation of the weight fraction of extract y at any position in the tower at height h above the base is given by the expression:

$$y = \frac{m z_0}{w(\alpha - \lambda)} \left\{ \exp(-\beta H) + \left[\frac{\exp(-\beta H) - \lambda}{\alpha - \exp(-\gamma H)} \right] \exp(-\gamma H) + \left[\frac{\lambda \exp(-\gamma H) - \alpha \exp(-\beta H)}{\alpha - \exp(-\gamma H)} \right] \right\}. \tag{10.69}$$

Example 2. In a particular hydrolysis of animal fat, the following set of operating parameters was recorded, and the effective value of $K_E a$ deduced:

$$m = 10{\cdot}32$$

$$\frac{z_0}{w} = 0{\cdot}0853$$

$$\alpha = \frac{mG}{L} = \frac{10{\cdot}32 \times 3760}{8540} = 4{\cdot}544$$

$$\beta = \frac{S\rho k}{L} = \frac{3{\cdot}688 \times 45{\cdot}05 \times 10{\cdot}2}{8540} = 0{\cdot}198$$

$$\gamma = \frac{K_E a S}{G}(\alpha - 1) = \frac{K_E a \times 3{\cdot}668}{3760} \times 3{\cdot}544 = 0{\cdot}00348\, K_E a$$

$$\lambda = 1 + \frac{3{\cdot}544 \times 0{\cdot}198}{0{\cdot}00348\, K_E a} = 1 + \frac{201{\cdot}6}{K_E a}$$

From equation (10.67):

$$0{\cdot}188 = \frac{10{\cdot}32 \times 0{\cdot}0853}{4{\cdot}544 - \exp(-0{\cdot}256\, K_E a)} \left[1 + \left(\frac{201{\cdot}6/K_E a}{3{\cdot}544 - \dfrac{201{\cdot}6}{K_E a}} \right) \right.$$

$$\left. \times \exp(-0{\cdot}256\, K_E a) - \left(\frac{3{\cdot}544}{3{\cdot}544 - \dfrac{201{\cdot}6}{K_E a}} \right) \exp(-14{\cdot}6) \right] .$$

Hence $4{\cdot}544 - \exp(-0{\cdot}256 K_E a) = 4{\cdot}682 \left[1 + \dfrac{\exp(-0{\cdot}256 K_E a)}{0{\cdot}0176 K_E a - 1} \right]$

and $$\exp(0{\cdot}256 K_E a) = \frac{3{\cdot}682 + 0{\cdot}0176 K_E a}{0{\cdot}138(1 - 0{\cdot}0176 K_E a)}$$

This equation, solved by trial and error, yields an answer of $K_E a = 14{\cdot}2$.

If a second order reversible reaction occurs in the continuous hydrolysing column described above, then the analysis would be as follows (Jenson, Jeffreys and Edwards, 1967):

Rate of removal of fat by reaction is

$$k_1 S \rho^2 \delta h \left[uz - \frac{1}{k} \times (1 - z - x - u) \right] = \gamma w \delta h \qquad (10.70)$$

Rate of transfer of glycerol from fat to water is

$$K_G a S(mx - y)\,\delta h = \alpha S h \qquad (10.71)$$

Rate of transfer of water from water to fat is

$$K_w a S(1 - y - nu)\,\delta h = \beta \delta h \qquad (10.72)$$

where the above expressions have been used to define new variables α, β, and γ.

Taking a glycerol balance in the form 'glycerol lost by fat phase = glycerol gained by water phase = glycerol transferred'

gives

$$\gamma - \frac{d}{dh}(Lx) = -\frac{d}{dh}(Gy) = \alpha \qquad (10.73)$$

A similar water balance gives

$$0{\cdot}587\gamma + \frac{d}{dh}(Lu) = \frac{d}{dh}[G(1 - y)] = \beta \qquad (10.74)$$

The equilibrium distribution ratio for water (n) is related to the solubilities of water in fat and fatty acid by

$$1/n = S_1 - (S_1 - S_0) z \qquad (10.75)$$

A total material balance between the element and the bottom of the column gives

$$L_0 + G = L + G_0 \tag{10.76}$$

and an equivalent glycerol balance over the same section gives

$$L_0\left(\frac{z_0}{w} + x_0\right) + Gy = L\left(\frac{z}{w} + x\right) + G_0 y_0 \tag{10.77}$$

The above equations with appropriate boundary conditions are sufficient to completely define the problem mathematically but equations (10.73) and (10.74) can be rearranged to a more useful form as follows. Subtracting the second half of equation (10.73) from (10.74) gives

$$\frac{dG}{dh} = \beta - \alpha \tag{10.78}$$

Differentiating the products on the left hand sides of equations (10.74) and (10.73) and using equations (10.76) and (10.78) to remove the derivatives of L gives

$$\frac{du}{dh} = \frac{\beta - 0{\cdot}587\gamma - (\beta - \alpha)u}{L} \tag{10.79}$$

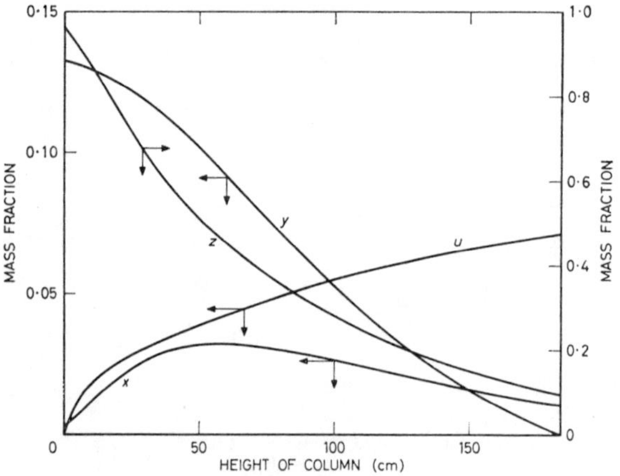

$T = 240°$ C
$R = 0{\cdot}69$
ZnO $= 0{\cdot}3\%$

FIG. 10.7. Computed composition profiles in a continuous hydrolysis

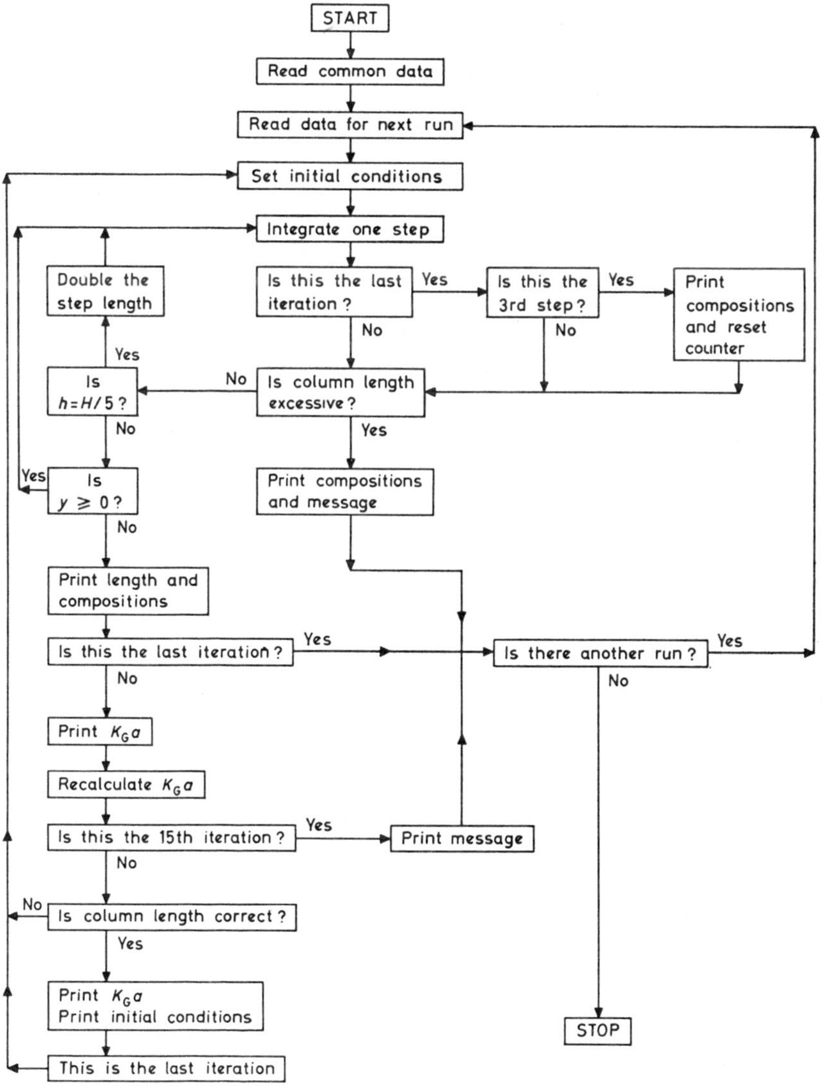

FIG. 10.8. Computer flow diagram for the analysis of a continuous hydrolyser

$$\frac{dx}{dh} = \frac{\gamma - \alpha - (\beta - \alpha)x}{L} \tag{10.80}$$

From equations (10.73) and (10.78)

$$\frac{dy}{dh} = \frac{-\alpha - (\beta - \alpha)y}{G} \tag{10.81}$$

Equations (10.78, 10.79, 10.80, 10.81) are now in a form ideally suited for a numerical step-by-step integration procedure provided the right hand sides can be evaluated at the start of each step. Assuming G, u, x, y calculated from the previous step, and G_0, u_0, x_0, y_0, L_0, z_0 given at the bottom of the column, equations (10.76, 10.77, 10.75, 10.70, 10.71, 10.72) can be solved consecutively for L, z, n, γ, α, β respectively and the integration can proceed. Thus only the flow rates and compositions of both streams at the bottom of the hydrolyser are needed to initiate the numerical procedure. The calculation stops when $y = 0$ corresponding to the pure water feed to the top of the column.

For a tallow containing 0·13 mole fraction, fed to a column operating at 240°C with a fat phase to extract phase ratio of 0·69, the graphical presentation of results is shown in Fig. 10.7 and the computer flow diagram is shown in Fig. 10.8.

REFERENCES

JEFFREYS, G. V., JENSON, V. G., and MILES, F. R. 1961. The analysis of a continuous fat-hydrolysing column. *Trans. Inst. Chem. Engrs,* **39**, 389.

JENSON, V. G., and JEFFREYS, G. V. 1967. The development and solution of mathematical models describing two-phase mass-transfer with chemical reaction during the hydrolysis of fats. *Inst. Chem. Engrs Symp. Ser.*, No. 23. , Institution of Chemical Engineers, London.

—, —, and EDWARDS, R. E. 1967. The continuous hydrolysis of fats and oils. *Proc.* 36th *internat. Congr. Ind. Chem., Brussels.* Special number T32, Vol. 3.

PIRET, E. L., PENNEY, W. H., and TRAMBOUZE, P. J. 1960. Extractive reaction: Batch and continuous-flow chemical reaction systems dilute case. *A.I.Ch.E.Jl,* **6**, 394.

INDEX